AF388949

Monitoring Lakes Water Based on Multisource Remote Sensing and Novel Modeling Techniques

Monitoring Lakes Water Based on Multisource Remote Sensing and Novel Modeling Techniques

Editors

Fei Zhang
Ngai Weng Chan
Xinguo Li
Xiaoping Wang

MDPI • Basel • Beijing • Wuhan • Barcelona • Belgrade • Manchester • Tokyo • Cluj • Tianjin

Editors
Fei Zhang
Xinjiang University
China

Ngai Weng Chan
Universiti Sains Malaysia
Malaysia

Xinguo Li
Xinjiang Normal University
China

Xiaoping Wang
Northwest A&F University
China

Editorial Office
MDPI
St. Alban-Anlage 66
4052 Basel, Switzerland

This is a reprint of articles from the Special Issue published online in the open access journal *Water* (ISSN 2073-4441) (available at: https://www.mdpi.com/journal/water/special_issues/lake_arid_remotesensing_modeling).

For citation purposes, cite each article independently as indicated on the article page online and as indicated below:

LastName, A.A.; LastName, B.B.; LastName, C.C. Article Title. *Journal Name* **Year**, *Volume Number*, Page Range.

ISBN 978-3-0365-7494-3 (Hbk)
ISBN 978-3-0365-7495-0 (PDF)

Cover image courtesy of Xiaoping Wang

Contents

About the Editors

Fei Zhang

Before January 2023, Fei Zhang was a professor at the College of Geography and Remote Sensing Sciences, Xinjiang University. February 2023-present, he is a professor at the College of Geography and Environmental Sciences, Zhejiang Normal University. He is mainly engaged in remote sensing research on eco-environments in arid zones, with a focus on remote sensing research on the water environment, remote sensing image classification, urbanization dynamic detection, and image processing. To date, he has published more than 100 journal papers as the first author or corresponding author. He is a Guest Editor of journals such as *Water* and *Remote Sensing*, and he is also an editorial board member of journals such as the *Journal of Lake Science* and *Spectroscopy and Spectral Analysis*.

Ngai Weng Chan

Ngai Weng Chan is a Professor at the GeoInformatic Unit, Geography Section, School of Humanities, Universiti Sains Malaysia. He is mainly engaged in human–environment relationships, with a focus on hydroclimatology, water resources, and hazard management. He served as Vice-President of the International Water Resources Association (IWRA) (2013-2015) and is a member of the International Association of Hydrological Sciences, International Water Association and Malaysian Water Association. He is active in civil society work in the area of water resources and environmental management. Currently, he is President of Water Watch Penang (WWP) (1997-Present) and a member of Malaysian Environmental NGOs (MENGOs) and the Malaysian Water Partnership. In 2010, he was announced as the Winner of the Asia Water Management Excellence Award. He has completed 67 research/consultancy projects and has published 30 books, 110 book chapters and 128 papers in professional journals. In the civil society arena, he is fondly referred to as "Malaysia's Waterman"!

Xinguo Li

Xinguo Li is a professor at the College of Geographic Sciences and Tourism, Xinjiang Normal University. He is mainly engaged in remote sensing research on changes in soil and water resources in arid zones, with a focus on the remote sensing mapping of soil salt content, soil organic carbon content, soil moisture and soil landscape. To date, he has published more than 40 journal papers. He is a reviewer of journals such as *Bulletin of Soil and Water Conservation*. in addition, he is a Guest Editor of *Water*.

Xiaoping Wang

Xiaoping Wang is an associate professor at the College of Resources and Environment, Northwest A&F University. He is mainly engaged in remote sensing research on terrestrial ecosystem carbon cycles, with a focus on the remote sensing mapping of vegetation canopy structures and biochemical parameters, ecological model (BEPS) ecosystem water and carbon cycle simulation and data assimilation. To date, he has published more than 20 journal papers. He is a Guest Editor of journals such as *Water* and the *International Journal of Environmental Research and Public Health*.

Editorial

Monitoring Lakes Water Using Multisource Remote Sensing and Novel Modeling Techniques

Xiaoping Wang [1], Fei Zhang [2,*], Ngai Weng Chan [3] and Xinguo Li [4]

[1] College of Natural Resources and Environment, Northwest A&F University, Yangling, Xianyang 712100, China
[2] College of Geography and Remote Sensing Sciences, Xinjiang University, Urumqi 830017, China
[3] School of Humanities, University Sains Malaysia, George Town 11800, Malaysia
[4] School of Geography and Tourism, Xinjiang Normal University, Urumqi 830046, China
* Correspondence: zhangfei3s@xju.edu.cn or zhangfei3s@163.com

Citation: Wang, X.; Zhang, F.; Chan, N.W.; Li, X. Monitoring Lakes Water Using Multisource Remote Sensing and Novel Modeling Techniques. *Water* 2022, 14, 3904. https://doi.org/10.3390/w14233904

Received: 1 November 2022
Accepted: 17 November 2022
Published: 1 December 2022

Publisher's Note: MDPI stays neutral with regard to jurisdictional claims in published maps and institutional affiliations.

1. Introduction

Inland lakes are indicators of climate change and environmental deterioration [1,2]. As a unique ecosystem unit, an inland lake is one of the basic places for human survival and development. In recent years, with the rapid development of regional societies and economies, the ecological environment of inland lakes has been rapidly degraded by human activities under the influence of large-scale water and soil exploitation activities, leading to massive deterioration of the ecological environment of lakes [3]. Therefore, lake ecological restoration and water quality monitoring under the coupling effect of climate change and human activities are the key to lake protection and management. In recent years, remote sensing has played an increasingly important role in the monitoring of the terrestrial water cycle.

Remote sensing technology has been applied in many fields, such as the monitoring and management of water storage, water quality, water levels, and hydrodynamics [4]. Remote sensing technology has been applied to water bodies since the 1970s [5]. Over time, the technology and theory of lake water color remote sensing monitoring have gradually matured, and an integrated lake remote sensing monitoring system of "Satellite-UAV-Ground" has been developed. In addition, the rapid development of computer and artificial intelligence technology in recent years provides a powerful algorithm support for the intelligent remote sensing observation of lakes. Therefore, the explosive growth of remote sensing data applications is driven by the coupling of multisource remote sensing data and the expansion of new modeling technology.

This Special Issue presents a review and recent advances of general interest in the use of remote sensing (RS) and geographic information systems (GIS) on inland lakes, with a focus on monitoring inland lakes (e.g., water storage, water quality, water levels, and hydrodynamics) and water resource management.

2. Overview of the Contributions

The call for papers was announced in July 2021, and after a rigorous peer-review process, a total of 11 papers were published [6–16]. To gain a better insight into the essence of the Special Issue, we will focus on the summary and analysis of these articles that mainly include four themes: (1) remote sensing monitoring of lake water quality; (2) remote sensing extraction and analysis of water area and water volume based on novel algorithms; and (3) remote sensing simulation and analysis of the watershed water environment.

2.1. Remote Sensing Monitoring of Lake Water Quality in Lakes and Reservoirs

Lakes (including reservoirs) have attracted more and more attention as the main drinking water source for more than 85% of the population in China. Remote sensing, as the only means to achieve large-scale, periodic, and operational monitoring, has played an important role in lake monitoring and research. Lake remote sensing, as a branch and cross-subject of lake science, remote sensing science, and other disciplines, enables researchers to learn from and promote each other. Zhang et al. [6] analyzed the response relationship between the water quality index and water reflectance, and used remote sensing technology to establish a water quality index monitoring model to monitor water quality in the Ebinur Lake watershed, producing a demonstration project for the use of remote sensing technology in lake monitoring in arid areas. Aranha et al. [7] used Sentinel-2 MSI TOA Level 1C reflectance images and analyzed the concentration of chlorophyll-a (Chl-a) in water bodies of five reservoirs located in the semi-arid region of northeastern Brazil. The model has a strong observation ability and high accuracy. Luo et al. [8] developed an online water quality assessment early warning system that integrates a high-frequency monitoring system (HFMS) and data quality control technology, which was applied in the Qiandao Lake region, China. The Early Warning System (EWS) focuses on data availability, quality control methods, statistical analysis methods, and data application, but not on the technical aspects of the detector, wireless data transmission, and interface software development. The development of this system provides a strong support for the automatic monitoring of lake water quality and three-dimensional lake hydrodynamic and ecosystem prediction. Together, these papers contribute to the development of the continuous monitoring of water quality in small and large reservoirs based on remote satellite-based analysis. Such an analysis is a strategic resource for promoting regional water security, and the future goal is to implement large-scale, intelligent remote sensing technology for the observation of water quality in lakes and rivers.

2.2. Remote Sensing Extraction and Analysis of Water Area and Water Volume Based on Novel Algorithms

The observation and monitoring of surface water area is of great significance for water resource management as well as ecological protection in a basin. Using remote sensing or hydrological model estimation methods can quickly obtain long time series of a water area, make up for the lack of data in a scarce-data area, and provide a basis for further research on surface water.

Li et al. [9] based their work on the GEE (Google Earth Engine) cloud platform and studied the effect of nine kinds of water indexes on the surface water extraction in Bosten Lake Basin by adding a slope mask to remove misclassified pixels to find the best extraction method for surface water extraction in the basin by means of accuracy verification and visual discrimination through a continuous iteration of the index threshold and slope mask threshold. The results show that when the threshold value is −0.25 and the slope mask is 8 degrees, the index WI2019 has the best effect on the surface water information extraction of Bosten Lake Basin, effectively eliminating the interference of shadow and snow. The extraction accuracy of surface water by remote sensing is improved, and provides a more accurate and convenient method for the extraction of surface water area under complex terrain. Chen et al. [10] adopted a spatial downscaling model for mapping lake water at a 10 m resolution by integrating Sentinel-2 and Landsat data, which was applied to map the water extent of Qinghai Lake from 1991 to 2020. This was further combined with the Hydroweb water-level dataset to establish an area-level relationship to acquire the 30-year data on water levels and water volumes. Then, the driving factors of the water dynamics were analyzed based on the grey system theory. The results were of great significance for local sustainable development and ecological protection. Zhang et al. [11] used DYRESM to estimate the water volume entering Waihai, part of Lake Dianchi, from 2007 to 2019 without historical hydrological observation data. Then, they combined this information with the monthly monitoring data of water quality to calculate the annual external loading. This method effectively solves the problem of the limited accuracy in the statistical results of lake water volume and external load estimation caused by a lack of data. Salama et al. [12] used remote sensing techniques and a geographic information system to analyze different satellite images, including multi-looking Sentinel-2, Landsat-9, and Sentinel-1 (SAR), to monitor the changes in the volume of water from 21 July 2020 to 28 August 2022. The volume of Nile water during and after the first, second, and third filling was estimated for the Grand Ethiopian Renaissance Dam's (GERD) reservoir lake, with comparisons for future hazards and environmental impacts. There are great challenges in the extraction of fine water based on remote sensing images. Future research will focus on developing water extraction algorithms suitable for multiple complex scenes (including highly heterogeneous urban scenes, cloudy and foggy scenes, and high-altitude mountain scenes), developing artificial intelligence algorithms with high accuracy, and developing fully automated extraction algorithms.

2.3. Remote Sensing Simulation and Analysis of Watershed Water Environment

The combination of the space in which people live in and the water body that can directly or indirectly affect human life and development is called the water environment. This water environment is applied to all kinds of natural factors and related social factors. Increasingly, the global watershed environment is facing more and more destruction, the inherent allocation mechanism of various elements of the natural environment is being maladjusted, and the environmental quality is deteriorating. Therefore, the observation and simulation methods and technologies of the watershed water environment need to be improved urgently, and remote sensing technology fills this gap fittingly. Tripp et al. [13] demonstrated their ability to monitor spatial and temporal changes in the playa water inundation area on sub-monthly time scales in West Texas, USA, using 10 m spatial resolution imagery from the Sentinel-2A/B satellites. The study developed a faster and more accurate method to cover a relatively small area compared with traditional monitoring methods. The methods provide a strong support for identification of small playas and ecological applications. Tang et al. [14] investigated the dynamics of the mid-channel bars (MCBs) in the Three Gorges Reservoir (TGR) using the Gravity Center Shifting Model. The number and area of MCBs changed dramatically with water-level changes, and the changes were dominated by MCBs. The study helped to reveal the mechanisms for the development of MCBs in the TGR. It also offers a scientific basis for the planning, optimal utilization, and ecological restoration of the MCBs in the TGR. Li et al. [15] used the Soil and Water Assessment Tool (SWAT) model in combination with the GCM model to address the separate and combined impacts of changes in climate and land use/land cover on the hydrological processes and sediment yield in the Xin'anjiang Reservoir Basin (XRB). The SWAT model simulation shows that climate change will exert a much larger influence on the sediment yield than land use/land cover (LULC) alteration in the XRB. These studies provide a deeper understanding of the sediment response to climate-driven forces and LULC changes in the XRB, which is beneficial for water quality protection and bloom prevention in the reservoirs in the East Asian monsoonal region. The watershed water environment is the main link to human activities in the basin. Large-scale, real-time remote sensing monitoring and simulation is the theme that needs undivided attention in the future.

3. Conclusions

The 11 papers summarized above contribute to the increasing interest in the study of monitoring lake water based on multisource remote sensing and novel modeling techniques. The Guest Editors hope that readers will be inspired by this Special Issue and will continue to study and innovate in the field of remote sensing observation of lake water color. In particular, the era of "big data" and "artificial intelligence (AI)" has arrived, which will usher in new development opportunities for the remote sensing observation of lake water color. In the future, with remote sensing and AI algorithms as the core methods, this field and others will focus on the miniaturization of spectral sensing devices, ease of use, lake water quality monitoring, and watershed water environment regulation. The dynamic monitoring of the integration of "heaven, earth, air and water" monitoring is a realistic requirement to promote the construction of an ecological civilization. Using space-based, ground-based, buoy-based, hand-held, and other methods to conduct all-weather and multidimensional monitoring and data analysis of water environments with a large spatial scope and long-time span and the ability to upload data analyses and results to online monitoring platforms through 4G/5G networks is anticipated to become the focal research topic in the future.

Author Contributions: Conceptualization, F.Z. and X.W.; formal analysis, N.W.C. and X.L.; writing—original draft preparation, X.W.; writing—review and editing, F.Z. and N.W.C.; funding acquisition, F.Z. and X.W. All authors have read and agreed to the published version of the manuscript.

Funding: This research was funded by the National Natural Science Foundation of China (42261006) and Fundamental Research Funds for Central Universities (2452022125).

Acknowledgments: The authors of this paper, who served as the Guest Editors of this Special Issue, wish to thank the journal editors, all contributing authors to the Special Issue, and all the referees who contributed to the revision and improvement of the 11 published papers.

Conflicts of Interest: The authors declare no conflict of interest.

References

1. Adrian, R.; O'Reilly, C.M.; Zagarese, H.; Baines, S.B.; Hessen, D.O.; Keller, W.; Livingstone, D.M.; Sommaruga, R.; Straile, D.; Van Donk, E.; et al. Lakes as sentinels of climate change. *Limnol. Oceanogr.* **2009**, *54*, 2283–2297. [CrossRef] [PubMed]
2. Woolway, R.I.; Kraemer, B.M.; Lenters, J.D.; Merchant, C.J.; O'Reilly, C.M.; Sharma, S. Global lake responses to climate change. *Nat. Rev. Earth Environ.* **2020**, *1*, 388–403. [CrossRef]
3. Yu, H.; Tu, Z.; Yu, G.; Xu, L.; Wang, H.; Yang, Y. Shrinkage and protection of inland lakes on the regional scale: A case study of Hubei Province, China. *Reg. Environ. Chang.* **2020**, *20*, 4. [CrossRef]
4. Sheffield, J.; Wood, E.F.; Pan, M.; Beck, H.; Coccia, G.; Serrat-Capdevila, A.; Verbist, K. Satellite Remote Sensing for Water Resources Management: Potential for Supporting Sustainable Development in Data-Poor Regions. *Water Resour. Res.* **2018**, *54*, 9724–9758. [CrossRef]
5. Huang, C.; Chen, Y.; Zhang, S.; Wu, J. Detecting, Extracting, and Monitoring Surface Water from Space Using Optical Sensors: A Review. *Rev. Geophys.* **2018**, *56*, 333–360. [CrossRef]
6. Zhang, F.; Chan, N.W.; Liu, C.; Wang, X.; Shi, J.; Kung, H.-T.; Li, X.; Guo, T.; Wang, W.; Cao, N. Water Quality Index (WQI) as a Potential Proxy for Remote Sensing Evaluation of Water Quality in Arid Areas. *Water* **2021**, *13*, 3250. [CrossRef]
7. Aranha, T.R.B.T.; Martinez, J.-M.; Souza, E.P.; Barros, M.U.G.; Martins, E.S.P.R. Remote Analysis of the Chlorophyll-a Concentration Using Sentinel-2 MSI Images in a Semiarid Environment in Northeastern Brazil. *Water* **2022**, *14*, 451. [CrossRef]
8. Luo, L.; Lan, J.; Wang, Y.; Li, H.; Wu, Z.; McBridge, C.; Zhou, H.; Liu, F.; Zhang, R.; Gong, F.; et al. A Novel Early Warning System (EWS) for Water Quality, Integrating a High-Frequency Monitoring Database with Efficient Data Quality Control Technology at a Large and Deep Lake (Lake Qiandao), China. *Water* **2022**, *14*, 602. [CrossRef]
9. Li, X.; Zhang, F.; Chan, N.W.; Shi, J.; Liu, C.; Chen, D. High Precision Extraction of Surface Water from Complex Terrain in Bosten Lake Basin Based on Water Index and Slope Mask Data. *Water* **2022**, *14*, 2809. [CrossRef]
10. Chen, Q.; Liu, W.; Huang, C. Long-Term 10 m Resolution Water Dynamics of Qinghai Lake and the Driving Factors. *Water* **2022**, *14*, 671. [CrossRef]
11. Zhang, Z.-M.; Zhang, F.; Du, J.-L.; Chen, D.-C. Surface Water Quality Assessment and Contamination Source Identification Using Multivariate Statistical Techniques: A Case Study of the Nanxi River in the Taihu Watershed, China. *Water* **2022**, *14*, 778. [CrossRef]
12. Salama, A.; ElGabry, M.; El-Qady, G.; Moussa, H.H. Evaluation of Grand Ethiopian Renaissance Dam Lake Using Remote Sensing Data and GIS. *Water* **2022**, *14*, 3033. [CrossRef]
13. Tripp, H.L.; Crosman, E.T.; Johnson, J.B.; Rogers, W.J.; Howell, N.L. The Feasibility of Monitoring Great Plains Playa Inundation with the Sentinel 2A/B Satellites for Ecological and Hydrological Applications. *Water* **2022**, *14*, 2314. [CrossRef]
14. Tang, Q.; Tan, D.; Ji, Y.; Yan, L.; Zeng, S.; Chen, Q.; Wu, S.; Chen, J. Dynamics of Mid-Channel Bar during Different Impoundment Periods of the Three Gorges Reservoir Area in China. *Water* **2021**, *13*, 3427. [CrossRef]
15. Li, H.; Yu, C.; Qin, B.; Li, Y.; Jin, J.; Luo, L.; Wu, Z.; Shi, K.; Zhu, G. Modeling the Effects of Climate Change and Land Use/Land Cover Change on Sediment Yield in a Large Reservoir Basin in the East Asian Monsoonal Region. *Water* **2022**, *14*, 2346. [CrossRef]
16. Zhang, R.; Luo, L.; Pan, M.; He, F.; Luo, C.; Meng, D.; Li, H.; Li, J.; Gong, F.; Wu, G.; et al. Estimations of Water Volume and External Loading Based on DYRESM Hydrodynamic Model at Lake Dianchi. *Water* **2022**, *14*, 2832. [CrossRef]

Article

High Precision Extraction of Surface Water from Complex Terrain in Bosten Lake Basin Based on Water Index and Slope Mask Data

Xingyou Li [1], Fei Zhang [1,2,*], Ngai Weng Chan [3], Jinchao Shi [4], Changjiang Liu [1] and Daosheng Chen [1]

1   College of Geography and Remote Sensing Sciences, Xinjiang University, Urumqi 830046, China
2   Key Laboratory of Oasis Ecology, Urumqi 830046, China
3   GeoInformatic Unit, Geography Section, School of Humanities, Universiti Sains Malaysia, 11800 USM, Penang 10790, Malaysia
4   Departments of Earth Sciences, The University of Memphis, Memphis, TN 38152, USA
*   Correspondence: zhangfei3s@163.com or zhangfei3s@xju.edu.cn

Abstract: The surface water extraction algorithm based on satellite remote sensing images is advantageous as it is able to obtain surface water information in a relatively short time. However, when it is used to extract information on surface water in large-scale, long-time series and complex terrain areas, there will be a large number of misclassified pixels, and a large amount of image preprocessing work is required. The accuracy verification is time-consuming and laborious, and the results may not be accurate. The complex climatic and topographic conditions in Bosten Lake Basin make it more difficult to monitor and control surface water bodies. Therefore, based on the GEE (Google Earth Engine) cloud platform, and the studies of the effect of nine kinds of water indexes on the surface water extraction in Bosten Lake Basin, this paper adds a slope mask to remove misclassified pixels and finds the best extraction method of surface water extraction in the basin by means of accuracy verification and visual discrimination through continuous iteration of index threshold and slope mask threshold. The results show that when the threshold value is −0.25 and the slope mask is 8 degrees, the index WI2019 has the best effect on the surface water information extraction of Bosten Lake Basin, effectively eliminating the interference of shadow and snow. The effect of water extraction in the long-time series is discussed and it was found that the precision of water extraction in the long-time series is also better than other indexes. The effects of various indexes on surface water extraction under complex terrain are compared. It can quickly and accurately realize the long-time series of surface water extraction under large-area complex terrain and provides useful guiding significance for water resources management and allocation as well as a water resources ecological assessment of Bosten Lake Basin.

Keywords: water extraction; water index; optimal threshold; Google Earth engine; slope mask

Citation: Li, X.; Zhang, F.; Chan, N.W.; Shi, J.; Liu, C.; Chen, D. High Precision Extraction of Surface Water from Complex Terrain in Bosten Lake Basin Based on Water Index and Slope Mask Data. *Water* **2022**, *14*, 2809. https://doi.org/10.3390/w14182809

Academic Editor: Frédéric Frappart

Received: 26 July 2022
Accepted: 6 September 2022
Published: 9 September 2022

Publisher's Note: MDPI stays neutral with regard to jurisdictional claims in published maps and institutional affiliations.

1. Introduction

There is no life without water. Water plays a vital role in the survival of human beings and other creatures as well as the rise of civilization and development of human society. Surface water generally includes rivers, lakes, glaciers, and swamps. It is the main component of freshwater resources on earth, which is irreplaceable in maintaining the ecological balance of river basins as well as meeting human demands, including power, water supply, irrigation, industrial needs and others. Interestingly, changes in surface water area can reflect and characterize the impact of climate change and human activities on surface water. Quickly and accurately extracting surface water information and grasping the spatial distribution of surface water have important practical significance for flood and drought disaster research, water resources monitoring research, water resources management research, etc. [1,2]. The surface water in arid and semi-arid environments is threatened by both natural and anthropogenic pressures. Mapping the distribution of surface water bodies is essential for managing and addressing degradation of both water quantity and quality [3]. Before the application of remote sensing technology in mapping water resources, most of the water was extracted based on manual measurement. This manual method has low precision, a massive heavy workload, a large cost and poor macro and continuous and real-time monitoring effect, all of which are impediments in meeting the requirements spatially and temporally. On the other hand, remote sensing technology has the advantages of macro scale data collection, dynamic monitoring and low cost. The use of remote sensing images to extract water information can accurately grasp the spatial distribution status and changing trend of water bodies in basins, and provide basic data for comprehensive management of basins, flood monitoring and water resources protection and conservation, all of which are of great importance [4–6].

At present, there are many algorithms for extracting water information from remote sensing images. In general, these methods can be roughly divided into several categories: single band method [7,8], spectral relationship method [4,9], image classification method [10] and water index method. Among them, the water index method is a popular index widely used by researchers. The most influential water index algorithms mainly include Normalized difference water index (NDWI), which weakens the influence of non-watery factors such as vegetation and soil. It is generally effective in extracting water from large lakes and reservoirs, but it still contains a lot of interference information in urban water extraction [11]. The modified normalized difference water index (MNDWI) is proposed on the basis of NDWI method, using Landsat TM short wave infrared (TM5) instead of near-infrared (TM4). MNDWI can weaken the impact of soil and buildings, but has a good effect on the removal of building shadows in urban areas [12]. The water index WI2006 uses the natural pairs of each band of landsat7 ETM+ images to reflect the reflection coefficient and interaction conditions and is used to extract wetlands covering eastern Australia [13]. The enhanced water index (EWI) is constructed by using the green light band (TM2), a near-infrared band (TM4) and mid-infrared band (TM5) of TM images, and this method is used to extract the water system information of semi-arid areas. This index allows the researcher to ignore the influence of atmospheric factors [14].

By analyzing the creation process of enhanced water index EWI, it is verified that the surface water can be extracted well whether the remote sensing image has been atmospherically corrected or not [15]. The modified normalized difference water index RNDWI (revised normalized difference water index) is constructed on the basis of analyzing the spectral characteristics of three ground feature types, viz. water, vegetation and soil. It can eliminate the influence of mountain shadows and accurately extract the water and land boundaries of Miyun Reservoir by using this index [16]. The new water index NWI (new water index) is proposed in combination with the strong absorption of water in the near-infrared and mid-infrared bands. NWI can partially eliminate the impact of solar altitude angle, terrain, shadow and atmospheric conditions, and its accuracy is very high [17]. The new water index NEW is a band ratio algorithm constructed by using the blue-green band (TM1) and mid-infrared band (TM7) of tm/etm+ images. This index can not only extract natural water but also eliminate the impact of terrain differences, thus solving the problem of shadow in water information [18].

In recent years, more new water indexes have been created with good verification results. For example, the automatic water extraction index AWEI is proposed based on TM image data. The main goal of AWEI is to separate water and nonwater pixels to the greatest extent by subtracting and adding between bands and assigning different coefficients to bands. It has been verified that AWEI has higher accuracy than MNDWI in extracting water information [19]. Another new index, Water index WI2015, is a water extraction algorithm based on linear discriminant analysis proposed on the basis of WI2006. The index uses linear discriminate anti-analysis classification (LDAC) to determine the coefficient of the best classification of the training area, which improves the classification accuracy [20]. The multi-band water index MBWI (multi-band water index) can weaken the impact of mountain shadows and dark pixels of buildings, and reduce the seasonal impact caused by changes in solar conditions [21]. Finally, the water index wi2019 (water index2019) is constructed on the basis of the analysis of the light break characteristics of water and snow, which improves the differentiation between water and snow in the classification process.

Among the numerous water remote sensing information extraction technologies, the one based on water remote sensing index is undoubtedly the most widely used. At present, the global and regional surface water distribution mapping is almost inseparable from the water remote sensing index. Some scholars carried out early high-resolution remote sensing mapping of global land surface water bodies [22], in which NDWI (normalized difference water index) and MNDWI (modified NDWI) water index were used as the main technologies. Taking MNDWI as a key algorithm, the global river distribution range and area can be calculated [23]. With the continuous construction and improvement of water index, many water indexes have been developed and often used to compare the effects of surface water extraction in the process of extracting surface water in different regions and are now widely used for surface water extraction in inland water bodies, wetlands, delta areas, coastal areas, dry, arid and semi-arid and other complex terrains [23–29]. These studies have achieved good water extraction results in the study area. Different water indexes have different advantages in surface water information extraction. The construction of new water indexes is based on the spectral information difference of typical ground object sample points in study area. They often achieve high extraction accuracy and differentiation effect within a study area. When selecting other study areas for verification, the extraction effect tends to decline, with the threshold value of extracted surface water also changing greatly [20,30,31]. When the threshold value is too small, it cannot effectively eliminate the misclassification of pixels, whereas too large a domain value will cause the loss of surface water information. In the specific application process, the automatic optimal threshold selection method often cannot achieve the best water extraction effect [32]. Therefore, it is necessary to optimize the water index and find the optimal threshold when using the water index to explore the change in water area in the study area [33].

Various water indexes have different effects when they are used to distinguish between water bodies and nonwater bodies. Shadows, ice, snow and clouds are the main misclassification types of water bodies. Through preliminary research, water indexes $AWEI_{nsh}$ and WI2019 can more effectively remove the effects of shadows and dark surfaces on surface water differentiation in the study area than other water indexes, but the effect on distinguishing ice, snow and water bodies is poor. Although WI2019 can effectively distinguish ice, snow and water bodies, the effect of distinguishing shadows is not good. Snow and ice in the region are mainly distributed in mountainous areas with high altitudes and large slopes. The shadow is also caused by the slope due to the land's topography. In places with a large slope, it is difficult to retain water bodies. Some scholars have tried to apply topographic factors to the process of surface water extraction and achieved good results [9,34]. Therefore, using slope data as a mask can effectively eliminate snow and shadow areas that are mistakenly classified as water pixels. Using the GEE cloud platform to call remote sensing images in the database can avoid a lot of image and processing work. Recently, many scholars have used the GEE platform to extract large-scale, long-time series surface water bodies and achieved good extraction results [35–40]. During the iteration of water index threshold and slope, the effect of the water extraction image and accuracy verification results can be observed synchronously to drive the final results. Theoretically, the GEE platform can be used to explore the optimal method of water extraction in any image area, and this greatly increases the work efficiency. Furthermore, it can realize the comparison of regional long-time water extraction effects under the GEE platform. The purpose of this paper is: (1) To realize the calculation and display of the water index under the GEE platform, and to explore the applicability of various water indexes under complex terrain; (2) By iterating the water index threshold and slope mask threshold, the most suitable water index method and the best threshold for water extraction in Bosten Lake

Basin are determined; (3) To realize the water extraction in the long-time series within the watershed and discuss the stability of different indexes in the long-time series.

2. Materials and Methodology

2.1. Study Area

In this paper, the boundaries of seven county and city level administrative divisions are merged as the research boundary of Bosten Lake Basin (Figure 1). Bosten Lake Basin, also known as the Kaidu River-Kongque River Basin, is located in the inland arid area of Xinjiang, China. Its geographical coordinates are $85°20'{\sim}87°25'$ E, $41°10'{\sim}43°30'$ N. Although Bosten Lake basin is mainly composed of Kaidu River Basin and Kongque River Basin, it also includes Yanqi Basin and its surrounding mountainous areas, and most areas to the north of the lower reaches of the Tarim River. Bosten Lake Basin is adjacent to Tianshan Mountains in the north and Tarim Basin in the south. The terrain is high in the northwest and low in the southeast. The geomorphic division belongs to the Tianshan Mountains region, including three small areas of Tianshan Mountains, Youledus basin and Yanqi Basin. The entire basin has a total area of 7.7×10^4 km^2, accounting for 45.06% of the drainage area. The basin is not totally mountainous as the plain area is 4.26×10^4 km^2, accounting for 55.32% of the drainage area. The landform in the area is complex as Bosten Lake Basin is surrounded by mountains on three sides. The overall terrain is high in the north and south, West and low in the East. The geomorphic units in the basin can be divided into intermountain basin landform, canyon landform and alluvial proluvial basin landform, of which the large and small Yudus basins belong to intermountain basin landform. The reach from the source of Kaidu River to the river canyon in the north of Yanqi basin is canyon landform, and the terrain height is obviously graded. The Kaidu River enters the Yanqi Basin from the east of Dashankou. The terrain is relatively flat and open, showing the geomorphic characteristics of an alluvial proluvial basin. The Yanqi basin is a local faulted basin formed between the main vein of the eastern Tianshan Mountains and its branches. Bosten Lake is in the southeast of the Yanqi Basin. There are a large number of relatively small wetland lakes in the southwest and northwest of Bosten Lake. Bayinbuluk grassland also contains many wetland water bodies, and a large number of snow mountains are distributed in the region, with the terrain fluctuating greatly. There are many seasonal rivers and lakes supplied by snow melt under the snow mountain terrain. As the longest river in the region, the Kaidu River flows through the main cities, mountains, deserts, wetlands, and other landforms in the region. These complex landforms jointly increase the difficulty of surface water monitoring and regulation in Bosten Lake Basin.

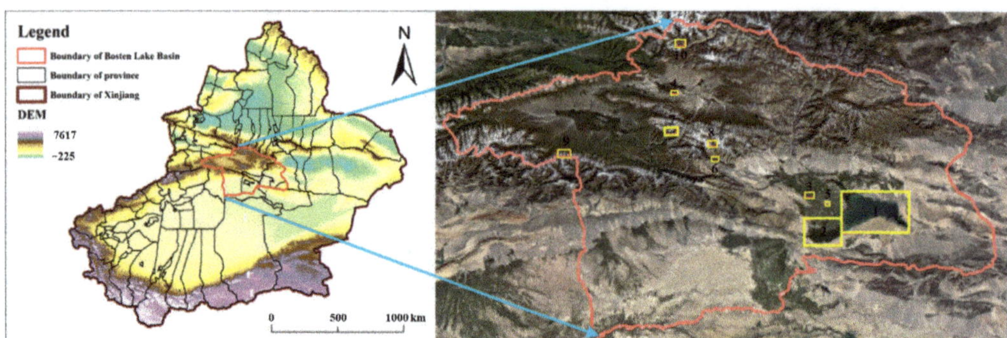

Figure 1. Schematic diagram of study area.

2.2. Data Resources

The data used in this paper mainly include Landsat8 OLI, Landsat7 ETM, Bosten Lake Basin vector boundary, and Google Earth high-resolution image and (JAXA/ALOS/AW3D30_V1_1) elevation data in GEE platform database.

The image used for the optimal water index threshold and slope and discussion selects the USGS Landsat 8 Collection 1 Tier 1TOA Reflection ("LANDSAT/LC08/C01/T1/TOA") data with cloud cover less than 8% from May to August 2021. In order to explore the surface water extraction effect of various indexes in the long-term academic column, a total of 108 images with less than 8% cloud cover in the USGS Landsat 8 Collection 1 Tier 1 TOA Reflection ("LANDSAT/LC08/C01/T1/TOA") from 2013 to 2021 in the Google Earth engine database are used, and 156 images with less than 8% cloud cover in the USGS Landsat 7 Collection 1 Tier 1 TOA Reflection ("LANDSAT/LE07/C01/T1/TOA") from 2000 to 2012 are used (Figure 2), and the elevation data call (JAXA/ALOS/AW3D30_V1_1) is used, including the latest 2021 Google Earth high-resolution image images provided by Google Earth Pro and Ovey Interactive Maps to compare the extraction effect.

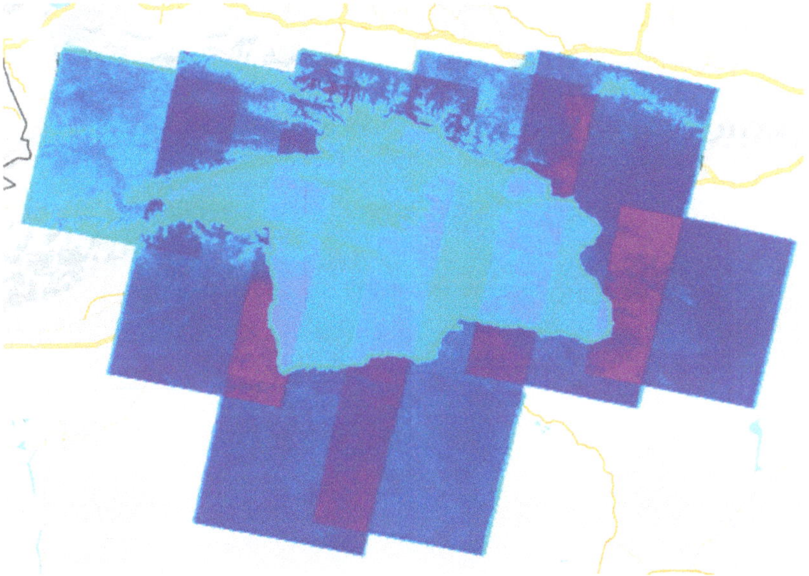

Figure 2. Image availability analysis in the study area.

2.3. Methodology

2.3.1. Remote Sensing Image

The ee.ImageCollection function calls the USGS Landsat 8 Collection1 Tier1 TOA Reflection dataset in the GEE database and sets the time interval from May to July 2021. The image BQA band is used for traffic screening and cloud removal. The image 'B6', 'B7', and 'B4' bands are set to the red, green and blue channels, respectively. Map.addlayer function performs false color synthetic display of images, highlighting the differences between water bodies and other ground objects, and serves as the basis for sample point selection and one of the bases for surface water extraction effect.

2.3.2. Selection of Sample Points for Accuracy Verification

Based on the GEE platform, the Configure geometry import tool was used to create six categories of ground objects. In combination with Google Earth satellite images and landsat8 false color composite images, 1283 water sample points (Figure 3) and 1538 nonwater sample points were selected in the study area, including 436 vegetation sample points, 228 building sample points, 189 wetland sample points, 218 bare land sample points and 220 snow sample points. Additionally, 247 shadow sample points are used as the basis for verification of water extraction accuracy and supervision and classification.

Figure 3. Selection of typical feature sample points in the study area.

In order to reflect the correctness of sample point selection, the sampleRegions function is used on the GEE platform to extract the reflectance values of all bands under each sample point of various ground objects, and the average value is obtained to prepare the spectral characteristic curve of typical ground objects in Bosten Lake Basin (Figure 4). It can be seen from the figure that the spectral reflectance characteristics of ground objects in the flow domain conform to the spectral characteristics of typical ground objects, which proves the correctness of sample point selection.

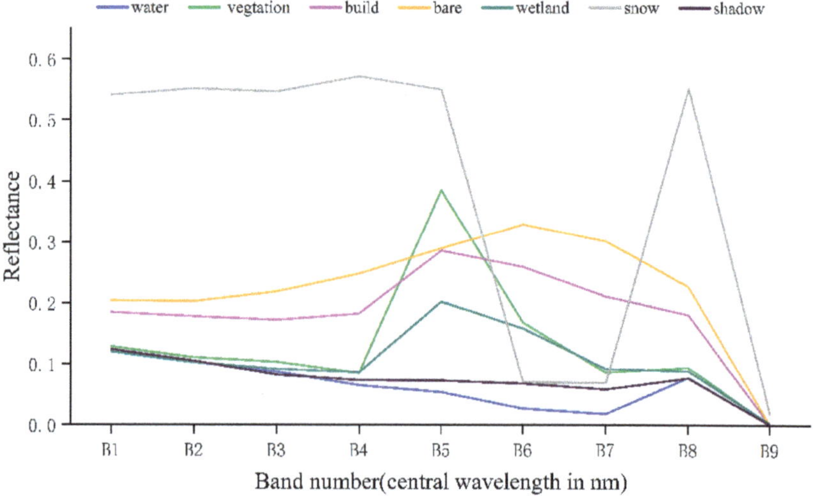

Figure 4. Spectral reflectance characteristics of typical ground objects in the study area.

2.3.3. Selection of Water Index and Realization of Index Calculation

By consulting the relevant literature on the construction of the water index, the water index suitable for Landsat series images is selected (Table 1). The normalized difference and expression functions are used to calculate the index band on the GEE platform, and Map.addLayer function shows the result of the exponential operation. The reclassification judgment formula is used to calculate the water index. A pixel greater than 0 in the grid image is assigned as 1 as the surface water pixel, whereas a pixel less than 0 is assigned as 0 as the nonsurface water pixel, so as to distinguish the surface water and nonsurface water with 0 as the threshold. The Map.addLayer function displays the exponential operation results by using Export.image. to drive function exports of the final water extraction grid image to Google cloud disk and downloads it locally. Arcmap10.8 is used to further display and analyze the results.

Table 1. Main water index calculation formula.

Index Name	Index Formula	Reference
NDWI	$\dfrac{\rho_{GREEN}-\rho_{NIR}}{\rho_{GREEN}+\rho_{NIR}}$	[11]
MNDWI	$\dfrac{\rho_{GREEN}-\rho_{SWIR1}}{\rho_{GREEN}+\rho_{SWIR1}}$	[12]
AWEI$_{nsh}$	$4 \times (\rho_{GREEN} - \rho_{SWIR1}) - (0.25 \times \rho_{NIR} + 2.75 \times \rho_{SWIR2})$	[19]
AWEI$_{sh}$	$\rho_{BLUE} + 2.5 \times \rho_{GREEN} - 1.5 \times (\rho_{NIR} + \rho_{SWIR1}) - 0.25 \times \rho_{SWIR2}$	[19]
EWI	$\dfrac{\rho_{BLUE}-\rho_{RED}-\rho_{NIR}}{\rho_{BLUE}+\rho_{RED}+\rho_{NIR}}$	[14]
ANWI	$\dfrac{\rho_{BLUE}+\rho_{GREEN}+\rho_{RED}-\rho_{NIR}-\rho_{SWIR1}-\rho_{SWIR2}}{\rho_{BLUE}+\rho_{GREEN}+\rho_{RED}+\rho_{NIR}+\rho_{SWIR1}+\rho_{SWIR2}}$	[41]
NWI	$\dfrac{\rho_{BLUE}-\rho_{NIR}-\rho_{SWIR1}-\rho_{SWIR2}}{\rho_{BLUE}+\rho_{NIR}+\rho_{SWIR1}+\rho_{SWIR2}}$	[17]
WI2015	$1.7204 + 171\rho_{GREEN} + 3\rho_{RED} - 70\rho_{NIR} - 45\rho_{SWIR1} - 71\rho_{SWIR2}$	[20]
WI2019	$\dfrac{1.75\rho_{GREEN}-\rho_{RED}-1.08\rho_{SWIR1}}{\rho_{GREEN}+\rho_{SWIR1}}$	[42]

Note: where ϱ Represents the band reflectance value, the band reflectance subscript corresponds to the corresponding band of different remote sensing images.

2.3.4. Use of Slope Mask and Determination of Optimum Threshold of Surface Water

In order to further eliminate the interference of snow and shadow and improve the accuracy of surface water extraction, slope factor judgment conditions are added to the index calculation results, and areas with excessive slope are divided into nonwater bodies. The constructed sample points are used as the verification basis. The Terrain.slope function converts (JAXA/ALOS/AW3D30_V1_1) DEM data into slope data, sets the data type of sample points to FeatureCollection, sets the Property of water extraction sample points to Class1 and the value to 1 and sets various nonwater sample points to FeatureCollection after fusion, with the Property set to class2 and the value set to 0. The Validation.filter function, which outputs the confusion matrix, overall accuracy, user accuracy, producer accuracy and kappa coefficient, are used as the basis for accuracy determination. The change of water extraction accuracy was observed by iterating the index threshold and slope mask. On the basis of threshold iteration, various indices are set at t (threshold) = 0, s (slope) = 0; t = 0, b = 10; t = b (best value), s = 0; water extraction is carried out in the four cases of t = b and s = b, and compared with Google Earth HD image to further identify the effect of water extraction.

2.3.5. Discussion on Extraction Methods of Other Water Bodies

In the process of extraction by the GEE window shows and satellite image layer, preliminary extraction results found that although the various indexes of regional water extraction effects are not the same, the water main area can be extracted by a large number of falsely divided bodies of water feature category, mainly for the mountain shadow and snow body city shadow, wetlands and other false points such as pixels compared to the previous two kinds or types. In order to discuss the classification details, LibSVM, SmileCart and MininumDistance classifiers were used for supervised classification using water sample points and non-water sample points on the GEE platform, and 10 typical water areas were selected in the study area for further visual discrimination of water extraction effect.

2.3.6. Validation of Water Extraction Accuracy in Long-Time Series

The third method of accuracy verification was to verify the stability of the water extraction effect in a long-time series by using the difference between water areas in dry and wet seasons. Therefore, to further explore the water extraction effect of various water indexes in the long-time series under the optimal water index and the optimal slope mask threshold, May–August and September–October were taken as the dry season and the wet season, respectively, in Bosten Lake Basin, and Landsat7 ETM and Landsat8 OLI were used to calculate the water area of various water indexes in the dry season and wet season from 2000 to 2021. The pixel value of surface water is set as 1 and that of non-surface water is set as 0. The pixel value of 1 multiplied by them is defined as the permanent surface water. Formula (1) is used to calculate the misdivided area of surface water in a long-time series. The surface water existing in both the dry season and wet season should be close to the one with the smaller area of the two. The reason why the permanent surface water is smaller is that the misdifferentiated pixels are removed during the superposition operation of surface water results in the dry and wet seasons. Therefore, the smaller the difference value is, the fewer misdifferentiated pixels are and the higher the surface water extraction accuracy is, which can be used as the basis for judging the surface water extraction effect.

$$E = \frac{\sum\limits_{i=1}^{n} \min(A_{ds}, A_{ws}) - A_p}{n} \tag{1}$$

Note: in the formula, E represents the annual average misclassified area, A_{ds} represents the extracted area of water in dry season, A_{ws} represents the extracted area of water in wet season, and A_p represents the permanent water area.

3. Results and Analysis

3.1. Extraction Effect of Surface Water with Water Index 0 as Threshold Value

It can be seen from Figures 5 and 6 that a large number of non-water pixels are misclassified into water pixels when various indices are used as the threshold of 0 for water extraction. According to the results, WI2019 and EWI have the best water extraction effect, whereas AWEI$_{sh}$, MNDWI and WI2015 have poor extraction effects. The water index extraction effect is compared in detail with Site1-lake, Site2-wetland, Site3-river, Site4-small surface water, Site5-city surface water, Site6-mountain surface water and the snow mountain surface water region corresponding to Figure 1 in the region. Results showed that the water index in lakes, rivers and the urban area obtained good results of water extraction, but in complex areas, WI2019 distinguishes better between snow and water effect. However, AWEI$_{nsh}$ is better than that of WI2019 for distinguishing shadow and water bodies, as most of the shadow is divided into surface water and the snow pixels, and the slope is the main cause of shadow, where a large amount of snow is also distributed on hills at higher altitudes. Therefore, a slope mask based on index WI2019 can further remove the misclassification pixels caused by mountain snow and mountain shadows.

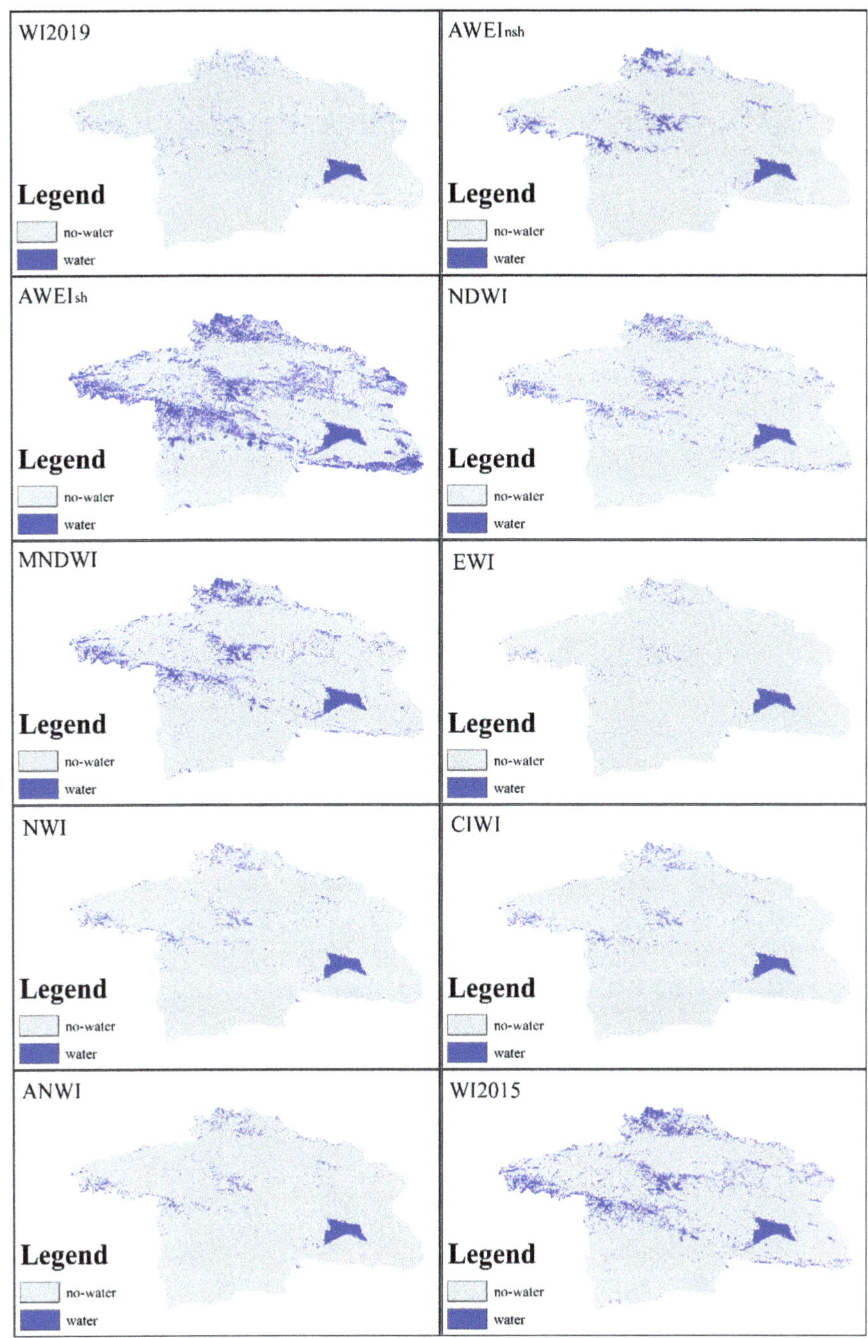

Figure 5. Each water index takes 0 as the threshold value to extract surface water results.

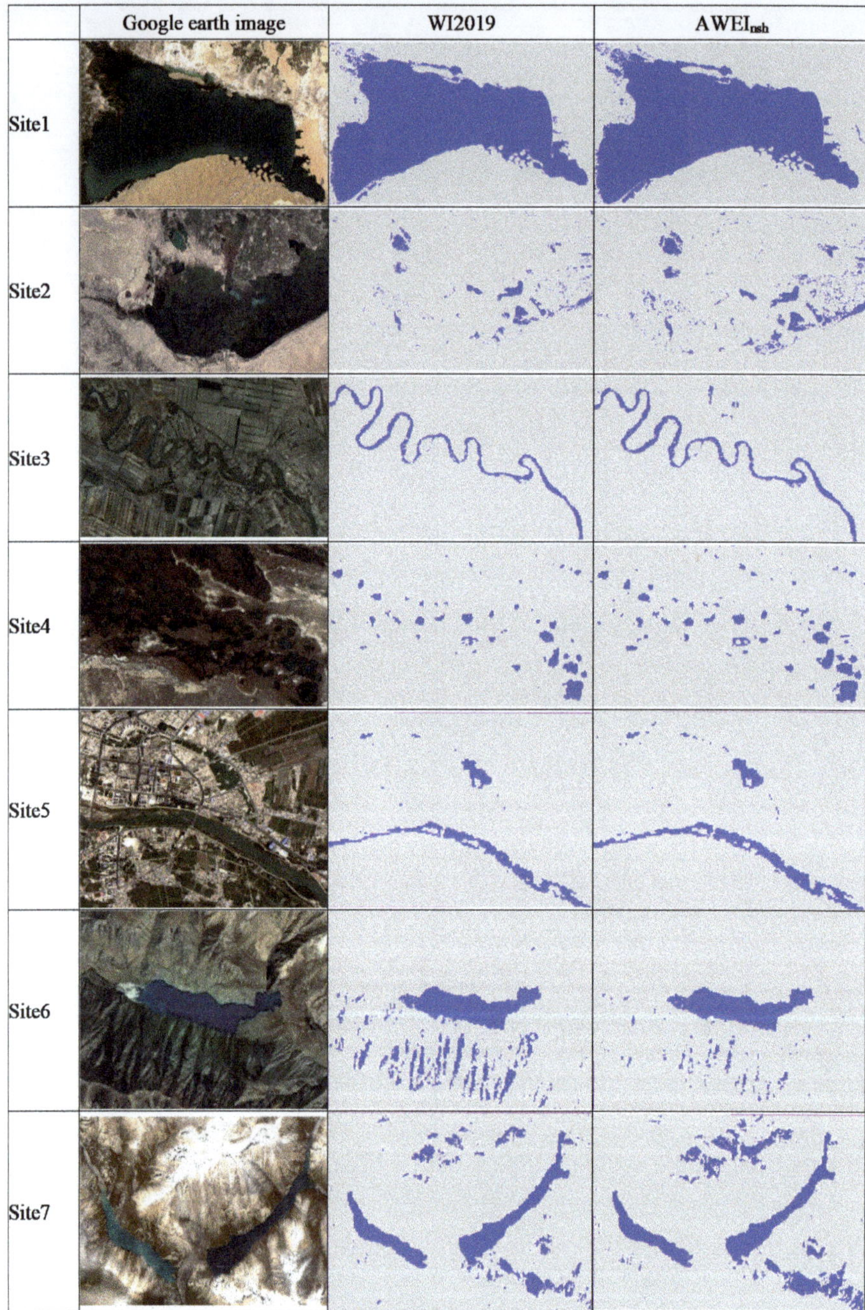

Figure 6. WI2019 and AWEI$_{nsh}$ surface water extraction details.

3.2. Changes of Extraction Accuracy under Water Index Threshold and Slope Iteration

The slope mask is added on the basis of the optimization of water index threshold. Through the continuous iteration of nine water index thresholds and slope thresholds, the accuracy of water extraction results by different methods is evaluated based on the overall accuracy, user accuracy, producer accuracy and kappa coefficient generated by the selected surface feature sample points in the study area, as shown in Table 2.

Table 2. Evaluation table of accuracy of water extraction results of different methods.

Classification Method	Threshold	Slope	Land Cover Class	Overall Accuracy	User Accuracy	Producer Accuracy	Kappa
WI2019	0.00	0.00	water	0.669	0.274	0.994	0.29
			nonwater		0.999	0.623	
	−0.25	0.00	water	0.679	0.299	0.984	0.31
			nonwater		0.996	0.630	
	0.00	10.00	water	0.908	0.807	0.989	0.81
			nonwater		0.992	0.861	
	−0.15	8.00	water	0.940	0.912	0.954	0.89
			nonwater		0.964	0.929	
AWEInsh	0.00	0.00	water	0.680	0.299	0.990	0.31
			nonwater		0.997	0.630	
	−0.1	0.00	water	0.682	0.306	0.987	0.319
			nonwater		0.997	0.632	
	0.00	10.00	water	0.901	0.944	0.853	0.80
			nonwater		0.865	0.949	
	−0.09	5.00	water	0.937	0.926	0.935	0.87
			nonwater		0.946	0.939	
AWEIsh	0.00	0.00	water	0.676	0.305	0.949	0.31
			nonwater		0.986	0.630	
	0.15	0.00	water	0.668	0.274	0.986	0.288
			nonwater		0.997	0.622	
	0.00	10.00	water	0.882	0.973	0.807	0.77
			nonwater		0.806	0.973	
	0.08	5.00	water	0.922	0.899	0.926	0.84
			nonwater		0.940	0.918	
MNDWI	0.00	0.00	water	0.681	0.307	0.973	0.32
			nonwater		0.993	0.632	
	0.15	0.00	water	0.682	0.305	0.989	0.32
			nonwater		0.997	0.632	
	0.00	10.00	water	0.903	0.976	0.837	0.81
			nonwater		0.841	0.977	
	0.00	5.00	water	0.931	0.935	0.915	0.86
			nonwater		0.928	0.945	
NDWI	0.00	0.00	water	0.675	0.289	0.987	0.30
			nonwater		0.997	0.627	
	0.15	0	water	0.674	0.300	0.946	0.30
			nonwater		0.986	0.268	
	0.00	10.00	water	0.907	0.880	0.913	0.81
			nonwater		0.930	0.903	
	0.03	9.00	water	0.916	0.859	0.951	0.83
			nonwater		0.963	0.891	
EWI	0.00	0.00	water	0.657	0.248	0.991	0.26
			nonwater		0.998	0.614	
	−0.15	0.00	water	0.669	0.275	0.989	0.290
			nonwater		0.997	0.623	
	0.00	10.00	water	0.845	0.694	0.954	0.68
			nonwater		0.972	0.792	
	−0.35	4.00	water	0.927	0.894	0.942	0.85
			nonwater		0.954	0.915	
ANWI	0.00	0.00	water	0.657	0.248	0.991	0.26
			nonwater		0.998	0.614	
	−0.15	0.00	water	0.667	0.272	0.989	0.287
			nonwater		0.997	0.622	
	0.00	10.00	water	0.839	0.670	0.964	0.67
			nonwater		0.979	0.781	
	0.1	6.00	water	0.930	0.901	0.943	0.86
			nonwater		0.954	0.920	

Table 2. *Cont.*

Classification Method	Threshold	Slope	Land Cover Class	Overall Accuracy	User Accuracy	Producer Accuracy	Kappa
NWI	0.00	0.00	water	0.657	0.248	0.991	0.26
			nonwater		0.998	0.614	
	−0.4	0.00	water	0.682	0.305	0.985	0.319
			nonwater		0.996	0.632	
	0.00	10.00	water	0.839	0.670	0.964	0.67
			nonwater		0.979	0.781	
	−0.40	4.00	water	0.930	0.901	0.943	0.86
			nonwater		0.954	0.920	
WI2015	0.00	0.00	water	0.682	0.306	0.980	0.32
			nonwater		0.995	0.632	
	0.05	0	water	0.681	0.306	0.980	0.319
			nonwater		0.995	0.632	
	0.00	10.00	water	0.904	0.973	0.840	0.81
			nonwater		0.973	0.975	
	0.00	5.00	water	0.932	0.932	0.920	0.86
			nonwater		0.932	0.943	
SmileCart				0.931			0.88
LibSVM				0.894			0.79
MinimumDistance				0.864			0.87

It can be seen from Table 2 that when the threshold value is 0 and slope is 0 for exponential water extraction, the overall accuracy is between 0.6–0.7 and the kappa coefficient is between 0.25–0.35. The user accuracy and producer accuracy of surface water and non-surface water are very different, with one being higher and the other being lower. With 0 as the threshold index calculation, results for the distinction between water and the water effect is poorer. However, with zero as the slope, and through iteration to find the best threshold value of various index, it was found that all kinds of indexes under the best threshold and under the extraction accuracy, in comparison with 0 as the threshold of the extraction, will only slightly improve accuracy when the threshold value is 0, and the slope for water extraction is 10. When each index of the extraction of the overall accuracy reached 0.8 or more, the user accuracy and producer accuracy exceeded 0.8, and when the gap is not big, the kappa coefficient reached 0.8 above, which can greatly improve the wetland information extraction effect and can effectively remove the water pixels. On the basis of the water index, iteration threshold and the slope, it was found that WI2019 and $AWEI_{nsh}$ achieved the highest accuracy with the best threshold and slope threshold. When the threshold of WI2019 was −0.15 and the slope mask threshold was 8, the overall accuracy reached 0.94 and the kappa coefficient reached 0.89. When the $AWEI_{nsh}$ threshold is −0.09 and the slope is 5, the overall accuracy reaches 0.937 and the kappa coefficient is 0.87. The three supervised classification methods also achieve high extraction accuracies. Therefore, complex terrain is further selected for visual discrimination of the water extraction effect.

3.3. Comparison of Water Extraction Effects under Complex Terrains

It can be seen from Figure 7 that although water pixels can be effectively extracted under complex terrain (site7–10), a large number of snow and mountain shadows are misclassified into water pixels. When water is extracted with the optimal slope threshold of 0, the misclassified pixel area is further increased. WI2019, $AWEI_{nsh}$ and other indexes show the same results. The reason is that although the overall accuracy and kappa coefficients have achieved great results using sample points for accuracy evaluation, there is a large difference between the user accuracy and producer accuracy of water and nonwater bodies, and the threshold value is small. Therefore, the classification results using this threshold value have produced many misclassification pixels. When WI2019 optimal threshold value and slope were used for water extraction, the best water extraction effect was achieved, and snow and mountain shadow misclassification pixels were excluded to the maximum extent, compared with the supervised classification results, and there are still a small number of snow and shadow pixels that are mistakenly classified as water bodies, and the effect is poor.

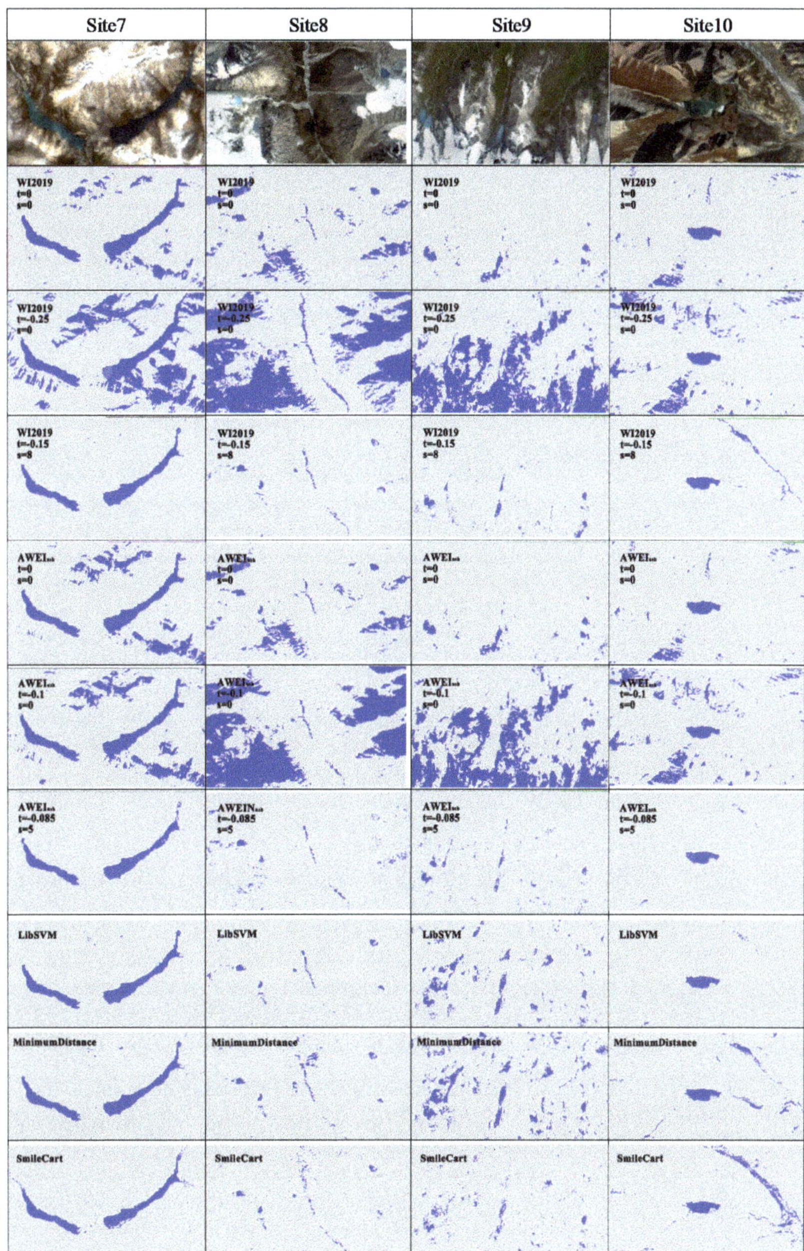

Figure 7. Comparison of extraction effects of different water extraction methods in more complex terrains.

3.4. Effect Analysis of Long Time Series Water Extraction

Landsat series remote sensing images of GEE platform from 2000 to 2021 were used to extract surface water in Bosten Lake Basin using various water indexes under the optimal index threshold and slope mask threshold, and the annual average water error area under different methods was respectively counted, as shown in Table 3.

Table 3. Annual average extraction error area of each water index from 2000 to 2021.

Water Index	WI2019	AWEI$_{nsh}$	AWEI$_{sh}$	MNDWI	NDWI	EWI	ANWI	NWI	WI2015
Error area (km^2)	140.5183	183.524	145.4784	391.2709	194.1482	266.785	177.5806	210.056	350.6671

The average annual extraction error of WI2019 under the optimal index threshold and slope mask threshold is 140.5183 km^2, which is still the optimal extraction effect. The error area of other water indices on long-term series is also small. For example, the average annual error area of AWEI$_{sh}$, ANWI and NDWI is only 145.478 km^2, 177.5806 km^2 and 194.1482 km^2, respectively, indicating that although its water extraction effect is not as good as WI2019, its extraction effect has good stability, which can be considered as a backup scheme for water extraction in other regions.

4. Discussion and Conclusions

4.1. Discussion

4.1.1. The Relationship between the Optimal Threshold of Water Index and Water Extraction Effect

When slope mask is not used to search for the optimal threshold of surface water, the optimal threshold obtained by using sample points to verify the accuracy of surface water extraction is often too small. In a visual interpretation, it is found that a large number of non-surface water pixels were misclassified into surface water pixels and that surface water extraction was carried out only with the optimal water index threshold. Good results can be achieved in areas with large continuous water areas or relatively flat terrain [3,43–45]. However, the results may be unreliable in large areas or areas with a complex geographical pattern. In the process of water extraction with water index, it is found that although the water index has its own advantages, the threshold value is the most critical factor to determine the effect of water extraction, whereas the structure of water index is secondary. Similar conclusions have been drawn in related studies on the optimal threshold value of water index [46,47].

4.1.2. The Optimal Threshold Value of Slope Mask Can Reflect the Effect of Water Index to Distinguish Shadows

The optimal slope mask of index wi2019 is 8, which is greater than the optimal slope mask threshold of water indexes such as AWEI$_{sh}$ and AWEI$_{nsh}$. In the actual process of surface water, the distinguishing effect of water indexes such as AWEI$_{sh}$ and AWEI$_{nsh}$ on shadow and surface water without slope mask is better than that of wi2019, indicating that the slope in the optimal threshold slope can reflect the shadow removal effect of water index. This is consistent with the comparison result that the AWEI index is better than other water indexes in the practical application of water index in relevant studies [48,49].

4.1.3. Commonality of Water Extraction Methods

This paper does not validate the water extraction method in other areas because the water extraction method proposed in this paper does not have a fixed index or threshold and can carry out the same workflow in different areas to find the optimal results. When judging the effect of water extraction in the long-time series, the water area in dry seasons in very few years is larger than that in wet seasons. According to the display effect of GEE in the extraction process, the reason is not only the extraction error but also the complex terrain and climate conditions in the region, which have little impact on the water extraction in the long-time series. The slope mask has achieved a good effect in distinguishing water bodies in areas with mountainous shadows and more snow, but it may not be applicable in areas with flat terrain. Although this method has achieved a good effect in extracting water bodies, there are still pixels incorrectly divided into water bodies, and the image resolution has a great impact on the extraction of small water bodies. Therefore, it is necessary to explore the applicability of this method in high-resolution images in future work and increase the comparison of more water extraction methods. The results of this study with a slope of 8 as a threshold are only applicable to specific RS imagery (Landsat series for this study) restricted by specific transit times and specific scanning angles. For example, when the hill slope and aspect are stable, the shadow area at noon must be smaller than that in the morning and evening. Therefore, there may be errors when using this method to extract water area from different remote sensing images. In addition, there is a certain difference in the spectral characteristics between the water pixels under shadow and those under light, and the spectral difference between them is smaller than that between the two. Although the difference between different bands enables the shadow water pixels to be extracted to a certain extent by the exponential method, water pixels, however, cannot be extracted in areas where the reflectance is too low or the water is completely shaded. Although there are few such water pixels, the loss of water pixels will still occur.

Although this method aims to get the higher precision of water extraction, and implements the long time series of water extraction, the prediction about the future of the water area of change need comprehensive consideration of many driving factors. Hence, under the premise of remote sensing image in the future, the lack of basic data is difficult to achieve, but the set of methods can be used in the future the extraction of remote sensing data for water. Previous water extraction results can provide historical data for water resource management and decision making in the region.

4.2. Conclusions

In this study, on the premise of preliminary discussion on various water indexes, the water index is used to extract the surface water of Bosten Lake Basin under the GEE platform. The surface water extraction effects of different water indexes are compared in three ways: sample accuracy verification, visual discrimination, and misclassification of area under long-time series. A method for surface water extraction in complex terrain areas is proposed by adding a slope mask. The interference of slope and snow cover is effectively removed, and the high-precision extraction of surface water in the region is realized through threshold iterative optimization. This method can be applied to the extraction of long-time series water in the region.

The index threshold has been optimized in previous studies on water information extraction using the water index. In this paper, when trying to extract water in the Bosten Lake basin with 0 as the threshold, various water indexes have achieved good water extraction results in lakes, wetlands, cities, and rivers, but the extraction effect in the whole region is generally poor. Therefore, taking 0 as the threshold for water extraction can reflect the advantages and disadvantages of the water extraction effect to a certain extent, but the water extraction result is not reliable, and it is still necessary to improve the water extraction effect through threshold optimization. The accuracy of water extraction has been significantly improved after slope masking of various water indexes. In Bosten Lake Basin, the ground feature types that affect the water extraction effect are mainly shadow and snow. The index wi2019 has a better distinguishing effect on water and snow than other water indexes. Adding the slope mask can further remove the interference of shadow and mountain snow on water extraction. The water index WI2019 takes -0.15 as the threshold and the slope of 8 as the mask, achieving the highest water extraction accuracy and the best visual discrimination effect in the study area, and it is better than the water extraction results of supervised classification. The error area of water extraction in the long-time series is smaller than in other indexes. It can achieve high-precision water extraction in the region under the condition of large topographic relief and more snow. It is of certain significance for monitoring and managing the dynamic changes of surface water in the region.

Author Contributions: Conceptualization, X.L. and F.Z.; methodology, X.L.; software, X.L.; validation, C.L., D.C. and X.L.; formal analysis, X.L.; investigation, F.Z. and C.L.; resources, X.L.; data curation, X.L.; writing—original draft preparation, X.L.; writing—review and editing, N.W.C. and J.S.; visualization, X.L., F.Z. and D.C.; supervision, F.Z.; project administration, F.Z.; funding acquisition, F.Z. All authors have read and agreed to the published version of the manuscript.

Funding: This research was funded by the National Natural Science Foundation of China (U2003205, U1603241), the National Natural Science Foundation of China (Xinjiang Local Outstanding Young Talent Cultivation) (U1503302) and the Tianshan Talent Project (Phase III) of the Xinjiang Uygur Autonomous.

Data Availability Statement: The data presented in this study are available on request from the corresponding author. The data are not publicly available due to confidentiality.

Conflicts of Interest: The authors declare no conflict of interest.

References

1. Papa, F.; Crétaux, J.-F.; Grippa, M.; Robert, E.; Trigg, M.; Tshimanga, R.M.; Kitambo, B.; Paris, A.; Carr, A.; Fleischmann, A.S. Water resources in Africa under global change: Monitoring surface waters from space. *Surv. Geophys.* **2022**, *20*, 1–51.
2. Garrido-Rubio, J.; Calera, A.; Arellano, I.; Belmonte, M.; Fraile, L.; Ortega, T.; Bravo, R.; González-Piqueras, J. Evaluation of remote sensing-based irrigation water accounting at river basin district management scale. *Remote Sens.* **2020**, *12*, 3187. [CrossRef]
3. Malahlela, O.E. Inland waterbody mapping: Towards improving discrimination and extraction of inland surface water features. *Int. J. Remote Sens.* **2016**, *37*, 4574–4589. [CrossRef]
4. Aznarez, C.; Jimeno-Sáez, P.; López-Ballesteros, A.; Pacheco, J.P.; Senent-Aparicio, J. Analysing the impact of climate change on hydrological ecosystem services in Laguna del Sauce (Uruguay) using the SWAT model and remote sensing data. *Remote Sens.* **2021**, *13*, 2014. [CrossRef]
5. Bretreger, D.; Yeo, I.-Y.; Kuczera, G.; Hancock, G. Remote sensing's role in improving transboundary water regulation and compliance: The Murray-Darling Basin, Australia. *J. Hydrol. X* **2021**, *13*, 100112. [CrossRef]
6. Domeneghetti, A.; Schumann, G.J.-P.; Tarpanelli, A. Preface: Remote sensing for flood mapping and monitoring of flood dynamics. *Remote Sens.* **2019**, *11*, 943. [CrossRef]
7. Lu, Z.; Wang, D.Q.; Deng, Z.D.; Shi, Y.; Ding, Z.B.; Ning, H.; Zhao, H.F.; Zhao, J.Z.; Xu, H.L.; Zhao, X.N. Application of red edge band in remote sensing extraction of surface water body: A case study based on GF-6 WFV data in arid area. *Hydrol. Res.* **2021**, *52*, 1526–1541. [CrossRef]
8. Yang, Y.; Ruan, R.Z. Extraction of Plain Lake Water Body Based on TM Imagery. *Remote Sens. Inf.* **2010**, *3*, 60–64.
9. Zhang, Q.; Wu, B.; Yang, Y.K. Extraction of Open Water in Rugged Area with A Novel Slope Adjusted Water Index. *Remote Sens. Inf.* **2018**, *33*, 98–107.
10. Vinayaraj, P.; Imamoglu, N.; Nakamura, R.; Oda, A. Investigation on perceptron learning for water region estimation using large-scale multispectral images. *Sensors* **2018**, *18*, 4333. [CrossRef] [PubMed]
11. McFeeters, S.K. The use of the Normalized Difference Water Index (NDWI) in the delineation of open water features. *Int. J. Remote Sens.* **1996**, *17*, 1425–1432. [CrossRef]
12. Xu, H.Q. A Study on Information Extraction of Water Body with the Modified Normalized Difference Water Index (MNDWI). *J. Remote Sens.* **2005**, *9*, 589–595.
13. Danaher, T.; Collett, L. Development, optimisation and multi-temporal application of a simple Landsat based water index. In Proceedings of the 13th Australasian Remote Sensing and Photogrammetry Conference, Canberra, ACT, Australia, 20–24 November 2006.
14. Yan, P.; Zhang, Y.; Zhang, Y. A study on information extraction of water enhanced water index (EWI) and GIS system in semi-arid regions with the based noise remove techniques. *Remote Sens. Inf.* **2007**, *6*, 62–67.
15. Xu, H.Q. Comment on the Enhanced Water Index (EWI): A Discussion on the Creation of a Water Index. *Geo-Inf. Sci.* **2008**, *10*, 776–780.
16. Cao, R.L.; Li, C.J.; Liu, L.Y.; Wang, J.H.; Yan, G.J. Extracting Miyun reservoirs water area and monitoring its change based on a revised normalized different water index. *Sci. Surv. Mapp.* **2008**, *33*, 158–160.
17. Ding, F. Study on information extraction of water body with a new water index (NWI). *Sci. Surv. Mapp.* **2009**, *34*, 155–157.
18. Xiao, Y.-F.; Zhu, L. A study on information extraction of water body using band 1 and band 7 of TM imagery. *Sci. Surv. Mapp.* **2010**, *35*, 226–227.
19. Feyisa, G.L.; Meilby, H.; Fensholt, R.; Proud, S.R. Automated Water Extraction Index: A new technique for surface water mapping using Landsat imagery. *Remote Sens. Environ.* **2014**, *140*, 23–35. [CrossRef]
20. Fisher, A.; Flood, N.; Danaher, T. Comparing Landsat water index methods for automated water classification in eastern Australia. *Remote Sens. Environ.* **2016**, *175*, 167–182. [CrossRef]
21. Wang, X.B.; Xie, S.P.; Zhang, X.L.; Chen, C.; Guo, H.; Du, J.K.; Duan, Z. A robust Multi-Band Water Index (MBWI) for automated extraction of surface water from Landsat 8 OLI imagery. *Int. J. Appl. Earth Obs. Geoinf.* **2018**, *68*, 73–91. [CrossRef]
22. Yamazaki, D.; Trigg, M.A.; Ikeshima, D. Development of a global similar to 90 m water body map using multi-temporal Landsat images. *Remote Sens. Environ.* **2015**, *171*, 337–351. [CrossRef]
23. Allen, G.H.; Pavelsky, T.M. Global extent of rivers and streams. *Science* **2018**, *361*, 585–588. [CrossRef] [PubMed]
24. Acharya, T.D.; Subedi, A.; Huang, H.; Lee, D.H. Application of water indices in surface water change detection using Landsat imagery in Nepal. *Sens. Mater.* **2019**, *31*, 1429–1447. [CrossRef]

25. Colditz, R.R.; Troche Souza, C.; Vazquez, B.; Wickel, A.J.; Ressl, R. Analysis of optimal thresholds for identification of open water using MODIS-derived spectral indices for two coastal wetland systems in Mexico. *Int. J. Appl. Earth Obs. Geoinf.* **2018**, *70*, 13–24. [CrossRef]

26. Ogilvie, A.; Belaud, G.; Massuel, S.; Mulligan, M.; Le Goulven, P.; Calvez, R. Surface water monitoring in small water bodies: Potential and limits of multi-sensor Landsat time series. *Hydrol. Earth Syst. Sci.* **2018**, *22*, 4349–4380. [CrossRef]

27. Zou, Z.H.; Xiao, X.M.; Dong, J.W.; Qin, Y.W.; Doughty, R.B.; Menarguez, M.A.; Zhang, G.; Wang, J. Divergent trends of open-surface water body area in the contiguous United States from 1984 to 2016. *Proc. Natl. Acad. Sci. USA* **2018**, *115*, 3810–3815. [CrossRef]

28. Chen, J.; Mueller, V. Coastal climate change, soil salinity and human migration in Bangladesh. *Nat. Clim. Chang.* **2018**, *8*, 981–985. [CrossRef]

29. Müller, M.F.; Yoon, J.; Gorelick, S.M.; Avisse, N.; Tilmant, A. Impact of the Syrian refugee crisis on land use and transboundary freshwater resources. *Proc. Natl. Acad. Sci.* **2016**, *113*, 14932–14937. [CrossRef] [PubMed]

30. Wang, D.Z.; Wang, S.M.; Huang, C. Comparison of Sentinel-2 imagery with Landsat8 imagery for surface water extraction using four common water indexes. *Remote Sens. Land Resour.* **2019**, *31*, 157–165.

31. Verpoorter, C.; Kutser, T.; Tranvik, L. Automated mapping of water bodies using Landsat multispectral data. *Limnol. Oceanogr. Methods* **2012**, *10*, 1037–1050. [CrossRef]

32. Xu, H.Q. Development of remote sensing water indices:a review. *J. Fuzhou Univ.* **2021**, *49*, 613–625.

33. Ji, L.; Zhang, L.; Wylie, B. Analysis of dynamic thresholds for the normalized difference water index. *Photogramm. Eng. Remote Sens.* **2009**, *75*, 1307–1317. [CrossRef]

34. Li, X.D.; Ling, F.; Foody, G.M.; Boyd, D.S.; Jiang, L.; Zhang, Y.H.; Zhou, P.; Wang, Y.L.; Chen, R.; Du, Y. Monitoring high spatiotemporal water dynamics by fusing MODIS, Landsat, water occurrence data and DEM. *Remote Sens. Environ.* **2021**, *265*, 112680. [CrossRef]

35. Tobón-Marín, A.; Canon Barriga, J. Analysis of changes in rivers planforms using google earth engine. *Int. J. Remote Sens.* **2020**, *41*, 8654–8681. [CrossRef]

36. Ismail, M.A.; Waqas, M.; Ali, A.; Muzzamil, M.M.; Abid, U.; Zia, T. Enhanced index for water body delineation and area calculation using Google Earth Engine: A case study of the Manchar Lake. *J. Water Clim. Chang.* **2022**, *13*, 557–573. [CrossRef]

37. Deng, Y.; Jiang, W.G.; Tang, Z.G.; Ling, Z.Y.; Wu, Z.F. Long-Term Changes of Open-Surface Water Bodies in the Yangtze River Basin Based on the Google Earth Engine Cloud Platform. *Remote Sens.* **2019**, *11*, 2213. [CrossRef]

38. Jin, Y.L.; Xu, M.L.; Gao, S.; Wan, H.W. Analysis on the Dynamic Changes and Driving Forces of Surface Water in the Three-River Headwater Region from 2001 to 2018. *Remote Sens. Technol. Appl.* **2021**, *36*, 1147–1154.

39. Chai, X.R.; Li, M.; Wang, G.W. Characterizing surface water changes across the Tibetan Plateau based on Landsat time series and LandTrendr algorithm. *Eur. J. Remote Sens.* **2022**, *55*, 251–262. [CrossRef]

40. Yu, Z.Q.; An, Q.; Liu, W.S.; Wang, Y.H. Analysis and evaluation of surface water changes in the lower reaches of the Yangtze River using Sentinel-1 imagery. *J. Hydrol. Reg. Stud.* **2022**, *41*, 101074. [CrossRef]

41. Rad, A.M.; Kreitler, J.; Sadegh, M. Augmented Normalized Difference Water Index for improved surface water monitoring. *Environ. Model. Softw.* **2021**, *140*, 105030. [CrossRef]

42. Huang, Y.L.; Deng, K.Y.; Ren, C.; Yu, Z.W.; Pan, Y.L. New water index and its stability study. *Prog. Geophys.* **2020**, *35*, 829–835.

43. Chen, Y.H.; Mu, Y.C.; Xie, Y.W. Extraction of the wetland information in the middle reaches of Heihe River based on Landsat images. *J. Lanzhou Univ. Nat. Sci.* **2016**, *52*, 587–592, 598.

44. Nguyen, U.N.; Pham, L.T.; Dang, T.D. An automatic water detection approach using Landsat 8 OLI and Google Earth Engine cloud computing to map lakes and reservoirs in New Zealand. *Environ. Monit. Assess.* **2019**, *191*, 235. [CrossRef] [PubMed]

45. Tew, Y.L.; Tan, M.L.; Samat, N.; Chan, N.W.; Mahamud, M.A.; Sabjan, M.A.; Lee, L.K.; See, K.F.; Wee, S.T. Comparison of Three Water Indices for Tropical Aquaculture Ponds Extraction using Google Earth Engine. *Sains Malays.* **2022**, *51*, 369–378. [CrossRef]

46. Yue, H.; Li, Y.; Qian, J.X.; Liu, Y. A new accuracy evaluation method for water body extraction. *Int. J. Remote Sens.* **2020**, *41*, 7311–7342. [CrossRef]

47. Yulianto, F.; Kushardono, D.; Budhiman, S.; Nugroho, G.; Chulafak, G.A.; Dewi, E.K.; Pambudi, A.I. Evaluation of the Threshold for an Improved Surface Water Extraction Index Using Optical Remote Sensing Data. *Sci. World J.* **2022**, *2022*, 4894929. [CrossRef]

48. Acharya, T.D.; Subedi, A.; Lee, D.H. Evaluation of water indices for surface water extraction in a Landsat 8 scene of Nepal. *Sensors* **2018**, *18*, 2580. [CrossRef]

49. Zhao, R.; Fu, P.; Zhou, Y.; Xiao, X.M.; Grebby, S.; Zhang, G.Q.; Dong, J.W. Annual 30-m big Lake Maps of the Tibetan Plateau in 1991–2018. *Sci. Data* **2022**, *9*, 164. [CrossRef]

Article

Evaluation of Grand Ethiopian Renaissance Dam Lake Using Remote Sensing Data and GIS

Asem Salama [1,2,*], **Mohamed ElGabry** [1,2], **Gad El-Qady** [1] **and Hesham Hussein Moussa** [1,2]

[1] National Research Institute of Astronomy and Geophysics, Helwan 11421, Egypt
[2] African Disaster Mitigation Research Center (ADMIR), Helwan 11421, Egypt
* Correspondence: asem.mostafa@nriag.sci.eg; Tel.: +20-21224017195

Abstract: Ethiopia began constructing the Grand Ethiopian Renaissance Dam (GERD) in 2011 on the Blue Nile near the borders of Sudan for electricity production. The dam was constructed as a roller-compacted concrete (RCC) gravity-type dam, comprising two power stations, three spillways, and the Saddle Dam. The main dam is expected to be 145 m high and 1780 m long. After filling of the dam, the estimated volume of Nile water to be bounded is about 74 billion m^3. The first filling of the dam reservoir started in July 2020. It is crucial to monitor the newly impounded lake and its size for the water security balance for the Nile countries. We used remote sensing techniques and a geographic information system to analyze different satellite images, including multi-looking Sentinel-2, Landsat-9, and Sentinel-1 (SAR), to monitor the changes in the volume of water from 21 July 2020 to 28 August 2022. The volume of Nile water during and after the first, second, and third filling was estimated for the Grand Ethiopian Renaissance Dam (GERD) Reservoir Lake and compared for future hazards and environmental impacts. The proposed monitoring and early warning system of the Nile Basin lakes is essential to act as a confidence-building measure and provide an opportunity for cooperation between the Nile Basin countries.

Keywords: Grand Ethiopian Dam; GIS; the first, second and third storages; satellite data

Citation: Salama, A.; ElGabry, M.; El-Qady, G.; Moussa, H.H. Evaluation of Grand Ethiopian Renaissance Dam Lake Using Remote Sensing Data and GIS. *Water* **2022**, *14*, 3033. https://doi.org/10.3390/w14193033

Academic Editors: Fei Zhang, Ngai Weng Chan, Xinguo Li and Xiaoping Wang

Received: 30 August 2022
Accepted: 25 September 2022
Published: 27 September 2022

Publisher's Note: MDPI stays neutral with regard to jurisdictional claims in published maps and institutional affiliations.

1. Introduction

The construction of massive hydraulic infrastructures, such as big dams, has expanded to an unprecedented level around the world in the 20th century. With their influence on social and political relations, they are also shaped by political, social, and cultural conditions [1,2]. The downstream countries in the main world river system are generally opposed to the upstream project dams [3,4]. These dam projects cause many concerns in the downstream countries because of their possible social and environmental impacts, including droughts, water salinity, and water flow effects. In the Euphrates Basin, the downstream countries of Iraq and Syria were affected by four droughts in 2000, 2006, 2008, and 2009, which are a cascading effect of climate change and a large number of dams being constructed along the Euphrates River, which is known as the Southeastern Anatolia (GAP) Project [5,6]. The GAP project includes the construction of 22 dams and 19 hydraulic power plants for irrigation and the generation of electricity on the Euphrates and the Tigris rivers and their tributaries [2,5]. The Three Gorges Dam (TGD) was constructed in China in the Yangtze River, affecting the sediment discharge and regulation of the flow process in the downstream provinces, which resulted in severe scouring and changes in the hydrogeological regime [7]. Dam projects were established along the Mekong River from 1965 to 2019 in northeastern Thailand, China, Vietnam, Loas, and Cambodia for power electricity generation [8]. These dam projects have environmental, economic, river hydrogeology, biological, and sediment transfer effects in Myanmar, Laos, Thailand, China, Cambodia, and Vietnam [9].

In April 2011, Ethiopia started the construction of the Grand Ethiopian Renaissance Dam (GERD). Understanding the context of the dam and its position relative to other dams on the

Blue Nile is essential. The newly built dam is located downstream of the Tana Lake, a highland lake at an average altitude of 1800 m a.s.l., with a surface area of 3060 km^2 at an average lake level of 1786 m a.s.l. This lake has a maximum depth of 15 m [10]. Four major tributaries feed the Tana Lake sub-basin, the Gelgel Abay in the south, Rib and Gomera in the east, and Megech in the north (Figure 1). The GERD is a gravity roller-compacted concrete dam with a target height of 145 m and length of 1780 m. The dam's crest is supposed to be at a height of 655 m above sea level, with the prospective to impound a lake with a capacity of 74 billion m^3 [11]. About 116 km upstream of the GERD, the Rosaries Dam is located in Sudan, constructed in 1961 and heightened in 2013, with a current storage capacity of 7.4 billion m^3 (Nile Basin Atlas Program) [12]. About 100 km downstream of Rosaries, the Sennar Dam was constructed in 1926 with a capacity of about 390 million m^3 (Nile Basin Atlas Program) [12]. Further north in Sudan is the Meroe Dam, with an impoundment capacity of about 12.3 billion m^3. Further to the north in the very south of Egypt, the Aswan High Dam was constructed in 1970 and is considered to bee the last dam near the mouth of the Blue Nile. The total capacity of the Aswan High Dam is 164 billion m^3. It consists of dead storage of 31.6 billion m^3, active storage of 90 billion m^3 (BCM), and emergency storage for flood protection of 41 billion m^3 (Nile Basin Atlas Program) [12].

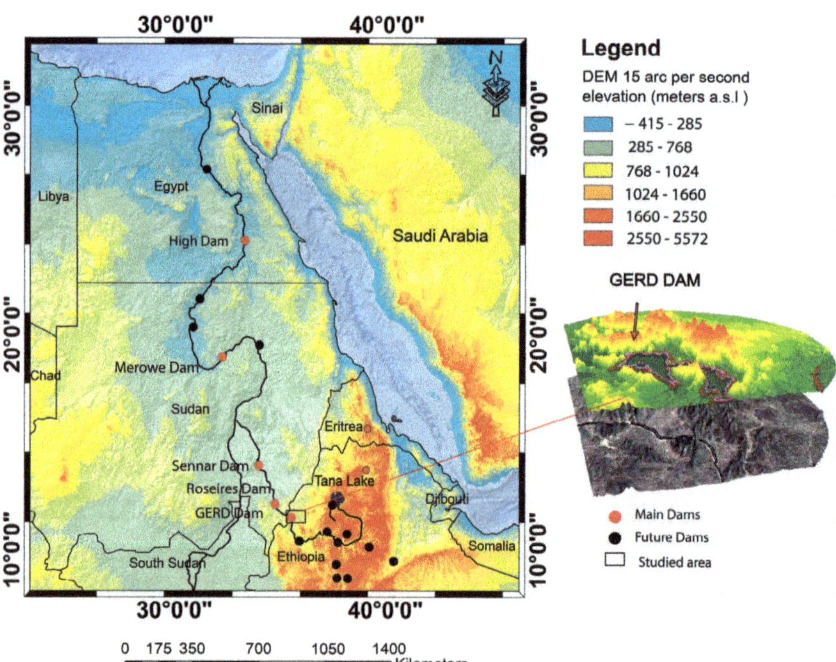

Figure 1. The location map of the studied area, which includes the location of the main artificial and proposed future dams in Africa. Data sourced from Wheeler et al. [13].

The Eastern Nile Basin is affected by historical complex hydropolitics over the use of the Nile water [14,15]. In the summer of 2020, the first phase of construction of the GERD was finished, and shortly after, the first filling of the GERD Lake started. During this season, the Sudanese dams, especially of Rosaries and Sennar, were confusedly operated due to a lack of prior information about the size and timing of the filling (reported by the Sudanese Minister of Irrigation Yasser Abbas on 26 August 2021 *Daily News* [16]. This may be due to the frozen agreement of the Eastern Nile Basin Initiative (NBI) activities [14,15]. This resulted in a shortage in freshwater during June and July, the filling months, in the capital Khartoum and many other cities after Sudanese water treatment stations went out of service due to low river levels. Later, in the same season, after the end of filling, Sudan faced a vast flood as the level of the Nile reached 17.48 m on 27 August 2020 at Khartoum, which was considered the second-highest level after the 1912 flood according to Prime Minister Abdalla Hamdouk [17] (*Guardian Journal*; date 5 September 2020). Ninety-nine people were killed in this flood as mentioned by the state of emergency in Sudan. Ramadan et al. [18] referred to the negative impacts, including environmental, economic, and social problems, on Egyptian countries by applying different scenarios along 2, 3, and 6 years of the filling of the GERD under different flow conditions. Omran and Negm [19] considered the different filling scenarios and indicated that Egypt and Sudan would experience severe impacts during the filling phase of GERD in some scenarios.

Remote sensing has been used to estimate and monitor the volume of lakes worldwide in various case studies. The key parameters controlling the water quantity of small or large lakes are the area and top level [20–22]. The spatial and temporal changes in the volume of water bodies can be calculated by several methods depending on the availability of morphometric and areal data. Amitrano et al. [21] used DEM (9 to 15 m resolution obtained from SAR data) to estimate the depth. They analyzed both Sentinel-1 and COSMO-SkyMed imagery to obtain more accurate results to extract the boundary of the basin as the water level increased, reflected by increases in the contour, to estimate the reservoir surface volume and retained water volume of the reservoir in the Labaa Basin in Ghana region. Xiaoqi et al. [23] used STRM DEM of the above lake level to construct the relationship between the elevation and the area to estimate the volume of the Namsto Lake in China. Pipitone et al. [22] used both optical (Landsat 5 TM, Landsat 8 OLI-TIRS and ASTER images) and synthetic aperture radar (SAR) images to monitor the water surface and the level of the Castello Dam Reservoir. They defined the displacement using the global navigation satellite system (GNSS) to detect the relationship between the water level and dam deformation in Castello Dam on Magazzolo Reservoir in south Italy. Ahmed et al. [24] used the time series of Landsat images of 2001, 2011, and 2019 to extract the modified normalized difference water index and combined it with field observation water level data to calculate the lake volume from 2001 to 2019 in Deeper Beel, which is situated in the southwestern part of Guwahati, Assam in India. Jiang et al. [25] used the average annual coefficients of the VH backscatter for Sentinel-1A and the normalized difference water index (NDWI) of Sentinel-2 to map small water bodies in the mountain region in China for water-related environment monitoring and resource management. In the Nile Basin, Hossen et al. [26] built bathymetric and water capacity relationships based on Sentinel-3 optical and radar data for Aswan High Dam Lake, Egypt. Kansara et al. [27] used an analysis of multi-source satellite imagery and Sentinel-1 SAR imagery to display the number of classified water pixels in the GERD from early June 2017 to September 2020, indicating a contrasting trend in August and September 2020 for all upstream/downstream water bodies using a Google Earth Engine (GEE). Their results show that upstream of the dam rises steeply while it decreases downstream.

In the last 20 years, multispectral remote sensing and Sentinel (SAR-1) data have been widely used for surface water monitoring to overcome the limitations and lack of field observations for monitoring of the storage volume of water reservoirs [19,21,23,28]. The dynamic volume change in GERD Lake is essential for all Blue Nile countries, including Ethiopia, Sudan, and Egypt, to understand the balance of the water security

2. Methods and Materials

2.1. Depth Estimations

The depth of the GERD Lake was estimated using Shuttle Radar Topography Mission (SRTM) data, which map the topography of the Earth's surface using radar interferometry. The Shuttle Radar Topography Mission (SRTM) is an international project spearheaded by the National Geospatial-Intelligence Agency and NASA, whose objective is to obtain the most complete high-resolution digital topographic database of the Earth. We downloaded the SRTM 1 arc per second data courtesy of the U.S. Geological Survey from https://

earthexplorer.usgs.gov/ accessed on 21 June 2020. It was measured on 11 March 2000. It was used in the study of the GERD Lake to obtain the elevation difference obtained through interferometry, which was transformed into a 3D digital elevation model (DEM), which was used as the GERD Basin Reservoir depth before the filling process.

2.2. Satellite Data Processing and Water Level Estimation

In this study, we tracked changes in the water capacity level boundary for the GERD Lake using the multi-optical satellite data and Sentinel 1A (SAR). We acquired the multi optical Sentinel 2A and Landsat-8 with time series from 21 July 2020 to 3 July 2021 courtesy of the U.S. Geological Survey, https://ers.cr.usgs.gov/ website accessed on 21 June 2020 while the Landsat-9 and Sentinel-1 (SAR) with time series were from 16 July 2021 to 28 August 2022. The sentinel-1 (SAR) was obtained from Copernicus Open Access Hub https://scihub.copernicus.eu/ website accessed on 29 August 2022. The Sentinel-2 data was characterized by higher spatial and spectral resolutions in the near-infrared region. The Sentinel-2 sensor, the EO satellite of the Copernicus program, has 12 bands with spatial resolutions of 10 (four visible and near-infrared bands), 20 (six red-edge and shortwave infrared bands), and 60 m (three atmospheric correction bands) [29]. Recently, the Landsat-9 satellite was launched on 27 September 2021. It is similar to Landsat-8 and characterized by four visible spectral bands, one near-infrared spectral band, three shortwave-infrared spectral bands at a 30 m spatial resolution, plus one panchromatic band at a 15 m spatial resolution, and two thermal bands at a 100 m spatial resolution. The problem of dense cloud cover is encountered in some optical satellite imagery, which masks the lake in rainy seasons, especially in June and July each year. We used a filter to remove cloud pixels, using the threshold to identify the pixel range as cloud using ArcGIS 10.8 software [30]. We found incomplete filter-out cloud in some multi-optical satellite images. We instead used Sentinel-1 SAR to obtain the water level boundary, especially in the cloud periods, which mask the GERD Lake boundaries. SAR sentinel-1 (Synthetic Aperture Radar (S-1 SAR) data are insensitive to cloud. However, Sentinel-1 SAR data are characterized by speckle noise and have some difficulties in detecting the water surface of water bodies. This can be solved by applying several techniques such as aggregation of the brightness pixels, which was proposed by Pipitone et al. [22].

The analysis scheme used to estimate the water volume in the GERD Lake is summarized in Figure 2 for optical multispectral and SAR data analysis, which was applied in this study. We used ArcMap 10.8 [30] for multioptical satellites (i.e., LandSat-8 and -9 and Sentinel-2) to separate the shape of the GERD Lake using the normalized difference water index (NDWI) as it enhances the presence of water bodies, a method introduced by Mcfeeders [31]. NDWI uses reflected near-infrared radiation and visible green light to enhance the presence of water bodies such as lakes and rivers. This method is characterized by its ability to eliminate the presence of soil and terrestrial vegetation features. The equation depends on the use of bands with a relatively high reflectance of the water green band (band-3) and one with low or no reflectance near-infrared (NIR) (i.e., band-8 in the case of multispectral Sentinel-2 and band-5 in the case of Landsat-8 and -9) as follows:

$$NDWI = \frac{Band\ 3 - Band\ 8\ or\ 5\ (NIR)}{Band\ 3 + Band\ 8\ or\ 5\ (NIR)} \tag{1}$$

The preprocessing of the workflow of Sentinel SAR-1 was applied by the Sentinel Application Platform (SNAP) [32], an open-source software version of 8.0.9 (http://step.esa.int/main/toolboxes/snap/ accessed on 1 October 2020), as follows: (a) a subset tool was used to delineate the area of the study. (b) The orbit file was applied, which allows updating of the orbit state vectors for each SAR scene, providing accurate satellite position and velocity information. (c) The thermal to noise removal algorithm was used to remove and reduce noise effects in the inter-sub-swath texture and normalize the backscatter for scenes in multi-swath acquisition modes. (d) Calibration equation was used to convert the image intensity values to sigma nought values in which the digital pixel was converted to radiometrically calibrated SAR backscatter concerning the nominally horizontal plane of Sentinel-1 GRD. (e) Terrain corrections were used to compensate for some distortions related to the side-looking geometry to be close to the real world. (f) We used coregistration with an average stack of two time series images per month to obtain a single image. We applied coregistration instead of a speckle filter to remove noise without affecting the resolution of the optical image of Sentinel SAR-1, which may result from temporal decorrelation effects. The final step was to convert it to linear transformations and apply the band math equation depending on the image histogram. The water lake was delineated using this equation in which the thresholding values range between -1, which refers to land, and $+1$, which refers to water bodies.

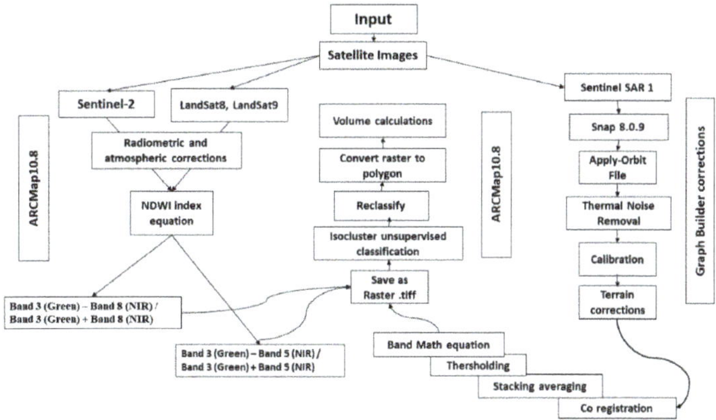

Figure 2. Workflow explaining the steps of satellite image analysis carried out in this study to calculate the volume of the GERD Lake.

2.3. Water Level Validations

The water levels were collected for Nasser, Tana, and GERD lakes in Egypt and Ethiopia, respectively. The in situ water level data was recorded by the gauging station at Nasser Lake and obtained from the Nile Research Institute (NRI) database. The other water level data was collected at virtual stations from a satellite altimetry set obtained from the "Global Reservoirs and Lakes Monitor (G-REALM) project" of the U.S. Foreign Agricultural Service [33] and the level contour was extracted by optical multispectral satellite data in this study. Then, we calculated the average water level uncertainty as shown in Table 1.

Table 1. Water level for the Aswan, Tana, and GERD lakes.

Name Lake	Water Level from In Situ Station	Water Level from Virtual Stations Obtained from Satellite Altimeter Data from G-REALM) Project "m"	Location of Virtual Water Level Station		Water Level Extracted by Sentinel-2 Boundary "m" in This Study	Differences "m"
			Long.	Lat.		
Nasser lake, Aswan Egypt	180.5	181.93	32.57	22.8	181	1.43
Tana Lake, Ethiopian	-	1789.31	37.3	12.0	1787.81	1.6
GERD Lake, Ethiopian	-	581.38	10.579	10.552	580	1.4
Average calculated water level uncertainty						±1.45

3. Results and Volume Calculation

The required parameters needed to compute the volume of the GERD Lake were as follows: (a) the input surface (i.e., 3D depth of the lake), which was established from the digital elevation model. (b) The second parameter required is the "Z" value, which was defined as the plane surface height of the water level top boundary in which the lake polygons were extracted from an optical satellite image or Sentinel 1A–SAR. The volume equation was calculated using the ARCGIS10.8 volume tool, which was dependent on the empirical formula of the volume. The volume equation is as follows:

The volume of water bodies = average depth (d) of the Lake X Area of the lake (A)   (2)

The computation of the DEM raster surface was evaluated using the extent of the center point of each cell as opposed to the extent of the entire cell area. The resulting analysis will decrease the data area of the raster by half a cell relative to the data area displayed for the raster according to the manual of ARCGIS10.8.

The average volume uncertainty was calculated with the average uncertainty in volume (Figures 3 and 4) depending on the water level uncertainty $\pm$ 1.45 calculated in the previous section. The lakes' polygons' boundaries were extracted from a multi-optical satellite image and Sentinel SAR-1 to reflect the water area storage morphology in the GERD Lake (Figure 3). A chart of the average volume for the GERD Lake with the time series obtained every month from 21 July 2020 to 28 August 2022 is shown in Figure 4.

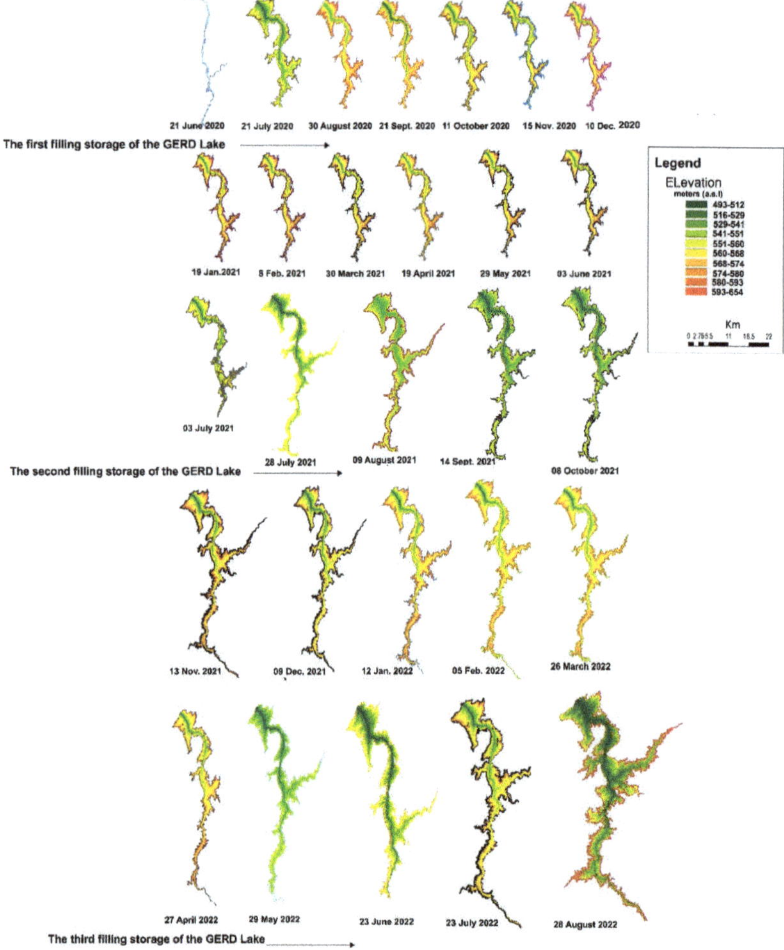

Figure 3. Digital elevation model (DEM) extracted by the GERD Lake polygons with time series from satellite images.

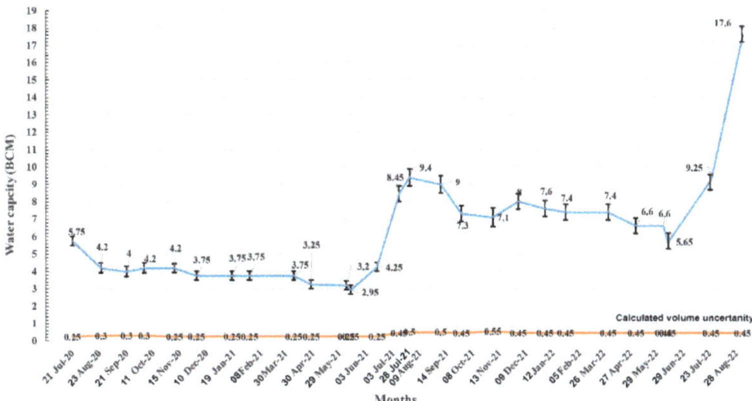

Figure 4. Chart showing the water volume capacity of the GERD Lake with the calculated volume uncertainty values in orange color from 21 July 2020 to 28 August 2022.

The volume of the GERD Lake in the first storage reached its maximum level, which appeared in the satellite images taken on 21 July 2020, with an area of 250.16 km^2 and a volume of 5.75 ± 0.25 billion m^3 (Figure 4). Although, there was a receding of the water in the GERD Lake in the next three months in August, September, and October in the year 2020, with an average water volume of 4.2 ± 0.3 billion m^3 (Figure 4). From November 2020 to 30 March 2021, the second receding of water storage in the GERD Lake reached an average volume of 3.75 ± 0.3 billion m^3, calculated from the satellite images. On 28 July 2021, the GERD Lake showed an increase in the polygon area extracted from the Sentinel SAR-1 satellite image of 316.54 km^2 and an average volume of 8.45 ± 0.45 billion m^3 (Figures 3 and 4). The average volume of storage of the GERD Lake increased in August and September 2021, with a maximum of 9.4 ± 0.5 billion m^3 during August 2021 after the second storage was carried out. Receding of the water of the lake was observed during October and November 2021 to an average volume of 7.3 ± 0.45 billion m^3 (Figure 4), with a slight increase in December 2021 to 8.0 ± 0.45 billion m^3. From January to 29 May 2022, the capacity of the reservoir lake decreased to 5.56 ± 0.45. Then, the third filling storage was reached by 23 July 2022, with an increase in the total capacity of 9.25 ± 0.25 billion m^3 and a significant large capacity of 17.4 ± 0.45 billion m^3 was reached on 28 August 2022 (Figures 3 and 4).

The Ethiopian government carried out the first storage in July 2020 while July 2021 and July 2022 represent the second and third storage stages. During the storage stages and closing of the GERD Dam gates, the GERD Lake was charged by rainfall and Tana Lake, which is considered the major source of the Blue Nile [10].

The water level was observed from satellite images to be on the lower limit of the saddle dam in the third filling on 28 August 2022. This saddle dam was built with a 5-km-long concrete face rockfall and 50 m high to maintain the required water surface elevation and depth at a relatively flat dam site. The saddle dam increases the natural features from 600 to 646 m asl, increasing the reservoir water level to the design level [34]. An emergency gated 300-m-wide spillway is located between the main dam and the saddle dam. The spillway, at a crest elevation of 624.9 m, is to be used for extreme flood conditions, releasing through a gully into the river downstream of the dam.

4. Discussion

The application of remote sensing and GIS to monitor GERD Lake volume changes provides critical information about the GERD Reservoir Lake water level and storage capacity. This will be very important for downstream countries in the case of a limitation or lack of information resulting from a stumble in negotiations between Ethiopia and the downstream countries Egypt and Sudan. Water safety is essential for both upstream and downstream countries. One of the most controversial debuts in GERD negotiations is the number of years for the initial reservoir filling, as a shorter filling time requires greater flow reduction and a higher investment return from the dam. A longer filling time requires lower flow reduction and lower investment return from the dam [34]. The water level shown by satellite data in this study was 600 ± 1.45 a.s.l on August 28 August 2022 in the lower level of the saddle dam. This level corresponds to 24.3% of the full storage capacity of 74 billion cubic meters. It was considered as more than the minimum reservoir fill rates, which is beneficial for hydroelectric generation without having an effect on stream flow into Egypt and Sudan as stated by Keith et al. [35]. King and Block [36] refer to the 25% filling policy, which can reduce the average downstream flow by more than 10 BCM per year. Hegay et al. [37] proposed numerous actions and mitigation strategies that could secure Egypt's water demands by minimizing the effects of the GERD project. These strategies should include the present-day operation of the AHD hydropower plant to mitigate imminent water shortages in combination with the increase in groundwater withdrawal as a backstop choice to quickly sustain the water demand. Water conservation strategies should additionally be integrated, mainly inside the agriculture sectors, by switching the countrywide production to crops that require less water.

Previous studies have investigated the possible future multi-environmental and hazard impacts on downstream countries. Wheeler et al. [38] described a post-filling period that includes severe multi-year droughts after filling of the dam with the uncertainty of the exact start and end time, which will require careful coordination to minimize possible harmful impacts on downstream countries. Donia and Negm [39] modeled three scenarios of the storage capacity of the GERD Lake. The storage capacity of the three models was estimated assuming 18 billion m^3 for the initial design storage capacity and 35 and 74 billion m^3 for the middle and final storage. Their results from scenario-3 of the full filling of GERD Lake in 5 years show a negative impact on agriculture due to the loss of silt, which is a result of restricting the water flowing to the Aswan High Dam in Egypt. Abulnaga's [40] study refers to scooping out accumulated mud and silts through dredging and the construction of onshore sediment ponds that are used for agricultural purposes due to the construction of the dam in Ethiopia. From an engineering point of view, EL Askary et al. [41] showed a deformation pattern associated with different sections of the GERD Lake and Saddle Dam (main dam and embankment dam) using 109 descending mode scenes from Sentinel-1 SAR imagery from December 2016 to July 2021. This may result in a dam failure flood, which will have harmful impacts in Sudan and Egypt.

In summary, the environmental impacts and other socio-political considerations of GERD extend across a diverse spectrum of issues from population growth, economic development, and water rights to sedimentation and/or changing flood regimes and the shock of climate change. It is necessary to examine the complex social and environmental values of water resources and the policies governing the use of water resources. A water cooperation policy is the best choice for the cooperative Nile basin initiative to overcome any debate on the remnant years of fillings [42]. Informal diplomacy has been successfully used to manage transboundary waters in a similar case in the Mekong River Dam [43]. The waterscape of the Mekong Dam issues has been extended to security actors that are not water experts within domestic politics. For this, the analysis could be extended to examine in more detail the knowledge channels within multiple tracks of diplomacy and how harms and inequalities are understood, beyond mere metrics of economic impacts and water quantities. This method of informal diplomacy can help change the frozen negotiation situations between Ethiopia and Egypt. Thus, understanding water diplomacy requires scrutiny of how power, knowledge, and the political economy of river basin development intersect.

5. Conclusions

The combination of open-source satellite optical and radar images with DEM provided a robust tool to estimate the water volume in the artificial GERD Lake during the initial phases of filling. The water level measured from satellite data refers to the consequent increase in the stored water volume of the GERD Reservoir Lake. Three stored water stages of the initial filling were considered for the lake, corresponding to volumes of 5.75 ± 0.25, 9.4 ± 0.5, and 17.4 ± 0.45 billion m^3 during 21 July 2020, 28 July 2021, and 28 August 2022, respectively.

Data collected from open sources combined with technical knowledge could provide very useful information that can be used to monitor the filling process and support informal diplomacy with transparent and trustful independent information that could possibly lead to a future agreement between all Nile basin countries. The authors believe that this work is a milestone in building a scientific initiative to utilize open-source data for the benefit of the community and to build a common agreement on the importance of investment in knowledge for sustaining water resources and their management. Further work is needed to extend this work to better understand the impact of the current filling process and its impact on the ecosystem and boost the knowledge and data exchange between riparian countries for integrated management plans for the Nile. An integrated database that combines ground- and satellite-based observations could utilize modern scientific techniques to integrate the dam's operation process and mitigate natural disasters and climate change's impact on the sustainable development in Nile Basin countries. Such an initiative could work as a confidence-building measure between Nile Basin countries and provide leveraging for science diplomacy to bridge cooperation and integration in an era of divergence and competition.

Author Contributions: Conceptualization, M.E. and G.E.-Q.; Data curation, A.S. and M.E.; Formal analysis, A.S.; Funding acquisition, M.E. and G.E.-Q.; Investigation, M.E.; Methodology, A.S., M.E., A.S. and H.H.M.; Writing—original draft, A.S. and M.E.; Writing—review and editing, H.H.M. and G.E.-Q. All authors have read and agreed to the published version of the manuscript.

Funding: This research received no external funding.

Institutional Review Board Statement: Not applicable.

Informed Consent Statement: Not applicable.

Data Availability Statement: Data and related results of this article are presented in the paper and can be asked from the first and second authors.

Acknowledgments: The authors are grateful to the Copernicus Sentinel-hub (https://scihub.copernicus.eu/) (accessed on 9 August 2022) and USGS https://earthexplorer.usgs.gov/ (accessed on 21 June 2020) for the availability of satellite data sources used in this research work. We would like to thank the National research institute of Astronomy and geophysics (NRIAG) for their financial and logistic support.

Conflicts of Interest: The authors declare no conflict of interest.

References

1. Hussein, H.; Conker, A.; Grandi, M. Small is beautiful but not trendy: Understanding the allure of big hydraulic works in the Euphrates-Tigris and Nile waterscapes. *Mediterr. Polit.* **2022**, *27*, 297–320. [CrossRef]
2. Conker, A.; Hussein, H. Hydraulic mission at home, hydraulic mission abroad? Examining Turkey's regional 'Pax-Aquarum' and its limits. *Sustainability* **2019**, *11*, 288. [CrossRef]
3. Mehtonen, K. Do the Downstream Countries Oppose the Upstream Dams? In *Modern Myths of the Mekong: A Critical Review of Water and Development Concepts, Principles and Policies*; Helsinki University of Technology: Espoo, Finland, 2008; pp. 161–173. ISBN 978-951-22-9102-1.
4. Conker, A.; Hussein, H. Hydropolitics and issue-linkage along the Orontes River Basin: An analysis of the Lebanon–Syria and Syria–Turkey hydropolitical relations. *Int. Environ. Agreem. Polit. Law Econ.* **2020**, *20*, 103–121. [CrossRef]
5. Giovanis, E.; Ozdamar, O. *The Transboundary Effects of Climate Change and Global Adaptation: The Case of the Euphrates-Tigris Water Basin in Turkey and Iraq*; Report no. WP 1517; Economic Research Forum (ERF): Giza, Egypt, 2021; pp. 1–60.
6. Özgüler, H.; Yıldız, D. Consequences of the Droughts in the Euphrates-Tigris Basin. *Water Manag. Dlomacy* **2020**, *1*, 29–49.
7. Zhou, Y.; Li, Z.; Yao, S.; Shan, M.; Guo, C. Case Study: Influence of Three Gorges Reservoir Impoundment on Hydrological Regime of the Acipenser sinensis Spawning Ground, Yangtze River, China. *Front. Ecol. Evol.* **2021**, *9*, 624447. [CrossRef]
8. Zeitoun, M.; Cascão, A.E.; Warner, J.; Mirumachi, N.; Matthews, N.; Menga, F.; Farnum, R. Transboundary water interaction III: Contest and compliance. *Int. Environ. Agreem.* **2017**, *17*, 271–294. [CrossRef]

9. Soukhaphon, A.; Baird, I.G.; Hogan, Z.S. The impacts of hydropower dams in the mekong river basin: A review. *Water* **2021**, *13*, 1–18. [CrossRef]

10. Ayana, E.K.; Philpot, W.D.; Melesse, A.M.; Steenhuis, T.S. Bathymetry, Lake Area and Volume Mapping: A Remote-Sensing Perspective. In *Nile River Basin: Ecohydrological Challenges, Climate Change and Hydropolitics*; Melesse, A.M., Abtew, W., Setegn, S.G., Eds.; Springer International Publishing: Cham, Switzerland, 2014; pp. 253–267. [CrossRef]

11. Melesse, A.M.; Abtew, W.; Moges, S.A. *Nile and Grand Ethiopian Renaissance Dam: Past, Present and Future*; Springer: Cham, Switzerland, 2021; p. 525. [CrossRef]

12. Nile Basin Initiative. Eastern Nile Subsidiary Action Program (ENSAP). 2010. Available online: http://entro.nilebasin.org/ensap (accessed on 20 July 2020).

13. Wheeler, K.G.; Basheer, M.; Mekonnen, Z.T.; Eltoum, S.O.; Mersha, A.; Abdo, G.M.; Zagona, E.A.; Hall, J.W.; Dadson, S.J. Cooperative filling approaches for the Grand Ethiopian Renaissance Dam. *Water Int.* **2016**, *41*, 611–634. [CrossRef]

14. Tvedt, T. *The River Nile in the Age of the British—Political Ecology and the Quest for Economic Power*; IB Tauris: London, UK, 2004; p. 464. ISBN 9789774160462.

15. Hussein, H.; Grandi, M. Dynamic political contexts and power asymmetries: The cases of the Blue Nile and the Yarmouk Rivers. *Int. Environ. Agreem. Polit. Law Econ* **2017**, *17*, 795–814. [CrossRef]

16. Available online: https://dailynewsegypt.com/2021/08/26/withholding-information-about-ethiopian-dams-filling-cost-us-dearly-sudan/ (accessed on 26 January 2022).

17. Slawson, N. Sudan Declares State of Emergency as Record Flooding Kills 99 People. Archived from the Original on 6 September 2020. Retrieved 6 September 2020–via The Guardian. Available online: https://www.theguardian.com/world/2020/sep/05/sudan-declares-state-of-emergency-record-flooding (accessed on 5 January 2021).

18. Ramadan, E.; Negm, A.; Elsammany, M.; Helmy, A. Quantifying the Impacts of Impounding Grand Ethiopian. In Proceedings of the Eighteenth International Water Technology Conference (IWTC 18), Sharm El Sheikh, Egypt, 12–14 March 2015. [CrossRef]

19. Omran, E.S.E.; Negm, A. Environmental Impacts of the GERD Project on Egypt's Aswan High Dam Lake and Mitigation and Adaptation Options. In *Grand Ethiopian Renaissance Dam versus Aswan High Dam*; Negm, A., Abdel-Fattah, S., Eds.; The Handbook of Environmental Chemistry; Springer: Cham, Switzerland, 2018; Volume 79. [CrossRef]

20. Kumar, R.; Bahuguna, I.M.; Ali, S.N.; Singh, R. Lake Inventory and Evolution of Glacial Lakes in the Nubra-Shyok Basin of Karakoram Range. *Earth Syst. Environ.* **2020**, *4*, 57–70. [CrossRef]

21. Amitrano, D.; Martino, G.D.; Iodice, A.; Mitidieri, F.; Papa, M.N.; Riccio, D.; Ruello, G. Sentinel-1 for Monitoring Reservoirs: A Performance Analysis. *Remote Sens.* **2014**, *6*, 10676–10693. [CrossRef]

22. Pipitone, C.; Maltese, A.; Dardanelli, G.; Lo Brutto, M.; La Loggia, G. Monitoring Water Surface and Level of a Reservoir Using Different Remote Sensing Approaches and Comparison with Dam Displacements Evaluated via GNSS. *Remote Sens.* **2018**, *10*, 71. [CrossRef]

23. Ma, X.; Lu, S.; Ma, J.; Zhu, L. Lake water storage estimation method based on topographic parameters: A case study of Nam Co Lake. *Remote Sens. Land Resour.* **2019**, *31*, 167–173.

24. Ahmed, I.A.; Shahfahad, S.; Baig, M.R.; Talukdar, S.; Asgher, M.S.; Usmani, T.M.; Ahmed, S.; Rahman, A. Lake water volume calculation using time series LANDSAT satellite data: A geospatial analysis of Deepor Beel Lake, Guwahati. *Front. Eng. Built Environ.* **2021**, *1*, 107–130. [CrossRef]

25. Jiang, H.; Wang, M.; Hu, H.; Xu, J. Evaluating the Performance of Sentinel-1A and Sentinel-2 in Small Waterbody Mapping over Urban and Mountainous Regions. *Water* **2021**, *13*, 945. [CrossRef]

26. Hossen, H.; Khairy, M.; Ghaly, S.; Scozzari, A.; Negm, A.; Elsahabi, M. Bathymetric and Capacity Relationships Based on Sentinel-3 Mission Data for Aswan High Dam Lake, Egypt. *Water* **2022**, *14*, 711. [CrossRef]

27. Kansara, P.; Li, W.; El-Askary, H.; Lakshmi, V.; Piechota, T.; Struppa, D.; Abdelaty Sayed, M. An Assessment of the Filling Process of the Grand Ethiopian Renaissance Dam and Its Impact on the Downstream Countries. *Remote Sens.* **2021**, *13*, 711. [CrossRef]

28. Condeça, J.; Nascimento, J.; Barreiras, N. Monitoring the Storage Volume of Water Reservoirs Using Google Earth Engine. *Water Resour. Res.* **2022**, *58*, e2021WR030026. [CrossRef]

29. Hedley, J.D.; Roelfsema, C.; Brando, V.; Giardino, C.; Kutser, T.; Phinn, S. Coral reef applications of Sentinel-2: Coverage, characteristics, bathymetry and benthic mapping with comparison to Landsat 8. *Remote Sens. Environ.* **2018**, *216*, 598–614. [CrossRef]

30. ESRI. *ArcGIS Desktop: Release 10*; Environmental Systems Research Institute: Redlands, CA, USA, 2011.

31. McFeeters, S.K. Using the Normalized Difference Water Index (NDWI) within a Geographic Information System to Detect Swimming Pools for Mosquito Abatement: A Practical Approach. *Remote Sens.* **2013**, *5*, 3544–3561. [CrossRef]

32. SNAP–ESA Sentinel Application Platform v2.0.2. Available online: http://step.esa.int (accessed on 1 October 2020).

33. Available online: https://ipad.fas.usda.gov/cropexplorer/global_reservoir/ (accessed on 5 June 2022).

34. Abtew, W.; Dessu, S.B. *The grand Ethiopian Renaissance Dam on the Blue Nile*; Springer Geography; Springer: Cham, Switzerland, 2019. [CrossRef]

35. Keith, B.; Ford, D.N.; Horton, R. Considerations in managing the fill rate of the Grand Ethiopian RenaissanceDam Reservoir using a system dynamics approach. *J. Def. Model. Sim.* **2017**, *14*, 33–43. [CrossRef]

36. King, A.D.; Block, P.J. An assessment of reservoir filling policies for the Grand Ethiopian Renaissance Dam. *J. Water Clim. Chang.* **2014**, *5*, 233–243. [CrossRef]

37. Heggy, E.; Sharkawy, Z.; Abotalib, A.Z. Egypt's budget deficit and suggested mitigations policies for the Grand Ethiopian Renaissance Dam filling scenarios. *Environ. Res. Lett.* **2021**, *16*, 074022. [CrossRef]
38. Wheeler, K.G.; Jeuland, M.; Hall, J.W.; Zagona, E.; Whittington, D. Understanding and managing new risks on the Nile with the Grand Ethiopian Renaissance Dam. *Nat. Commun* **2020**, *11*, 5222. [CrossRef] [PubMed]
39. Donia, N.; Negm, A. Impacts of Filling Scenarios of GERD's Reservoir on Egypt's Water Resources and Their Impacts on Agriculture Sector. In *Conventional Water Resources and Agriculture in Egypt*; Negm, A.M., Ed.; The Handbook of Environmental Chemistry; Springer: Cham, Switzerland, 2019; Volume 74, pp. 391–414. [CrossRef]
40. Abulnaga, B.E. Dredging the Clays of the Nile: Potential Challenges and Opportunities on the Shores of the Aswan High Dam Reservoir and the Nile Valley in Egypt. In *Grand Ethiopian Renaissance Dam versus Aswan High Dam*; Negm, A., Abdel-Fattah, S., Eds.; The Handbook of Environmental Chemistry; Springer: Cham, Switzerland, 2019; Volume 79. [CrossRef]
41. El-Askary, H.; Fawzy, A.; Thomas, R.; Li, W.; LaHaye, N.; Linstead, E.; Piechota, T.; Struppa, D.; Sayed, M.A. Assessing the Vertical Displacement of the Grand Ethiopian Renaissance Dam during Its Filling Using DInSAR Technology and Its Potential Acute Consequences on the Downstream Countries. *Remote Sens.* **2021**, *13*, 4287. [CrossRef]
42. Wheeler, K.G.; Hussein, H. Water research and nationalism in the post-truth era. *Water Int.* **2021**, *46*, 1216–1223. [CrossRef]
43. Mirumachi, N. Informal water diplomacy and power: A case of seeking water security in the Mekong River basin. *Environ. Sci. Policy* **2020**, *114*, 86–95. [CrossRef]

Article

Estimations of Water Volume and External Loading Based on DYRESM Hydrodynamic Model at Lake Dianchi

Rufeng Zhang [1], Liancong Luo [1,*], Min Pan [2], Feng He [2], Chunliang Luo [2], Di Meng [2], Huiyun Li [3,*], Jialong Li [1], Falu Gong [1], Guizhu Wu [1], Lan Chen [1], Jian Zhang [1] and Ting Sun [1]

[1]   Institute for Ecological Research and Pollution Control of Plateau Lakes, School of Ecology and Environmental Science, Yunnan University, Kunming 650504, China
[2]   Kunming Dianchi & Plateau Lakes Institute, Kunming 650500, China
[3]   Nanjing Institute of Geography and Limnology, Chinese Academy of Sciences, Nanjing 210008, China
*   Correspondence: billluo@ynu.edu.cn (L.L.); hyli@niglas.ac.cn (H.L.); Tel.: +86-(0)-181-1290-7530 (L.L.); +86-(0)-159-0515-7570 (H.L.)

Citation: Zhang, R.; Luo, L.; Pan, M.; He, F.; Luo, C.; Meng, D.; Li, H.; Li, J.; Gong, F.; Wu, G.; et al. Estimations of Water Volume and External Loading Based on DYRESM Hydrodynamic Model at Lake Dianchi. *Water* **2022**, *14*, 2832. https://doi.org/10.3390/w14182832

Academic Editors: Fei Zhang, Ngai Weng Chan, Xinguo Li, Xiaoping Wang and George Arhonditsis

Received: 29 June 2022
Accepted: 6 September 2022
Published: 12 September 2022

Publisher's Note: MDPI stays neutral with regard to jurisdictional claims in published maps and institutional affiliations.

Abstract: There are many rivers flowing from complex paths into Lake Dianchi. At present, there is a lack of inflow and water quality monitoring data for some rivers, resulting in limited accuracy of statistical results regarding water volume and external loading estimations. In this study, we used DYRESM to estimate the water volume entering Waihai of Lake Dianchi from 2007 to 2019 without historical hydrological observation data. Then, we combined this information with the monthly monitoring data of water quality to calculate the annual external loading. Our results showed that: (1) DYRESM could effectively capture the extreme changes of water level at Waihai, showing its reliable applicability to Lake Dianchi. (2) The average annual inflow of rivers entering Waihai was about $6.69 \times 10^8 \text{ m}^3$. The fitting relationship between river inflow and precipitation was significant on annual scale ($r = 0.74$), with a higher inner-annual fitting coefficient between them ($r = 0.98$), thus suggesting that precipitation and its caused river inflows are the main water source for Waihai. (3) From 2007 to 2010, the river loadings remained at a high level. They decreased to 2445.44 t (total nitrogen, TN) and 106.53 t (total phosphorus, TP) due to a followed drought in 2011. (4) The river loading had annual variation characteristics. The contribution rates of TN and TP loading in the rainy season were 63% and 67% respectively. (5) Panlong River, Daqing River, Jinjia River, Xinbaoxiang River, Cailian River and Hai River were the main inflow rivers. Their loadings accounted for 81.3% (TN) and 80.3% (TP) of the total inputs. (6) River loadings have gradually reduced and the water quality of Waihai has continually improved. However, Pearson analysis results showed that the water quality parameters were not significantly correlated with their corresponding external loading at Waihai, indicating that there might be other factors influencing the water quality. (7) The contribution rates of internal release to the total loads of TN and TP at Waihai were estimated to be 7.6% and 8.9% respectively, suggesting that the reductions of both external and internal loading should be considered in order to significantly improve the water quality at Waihai of Lake Dianchi.

Keywords: Lake Dianchi; eutrophication; DYRESM; inflow volume; external loading

1. Introduction

A lake is a key node at the intersection of terrestrial ecosystem elements, playing roles in freshwater supply, flood storage and species conservation in the geosphere. Lakes are a valuable resource that human beings depend on to improve productivity through functions such as regulating runoff, developing irrigation and conducting shipping [1,2]. With the rapid growth of the global population and the gradual advancement of industrialization, urbanization and modern agriculture, a large amount of anthropogenic pollutants have been discharged into lakes, increasing the harm of eutrophication [3,4]. Lake eutrophication refers to a large increase in essential plant nutrients, such as nitrogen (n) and phosphorus (p), in a lake, causing significant increases in the primary productivity of water ecosystems and resulting in the appearance of algal blooms, lower dissolved oxygen (DO) concentrations, reduced transparency, the death of aquatic animals, reduced biodiversity and damage to the normal habitat and function of the lake [5,6]. Studies [7,8] have found that eutrophication of a lake under natural conditions takes a long time to evolve, but disturbances from human activity can significantly accelerate the eutrophication process, shortening it from the original timeline of thousands of years to decades or even less. At the end of the 1990s, 61% of lakes worldwide were eutrophic [9], and the eutrophication rate of global inland waters had increased to 63% by 2012 [10]. The percentage of eutrophic lakes (reservoirs) among 110 important Chinese lakes (reservoirs) was 29% in 2020 [11]. The eutrophication of lakes has proven to be a threat to the sustainable development of human society.

Lake Dianchi is the largest freshwater lake on the Yunnan–Guizhou Plateau, playing key roles in the social and scientific development of Yunnan New Area [12]. However, it is in a moderate or severe eutrophic state all year round, and cyanobacterial blooms occur frequently, causing hidden dangers to the water environment and water safety of surrounding residents [13]. Guo et al. [14] found that n and P loadings from urban sewage and agricultural runoff are the main sources of pollution in Lake Dianchi. Ma et al. [15] considered the hydrological characteristics of highland lakes, and they concluded that the long retention time of water bodies, weak exchange capacity and excessive nutrient loading have led to a faster rate of eutrophication in Lake Dianchi. Dong [16] found that soil erosion highly contributed to non-point source pollution in Lake Dianchi Basin and pointed out that rainfall, agricultural structure or rural population changes were conducive to increases in non-point source pollution loadings; because it could be concluded that external loading is the main root of eutrophication in Lake Dianchi, controlling external loading may be the first step to address eutrophication. As the links between a lake and terrestrial ecosystem in a basin, rivers are the key intermediate links for external loading control because land-based pollutants enter lakes by rivers, causing the deterioration of lake water and ecosystem quality [17]. About 70–80% of the annual water supplementation to Lake Dianchi comes from river inflow [18], and the average annual river total nitrogen (TN) and total phosphorus (TP) input can account for 80.2% and 78.8%, respectively, of the total external loading in Lake Dianchi [19], so accurate statistics regarding external loading by rivers is important for eutrophication management. However, there are more than 120 rivers flowing through complex paths into Lake Dianchi, resulting in poor statistics regarding water input and external loading [20]. Therefore, how to effectively invert the missing water volume and calculate external loading was the focus of this study.

A water balance equation is constructed by describing different hydrological processes in a basin as continuous water saving and flow processes, mainly using relevant factors, such as precipitation, temperature and runoff, as input in the water quantity inversion [21]. Zhang et al. [22] combined water deficit measurements (small ditches and rivers leading into the lake, farmland drainage entering the lake and groundwater infiltration and exfiltration) into an uncertain incoming water term, and then they established a water balance equation for Lake Bosten with the incoming river flow, outgoing river flow and lake and evaporation consumption. Qin et al. [23] integrated incoming and outgoing flow, reservoir precipitation and evaporation and the loss of seepage from the reservoir to establish a water balance equation in Guanting Reservoir Station. Zan et al. [24] constructed a water balance equation for the Aral Sea based on regional rainfall, total evaporation and the amount of water entering and leaving the lake, and then they conducted a rough assessment of the total amount of groundwater data missing from monitoring. However, the authors of these articles mainly used historical data to build up simple mathematical equations, which are not computationally adequate for the dynamics of long-term time-series data. With the development of technology, researchers have effectively improved flood process forecasting accuracy by the machine learning method [25,26], and hydrological models have been widely applied to estimating variations of lake volume [27,28]. At present, the calculation principle of external loading is clear, mainly calculated through flow and water quality concentration [29]. Therefore, the authors of this study used the computationally powerful hydrodynamic model DYRESM (Dynamic Reservoir Simulation Model) to calculate the elements of water balance in Lake Dianchi [30]. The DYRESM is a one-dimensional hydrodynamic model with the advantages of simple profile and convenient parameter-rate determination; it has been used in many domestic and international hydrological studies [31–33]. In this study, we first inverted the incoming water volume of Lake Dianchi from 2007 to 2019 and then calculated the external loading by combining the volume information with river water quality monitoring data in order to provide a scientific basis and reasonable suggestions for the management and ecological restoration of Lake Dianchi, as well as to provide reference methods for similar studies.

2. Materials and Methods

2.1. Study Site

Lake Dianchi is located in the southwestern part of the urban area of Kunming, the capital of Yunnan Province (Figure 1). It is one of the four major fractured tectonic lakes in China, and the only sewage-receiving body of the lakes in Kunming [34]. The Lake Dianchi Basin covers an area of 2920 km^2, accounting for about 0.75% of the land area of Yunnan Province but carrying nearly 23% of the province's gross domestic product (GDP) and 8% of the population [35]. With the development of urbanization and agriculture in Kunming, the nutrient loading of basin has significantly increased, resulting in the perennial deterioration of Lake Dianchi's water quality. The water area of Lake Dianchi is 309.5 km^2 (at an elevation of 1887.4 m), with a storage capacity of 1.56×10^8 m^3 and an average depth of 5.3 m [36,37]. The southern part is called Waihai, which is the major part of Lake Dianchi, with a water area of 298.7 km^2 and average water resources that account for more than 90% of the total water resources of Lake Dianchi [37].

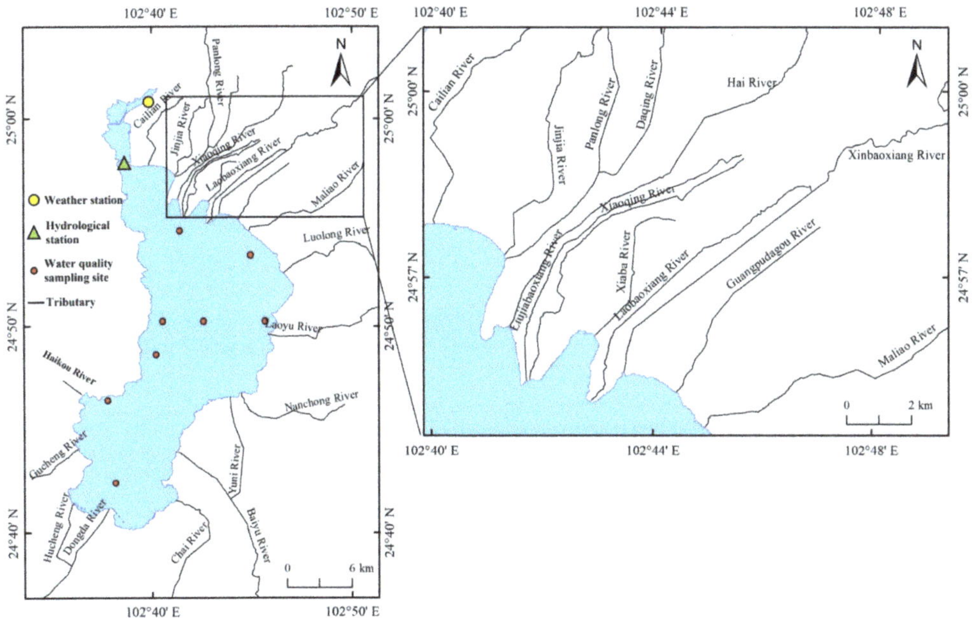

Figure 1. Distribution of meteorological stations, hydrological stations and main tributaries of Lake Dianchi.

Lake Dianchi Basin has a typical subtropical highland monsoon climate, with the mountains in the north blocking the northern cold streams in the winter, which allows the basin to have "four seasons like spring" all year round. The basin maintains a multi-year average temperature between 14.6 °C and 15.9 °C and an annual temperature difference of about 12 °C. The lowest annual temperature occurs in January, and the highest annual temperature occurs in July [38]. The average multi-year rainfall is 986 mm, which can be divided into distinct dry and rainy seasons. The rainy season occurs from May to October each year, with rainfall accounting for more than 85% of the annual total, while the dry season occurs from November to next April, with rainfall accounting for only 15% or less of the annual total. The average multi-year evaporation is about 1871 mm, which is significantly higher than the average annual rainfall [39]. There are many rivers entering Lake Dianchi with a characteristic of "short flow near the source", and the special functions of transport, migration and sink determine the prominent position and role of rivers in the Lake Dianchi ecosystem [19]. In order to meet the ecological water demand, partial tailwater of Kunming urban sewage is discharged into Lake Dianchi after purification and treatment [40]. At present, there are artificially controlled outlets for Caohai and Waihai, which are the Xiyuan Tunnel at the northwest water area and the Zhongtan Gate at the southwest water area, respectively [41]. In addition, to support the urban development of Kunming and to meet the production and living water needs in the basin, Lake Dianchi's water resources are developed and utilized to 90%; furthermore, the total water supply in the basin was 8.20×10^8 m^3 in 2015, of which 1.36×10^8 m^3 was supplied by Lake Dianchi, indicating that water supply is also an important outflow pathway [42]. Since Waihai is the major body of Lake Dianchi, the annual volume of water in and out of the lake and the external loading are absolutely dominant in the total volume of Lake Dianchi [43], so the authors of this paper selected Waihai as the study area. The multi-year distance level changes in chlorophyll a (Chl$-a$), TN and TP concentrations in Waihai are shown in Figure 2, indicating that the water quality has significantly improved after years of treatment.

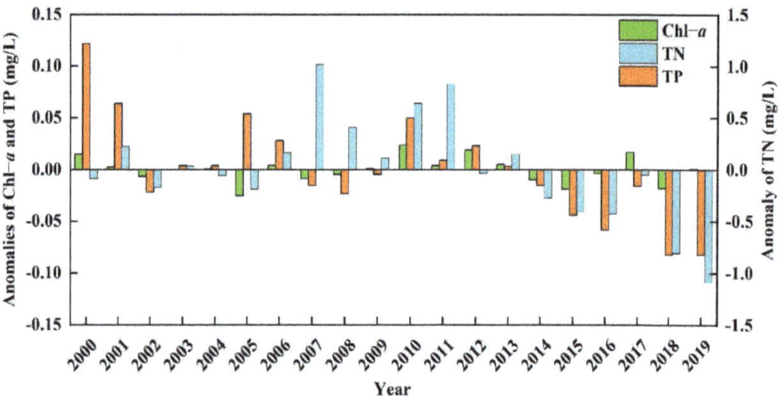

Figure 2. Annual anomalies of Chl$-a$ (green), TN (blue) and TP (brown) at Waihai from 2000 to 2019.

2.2. Data Source

To ensure the accuracy of our results regarding water inversion and external loading, data of detailed hydrological, water quality, meteorological, topographical and river channels at Waihai were collected in this study. They are presented in Table 1 below.

Due to equipment failure and condition restriction, some monthly monitoring data of TN and TP in river channels were missing. When the monitoring data were missing for less than three months, a linear interpolation method was used to supplement. When the data were missing for more than three consecutive months, they were supplemented by calculating the monthly mean data of the previous and following years.

Table 1. Data information.

Data Type	Data Period	Data Content	Data Source
Topography	\	Water contour (elevation of Lake Dianchi bottom-area generation)	Kunming Dianchi & Plateau Lakes Institute
Meteorology	2007–2019	Daily data of weather station at Kunming	The China Meteorological Data Service Center
Inflow river	2007–2019	Monthly monitoring values of TN and TP	Dianchi Administration Bureau of Kunming
	2007–2019	Monthly monitoring values of river flow	
Tail water inflow	2007–2020	Yearly data of municipal treated sewage in Kunming; proportion of tail water discharged into Waihai after treatment in 2020 (31.08%)	Kunming Environmental Statement, Dianchi Administration Bureau of Kunming
Urban water supply	2007–2019	Yearly urban water supply of Kunming; in 2015, urban water supply accounted for 54.24% of whole basin, and Lake Dianchi water supply accounted for 16.59% of basin water supply	Kunming Statistical Yearbook [42]
Water regimen	2007–2019	Average daily water level of Waihai	Dianchi Administration Bureau of Kunming
Water quality	2000–2019	Monthly monitoring values of water quality at Waihai	Kunming Municipal Ecology and Environment Bureau
Water outflow	2007–2019	Daily measured flow of Haikou River	Dianchi Administration Bureau of Kunming

There are 24 major input rivers around Waihai, and some of them presented a small amount of missing monthly water quality data that could be supplemented by using statistical methods. However, we found a large amount of missing data regarding instantaneous monthly river flow, which made it difficult to invert the water inflow volume. Therefore, the authors of this study collected measured data of river inflow, obtained the average values of historical flow for each river from January to December and then calculated the proportion of each river in the total annual flow after summing up the annual flow, which was used to allocate the inverse water volume. The percentage of missing measured data and each river flow in the total inflow volume from 2007 to 2019 are shown in Table 2.

Table 2. Information of rivers.

River	Ratio of Missing Data (%)	Proportion of River Flow in Total Volume (Before 2012)	Proportion of River Flow in Total Volume (After 2012)	River	Ratio of Missing Data (%)	Proportion of River Flow in Total Volume (Before 2012)	Proportion of in River Flow Total Volume (After 2012)
Cailian River	9.6	5.3	5.3	Luolong River	11.5	4.2	4.2
Jinjia River	59.6	8.6	8.6	Laoyu River	9.0	3.3	3.3
Panlong River	63.5	39.6	39.8	Nanchong River	19.9	0.6	0.6
Daqing River	12.2	11.2	11.3	Yuni River	24.4	1.4	1.4
Hai River	12.2	3.1	3.1	Chai River	15.4	1.5	1.5
Liujiabaoxiang River	33.3	0.4	Cutoff	Baiyu River	4.5	2.7	2.7
Xiaoqing River	72.4	1.1	1.1	Cixiang River	7.7	1.3	1.4
Wujiabaoxiang River	35.0	0.1	Cutoff	Dongda River	12.2	2.0	2.0
Xiaba River	62.2	1.8	1.9	Hucheng River	6.4	1.3	1.3
Laobaoxiang River	53.9	0.4	0.4	Gucheng River	1.9	0.4	0.4
Xinbaoxiang River	21.2	7.3	7.4	Guangpudagou River	23.1	0.9	0.9
Maliao River	20.5	0.9	0.9	Yaoan River	78.9	0.6	0.6

2.3. Model Description

DYRESM is a one-dimensional hydrodynamic model developed by the Centre for Water Research at the University of Western Australia that is mainly used for the simulation of lakes and reservoirs [30]. The model is capable of running alone to complete simulations of water temperature and salinity in the vertical direction of lakes and reservoirs, and it can be coupled with the CAEDYM (Computational Aquatic Ecosystem Dynamic Model) ecological model to simulate water quality and life processes of biological organisms in water areas, such as phytoplankton, fish and benthos, as well as the exchange of nutrients between water bodies and sediments [44].

The basic data required by DYRESM contained: (1) topographic basin data, such as the water surface area corresponding to different water depths that was calculated from the elevation–area relationship at the bottom of Lake Dianchi; we stratified the water body of Lake Dianchi at 0.1 m of water depth, with the maximum water depth being 11.5 m, and then separately calculated the water surface area at each depth. (2) The number of inflow channels, outflow channels and the elevation of the river entrance.

The DYRESM boundary conditions included: (1) meteorological files containing the daily data of solar short-wave radiation (W/m^2), air temperature (°C), water vapor pressure (hPa), average wind speed (m/s), cloudiness (0–1) or solar long-wave radiation (W/m^2), rainfall (m) and snowfall (m, set to 0 for areas without snowfall); (2) inflow and outflow files, with inflow files including daily inflow volume and water quality to the lake (m^3) and the outflow file mainly including daily outflow data (m^3).

The initial conditions of DYRESM were the water quality's distribution information in the vertical direction at the starting moment of simulation. The main physical parameters in the model and configuration files were debugged by drawing on the range of values provided in the literature for each parameter. The specific parameter values are shown in Table 3.

Table 3. Key parameters of DYRESM.

Parameter	Value Range	Unit	Value in This Paper
Bulk aerodynamic momentum transport coefficient	1.3×10^{-3}–1.9×10^{-3} [45,46]	\	1.3×10^{-3}
Mean albedo of water	0.07–0.084 [47,48]	\	0.075
Emissivity of water surface	0.94–0.96 [30,48]	\	0.96
Critical wind speed	3–6.5 [45,48]	m/s	5.00
Shear production efficiency	0.06–0.084 [45,48]	\	0.08
Potential energy mixing efficiency	0.15–0.29 [48,49]	\	0.2
Wind-stirring efficiency	0.06–0.9 [32,50]	\	0.2
Extinction coefficient	0.2–0.8 [32,49]	m^{-1}	0.8
Vertical mixing coefficient	200–2500 [32,51]	\	200

2.4. Calculation Principle of Lake Volume Variation

The heat consumed by the evaporation of a lake surface is calculated according to following equation [52]:

$$Q_{lh} = min\left[0, \frac{0.622}{P}C_L\rho_A L_E U_a(e_a - e_s(T_s))\Delta t\right] \tag{1}$$

where Q_{lh} (quantity of latent heat) refers to the heat (J/m^2) consumed by the evaporation of the water surface during Δt, P is the atmospheric pressure (hPa), C_L is the latent heat conduction coefficient (1.3×10^{-3}) of wind speed at a 10 m reference height, $_A$ is the air density (kg/m^3), L_E (2.453×10^6 J/kg) is the latent heat of water evaporation [47], U_a is the wind speed at a 10 m reference altitude (m/s), e_a is the vapor pressure of air (hPa), e_s (saturation vapor pressure) is the saturated vapor pressure (hPa) under the condition of water surface temperature (T_s) and Δt is the calculated time step of model, which is set to 3600 s [52].

The formula for calculating the mass change (kg) of the Nth-layer in a lake caused by evaporation is as follows [52]:

$$\Delta M_N^{(lh)} = \frac{-Q_{lh} A_N}{L_E} \tag{2}$$

where $\Delta M_N^{(lh)}$ represents the mass changes of water (kg) caused by evaporation in the Nth-layer ($N \geq 1$, $N = 1$ means the surface layer of the water column), A_N is the surface area of Nth-layer and other variables are as mentioned above.

The calculation formula of water level rise (m) caused by precipitation is as follows [52]:

$$r_h = R_h \frac{\Delta t}{N_d} \tag{3}$$

where r_h is the water level changes (m) of the Nth-layer caused by precipitation, R_h is the total daily rainfall (m) and N_d is the duration of daily rainfall (s).

The calculation formula of water mass change (kg) in different layers by precipitation is as follows [52]:

$$\Delta M_N^{(rain)} = \rho_N A_N r_h \tag{4}$$

where $\Delta M_N^{(rain)}$ is the mass changes (kg) caused by precipitation of the Nth-layer, ρ_N is the water density (kg/m^3) and other variables are as mentioned above.

The formula for calculating the total water mass change in the Nth-layer of the lake caused by evaporation and precipitation is as follows [52]:

$$\Delta M_N = \Delta M_N^{(lh)} + \Delta M_N^{(rain)} \tag{5}$$

According to the five above-described formulas, DYRESM can automatically calculate the daily evaporation of a lake surface and the corresponding water level variation.

2.5. Calculation Principle of Water Compensation Method

The principle of our water compensation method is shown in Figure 3. After configuring the original inflow and outflow files of DYRESM, considering the influence of lake precipitation and evaporation on the storage capacity, the daily simulated water level was obtained by model using the area and volume data corresponding to different depths provided in the underwater topographic map (scale 1:2000). The water level storage capacity curve of Lake Dianchi was constructed by linear fitting. The used fitting equation was y = 2.89x − 5435.77, where y is storage capacity ($\times 10^8$ m^3), x is water level (m) and correlation coefficient (r) = 0.99. Based on the water level–storage capacity curve, the model was able to calculate the daily simulated storage capacity and measured storage capacity, respectively. Then, the difference between the storage capacity of the next day and that of the previous day was calculated, allowing us to obtain daily simulated storage capacity difference and the measured storage capacity difference data.

The daily compensation value was obtained in the model by subtracting the difference between daily measured storage capacity difference and simulated storage capacity difference. If the compensation value was positive, simulated storage capacity was lower than measured storage capacity, meaning that inflow volume in the model needed to be increased; we set a virtual river channel in the inflow file to supplement increased inflow into the virtual river channel. A negative value indicated that the outflow of model needed to be increased. Taking the absolute value of the compensation value and adding it to the outflow flow to complete the primary water volume compensation calculation, this was followed by a comparison of the simulated water level results with the measured water level. If the error could be ignored, the calculation was stopped. If the error was obvious, water compensation began again. Then, the new inflow and outflow compensation values were calculated so that the inflow (outflow) could be accordingly modified. Following water compensation, the daily water volume of the virtual river was close enough to the daily total inflow of real rivers to be used in subsequent work.

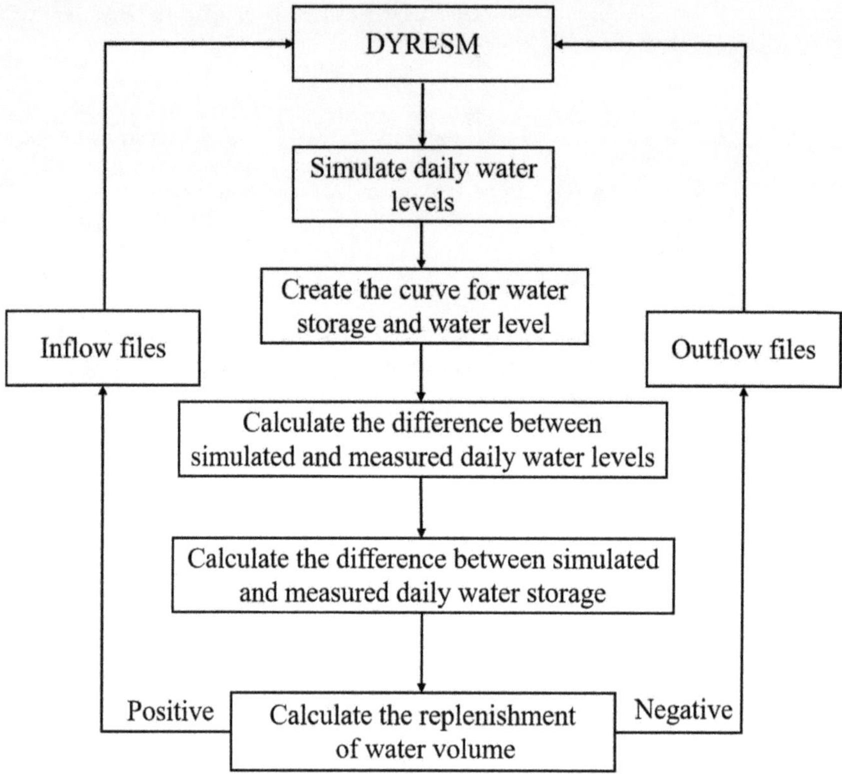

Figure 3. Flowchart of water compensation method for DYRESM.

2.6. Original Inflow and Outflow Files of DYRESM

According to Table 2, river inflow data were seriously missing and could not be configured for the model inflow file. Statistics regarding urban sewage treatment capacity over the years were obtained from the Kunming Environmental Statement. The annual tail water inflow could be calculated based on the proportion of tailwater flow into Waihai after sewage treatment in 2020. The tail water inflow was distributed every day to obtain original inflow profiles containing the tail water data. The original outflow document included the daily measured discharge of Haikou River and the daily water supply of Lake Dianchi. The calculation steps of daily water supply were as follows: First, the urban water supply of Kunming over the years was counted. Second, according to the proportion of urban water supply in Lake Dianchi Basin in 2015, the water supply of basin over the years was obtained. Finally, based on the proportion of Lake Dianchi water supply in the basin water supply in 2015, the water supply of Lake Dianchi over the years was calculated, and the daily water supply in the year was found to be equally distributed. At the same time, since only the water level change was considered, the water quality concentration in and out of the lake was set to 0.

2.7. Evaluation Standard of Model Error

The model error was verified by calculating the root mean square error (*RMSE*) between the measured and model-simulated values, and the Nash efficiency coefficient

(*NSE*) and correlation coefficient (*r*) between measured and simulation values [48]. *RMSE* and *NSE* were calculated as follows:

$$RMSE = \sqrt{\frac{1}{N}\sum_{i=1}^{N}(S_i - O_i)^2} \tag{6}$$

$$NSE = 1 - \frac{\sum_{i=1}^{N}(S_i - O_i)^2}{\sum_{i=1}^{N}(O_i - \overline{O})^2} \tag{7}$$

where O_i is the measured value, S_i is the simulated value, $\overline{O}$ refers to the arithmetic average of measurements and N is the number of data. *RMSE* results can explain the dispersion degree of samples; the smaller the value, the better the simulation effect. The *NSE* is a dimensionless statistical parameter that is commonly used to describe the fitting accuracy of models ($NSE \leq 1$); $NSE = 1$ indicates a complete fit, and $NSE \leq 0$ indicates that the fitting degree is very poor. When the *NSE* is positive, the simulated value can better express the law of the measured value than the average of the measured value. The closer the *NSE* value is to 1, the better the fitting degree and the better the simulation effect.

2.8. Calculation Method of External Loading by River

The dry and rainy seasons are distinct in Lake Dianchi Basin, and the discharge into Waihai of each month significantly varies. Therefore, we allocated the yearly retrieved water volume according to the annual inflow proportion of each river, distributing the annual water volume of each river to the month according to the proportion of historical monthly inflow. We used the monthly measured values of TN and TP of each river as the monthly concentration to obtain the external loading of each river channel, and the total external loading input by river channel was obtained by adding external loading.

3. Results

3.1. Waihai Water Level Simulation

We implemented a water balance analysis from January 2007 to December 2019. After this calculation, the simulated water level clearly agreed well with the observed water level (Figure 4). Before the calculation, the simulated water level at Waihai continued to decline because the annual evaporation in Lake Dianchi Basin is greater than its precipitation [39]. The DYRESM accurately reproduced the water level, with a high coefficient of determination and small relative error values (*RMSE* = 0.0072 m; *NSE* = 0.99; *r* = 0.99). The maximum measured water level was 1887.56 m on 11 August 2015, and the simulated water level on that day was 1887.57 m. The lowest water level occurred on 24 May 2010 (1886.35 m), and the simulated water level was also 1886.35 m. This showed that the DYRESM could reflect fine variations and extreme conditions in measured water levels well after calculation. From 2009 to 2010, the water level at Waihai significantly decreased, which was completely different from other periods and probably because of the drought in Yunnan Province in 2009 [53].

3.2. Retrieval Results of Water Inflow

We calculated the total annual inflow by river (Figure 5a). River flow is closely related to rainfall in the basin [54], so the annual total lake inflow was fitted with the annual total rainfall to verify the accuracy of the calculation results, which are shown in Figure 5b. The correlation coefficient between annual total runoff and annual total rainfall was 0.74 (Figure 5b), indicating a significant relationship between inversion water volume and precipitation, consequently demonstrating that the DYRESM's inversion water amount was feasible. From 2007 to 2019, the annual total lake inflow by river was consistent with the changing trend of the annual total precipitation (Figure 5a). During the study period, the annual average inflow volume of Waihai was about 6.69×10^8 m³, which was consistent with the annual average land water inflow of 6.97×10^8 m³ of Lake Dianchi [55]. The inflow volume significantly decreased after 2009, and it reached the lowest level of only 3.24×10^8 m^3 in 2011. According to the Kunming Statistical Yearbook, 2011 was the third consecutive year of drought relief in Kunming, with a total precipitation of 697.80 mm, 29 cut-off river channels and an accordingly decreased inflow volume. In 2017, the lake inflow was as high as 10.16×10^8 m^3, and the total annual precipitation was 1186.4 mm, both of which were the highest values from 2007 to 2019.

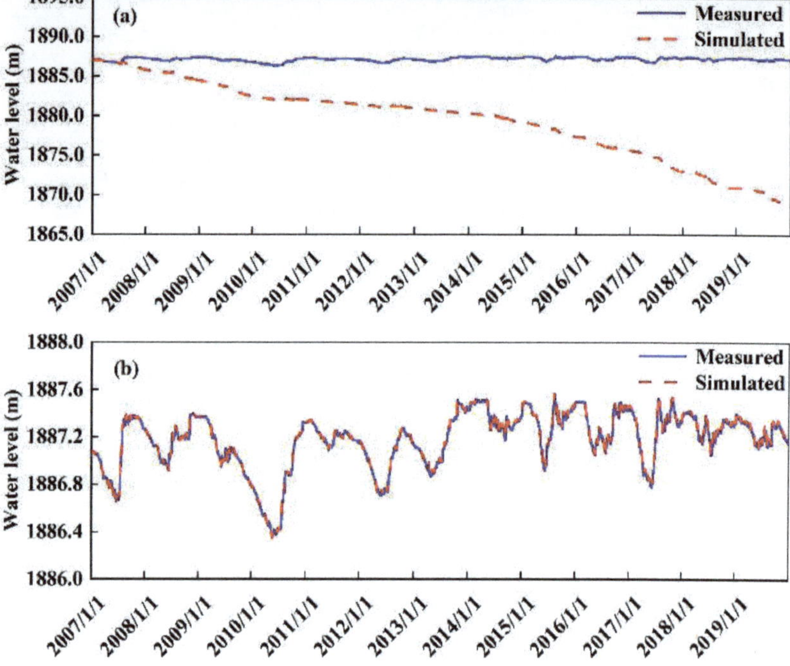

Figure 4. Simulated and measured daily water levels from 2007 to 2019: (**a**) before water compensating; (**b**) after water compensating.

3.3. Variation within the Year of Water Inflow

The simulated and calculated changes in inflow and precipitation with the year showed a single-peak trend of first rising and then decreasing. The water inflow was as high as 1.15×10^8 m^3 in July, and the average precipitation was 212 mm, both of which were first within the year. In the following August, the average inflow volume and precipitation were 1.01×10^8 m^3 and 200.44 mm, respectively. These results are consistent with the viewpoint summarized by Chen that "Flood season in Lake Dianchi Basin is mainly concentrated in July and August" [56]. Through fitting calculation, it was found that there was a close relationship between retrieval inflow and precipitation in the year (Figure 6b; $r = 0.98$), which proves the important significance of precipitation forecast in the flood control and waterlogging work of Lake Dianchi.

3.4. Calculation Results of External Loading by Riverway

To verify the accuracy of the river external loading calculated by simulated volume and measured water quality, we collected TN and TP loading data from inflow rivers during the study period and analyzed them with the calculated loading. The specific data are shown in the following table.

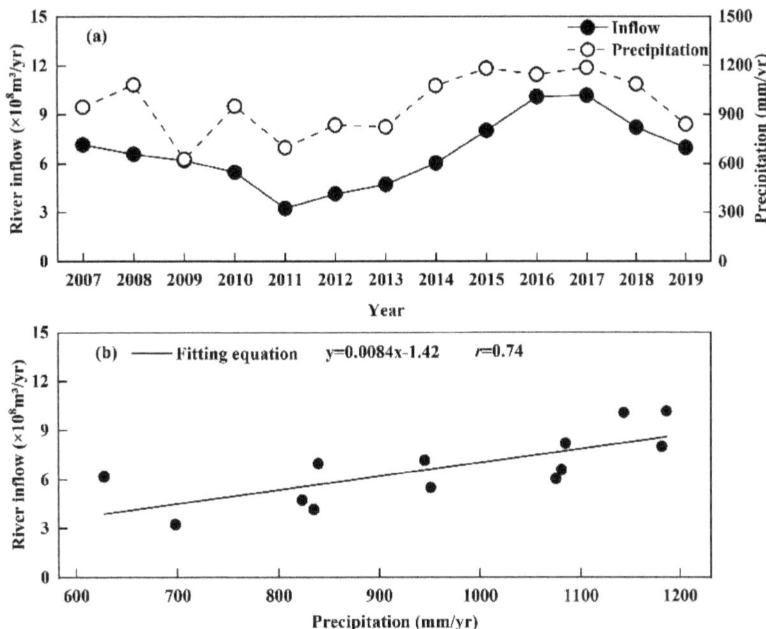

Figure 5. Yearly variations of total river inflow and precipitation (**a**) and their fitting relationship (**b**) for 2007–2019.

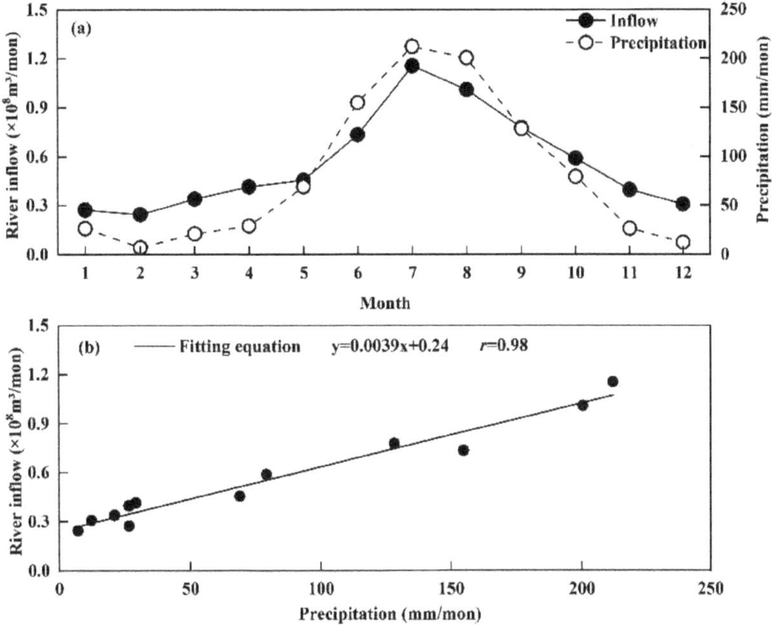

Figure 6. Intra-annual variations of total river inflow and precipitation (**a**) and their fitting relationship (**b**) for 2007–2019.

In some publications, only the basin emissions of TN and TP or the total amount of external loading of Lake Dianchi (including Caohai) have been studied. Therefore, in this study, we used the calculation coefficients of TN loading (the total external loading of Lake Dianchi accounted for 64% of the whole basin and the total amount of Waihai accounted for 73% of the total external loading of Lake Dianchi) and TP loading (the total amount of external loading of Lake Dianchi accounted for 60% of the whole basin and the total amount of Waihai accounted for 90% of the total external loading of Lake Dianchi) according to the specific values given in the "The 14th Five-Year Plan period for water environmental protection and governance in Lake Dianchi Basin". Additionally, the abovementioned coefficients were used to calculate the total amounts of TN and TP in Waihai.

From the data in Table 4, it can be seen that the calculation amount of river loading before 2011 was often higher than the actual amount. This was due to the fact that the loading calculation coefficients were based on Waihai data in 2009, resulting in a reduction effect in external loading before "The 12th Five-Year Plan period" being higher than in reality. However, the change trend of river loading and the total amount of Waihai remained roughly the same. During the study period, the average annual TN loading input by the river channel was 5480 t, and the average annual input loading of TP was 295 t. In 2011, due to a drought in the basin, the amount of water entering Waihai was significantly reduced, resulting in input loadings of TN and TP by the river channel of 2616 t and 107 t, respectively, which were minimum values in the calendar year. From the perspective of time scale, the river loadings of TN and TP declined year by year: the TN loading into Waihai in 2007 was 9080 t and the loading fell to 3728 t in 2019, with a reduction rate of 59%. The TP loadings into Waihai in 2007 and 2019 were 713 t and 115 t, respectively, and the reduction rate was as high as 84%.

Table 4. Annual external loading of Waihai from 2007 to 2019.

Year	TN Loading by Riverway (Calculated Value, Ton)	Total External Loading of TN (Literature Value, Ton)	TP Loading by Riverway (Calculated Value, Ton)	Total External Loading of TP (Literature Value, Ton)	Data Source
2007	9080	7452	713	782	[20]
2008	8290	4990	772	294	[57]
2009	8782	6231	512	697	[58]
2010	6265	4268	284	390	[59]
2011	2616	\	107	\	\
2012	3948	4299	161	370	[60]
2013	4145	3978	197	331	[61]
2014	4167	5358	207	538	[62]
2015	4235	5656	179	495	Lake Dianchi Protection and Governance Plan (2016–2020)
2016	5602	6590	203	566	Lake Dianchi Protection and Governance Three-year Tackling Action Implementation Plan (2018–2020)
2017	5906	3842	205	390	Lake Dianchi Protection Plan (2018–2035)
2018	4472	5109	183	450	[63]
2019	3728	3884	115	397	The 14th Five-Year Plan for Water Environment Protection and Management of Lake Dianchi Basin (2021–2025)

The TN and TP loadings of rivers showed regular changes with the year (Figure 7). TN loading by rivers reached high values of 1035 t and 687 t in July and August, respectively, accounting for 19% and 13% of the whole year's loading, which were similar to the proportions of total inflow volume in July and August. The TN loadings by rivers in the rainy season (May–October) and the dry season were 3473 t and 2007 t, respectively, accounting for 63% and 37% of the annual loading. The TP loadings in the rainy and dry seasons were 197 t and 98 t, respectively, accounting for 67% and 33% of the annual loading. These results show that the river loading has distinct annual distribution characteristics. In the rainy season, with the significant increase in river inflow, the external loading of the river significantly increases. Therefore, different measures should be taken to control the loading of river in the rainy and dry seasons.

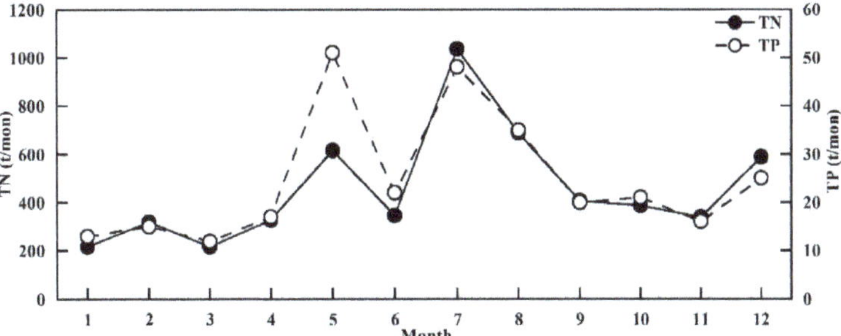

Figure 7. Intra-annual variation of TN and TP loadings for inflow rivers.

In order to clarify the focus of river management, the authors of this study calculated the multi-year proportion of 24 rivers' loading in the total pollutants and highlighted the main external input channels (Figure 8). Panlong River, Daqing River, Jinjia River, Xinbaoxiang River, Cailian River and Haihe River were found to account for more than 5% of the total input of TN and TP at Waihai. Panlong River was found to be the most important source of external loading, with TN and TP loadings accounting for 32.7% and 23.8%, respectively, due to the fact that the amount of water entering Panlong River is about 2.50×10^8 m^3 and the abundant water volume provides convenient conditions for receiving the basin's pollutants [64]. The proportions of the total TN and TP loading by river in Daqing River were second only to Panlong River at 18.1% and 20.4%, respectively, but the inflow volume of Daqing River was found to account for only 11.3%, indicating that the water quality of Daqing River is poor. This is because the upstream tributary called Mingtong River belongs to the sewage channel and the terminal sewage interception gate has the risk of overturning the weir in the rainy season, thus causing Daqing River to face risks of deteriorating water quality. Haihe River was shown to be similar to Daqing River, with an inflow rate of only 3.10%, but it was found to carry 6.1% of TN loading and 11.2% of TP loading, indicating that the water quality of Haihe River is worse than that of Daqing River. This water quality issue is possibly due to the incomplete diversion of rainwater and sewage in the drainage system of river basins, which allows the domestic sewage of villages to easily overflow into the river during the rainy season.

3.5. River Loading and Water Quality

From Figure 2, it can be seen that the water quality of Waihai was significantly improved after treatment. Compared to 2007, the improvement rates of TN and TP of Waihai in 2019 were 68.8% and 50%, so the pollution of the water body was mitigated. There was no obvious trend in the change in Chl−a concentration, which indicated that there are differences in the influencing factors of Chl−a and other water quality indicators and that more targeted treatment measures are needed. A Pearson correlation analysis of yearly data between river loading and Waihai water quality from 2007 to 2019 was performed, and the results are shown in Table 5.

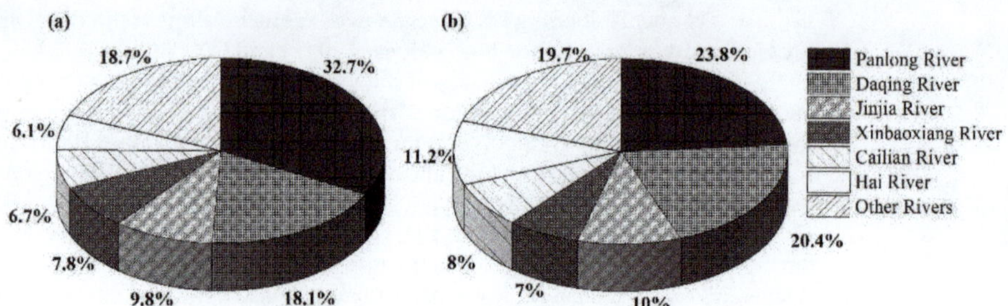

Figure 8. Proportion of external loadings of TN (**a**) and TP (**b**) in the main inflow rivers.

Table 5. Pearson's correlation between river loadings and water quality parameters from 2007 to 2019.

Indicators	Chl$-a$ (mg/L)	TN (mg/L)	TP (mg/L)	n
TN loading (t)	0.008	0.44	0.13	13
TP loading (t)	−0.13	0.54	0.12	13

This table shows that there was no significant correlation between external loading by river and water quality index of Waihai, indicating that there are other influencing factors of water quality besides river loading, implying that the influence of loading input by river is not yet possible without deeper research.

3.6. Effects of Different Pollution Sources

Lake pollution is divided into two types (internal source and external source), and external inputs are dominated by river loading, although atmospheric deposition also has a significant impact on lake pollution that is more significant in highland lakes [65]. The tailwater discharged after sewage treatment also carries certain pollutants. In this study, TN and TP data of Waihai regarding different pollution sources in 2014 were compiled based on the literature, and the results are shown in Table 6.

Table 6. TN and TP loadings from different sources at Waihai in 2014.

Data Type	TN Loading (t)	TP Loading (t)	Data Source
River loading	4167	207	Calculation result
Tail water loading	1953.18	65.11	Calculated by tail water inflow and water quality mission standards
Atmospheric deposition	407.73	34.51	[66]
Internal pollution	539.84	29.88	Basic investigation report on total volume control at Dianchi Basin

River loading was found to account for 59% and 61.5% of the total TN and TP, respectively, due to the significantly higher population density in Lake Dianchi Basin (which is close to twice that of Lake Chaohu) that has enabled the river loading and pollution absorption pressure in Lake Dianchi to become more prominent [67]. Over the years, in order to collect and treat point source sewage and surface source sewage, Kunming has vigorously promoted the construction of urban domestic sewage treatment plants, but there is still a large gap between sewage discharge standards and surface water environmental quality standards that has resulted in a large tailwater loading. Lake Dianchi Basin is also an important flower and vegetable production base, with fertilizer use nearly 2.5 times higher than the national level [68]. Because of low rainfall in the dry season, n and p particulates from biomass burning, industrial production emissions and fertilizer application losses are enriched over Lake Dianchi; in the rainy season, they enter the lake with precipitation and increase the lake pollution [66]. Lake characteristics largely determine the nutrient change pattern of Lake Dianchi, and dynamic disturbances and wind and wave processes in shallow lakes are likely to cause sediment suspension and internal pollution release [69], so the exchange of internal pollutants at the water–sediment interface will accelerate water quality deterioration or seriously affect nutrient loading reduction [70].

4. Discussion

4.1. Analysis of Water Quantity Retrieval and External Loading Results

The inversion of river volume by the DYRESM was a lake water balance model based on the core principle that the increase in lake water over a certain period is equal to all the water entering the lake minus all the water discharging from the lake, so the inversion data at Waihai also contained some groundwater-dominated uncertainty. Groundwater is often regarded as an important recharge source [22], but there are no major rivers transiting Kunming, and the regional water recharge mainly relies on seasonal rainfall, with a groundwater resource of about 1.98×10^8 m^3 [71]. With the development of society, groundwater levels in seven water-rich blocks in the Kunming area continued to decline from 2004 to 2013, accompanied by water level decreases ranging from 0.2 m to 12.6 m [72], so the groundwater recharge to Lake Dianchi could be ignored in this study. The fit verification between river inflow and Kunming precipitation on monthly or yearly scales showed that there was a strong connection between inverse inflow and precipitation, thus proving that precipitation is the fundamental water source of Lake Dianchi and providing a basis for flood prevention through precipitation forecasting. This analysis shows that the results of total inflow by the DYRESM inversion were reliable, and it is reasonable to consider all the total inflow as the river flow in Lake Dianchi Basin.

The measured flow data of river channels were seriously missing (Table 2), meaning the vacant values could not be replaced with statistical methods. Therefore, the authors of this study calculated the annual flow and monthly flow percentage of each river channel to allocate the inverse water volume. Although this method has errors, the real river flow should have distinctive monthly characteristics because precipitation is the main recharge source of the river channel (Figure 6), so the allocation method of this study reflected the monthly changes, and the error value was reduced. In addition, the authors of this paper used the monthly monitoring values of river water quality to represent the daily water quality in each month, which may have led to errors in external loading results. Because the main inflow rivers are located in the northeastern shore and pass through the main human activity area, the basin's pollutants tend to sink into rivers in the rainy season with short-term heavy rainfall, causing the temporary elevation of TN and TP in rivers. Therefore, the accuracy of using river water quality in extreme weather to represent the prevailing conditions in those months was limited. In summary, the frequency of water quality monitoring should be increased to solve this problem in the future.

4.2. Analysis of Water Quantity Retrieval and External Loading Results

Due to the difficulty of "Three rivers and Three lakes" in the key national governance, the governance process of Lake Dianchi is significant. After years of investment, the pollution degree of Lake Dianchi has been effectively alleviated. From 1993 to 2015, the water quality of Lake Dianchi was always deteriorating in the inferior V class, but the water quality changed to V class in 2016 and remained stable in IV class in 2018–2019 [43]. These changes demonstrate the gradual emergence of the treatment effect, a result that is consistent with the trends of water quality and river loading in Figure 2 and Table 3. Although TN and TP loadings declined from 2007 to 2010, they remained high. During the period of the "11th Five-Year Plan", Kunming began conducting engineering governance of inflow rivers, but the river situated at the north shore of Lake Dianchi flows through a main urban area with a large amount of urban sewage and rainwater pollution [18], resulting in limited reduction in river loading [18]. Due to the drought situation, the inflow volume was low in 2011, so external loading in that year was low. In addition, the local government implemented regulations on river management of Kunming in 2010, ensuring the effect of comprehensive regulation in the system while also increasing the investment in river treatment, which laid a solid economic foundation for comprehensive regulation [18]. In conclusion, effective lake environmental management requires long-term system and economic support. During the "12th Five-Year Plan period", Lake Dianchi Basin water pollution control and eutrophication comprehensive control technology were included in a special water project oriented to the whole basin. Accordingly, comprehensive control was enacted and industrial point source pollution was controlled. Attention to inflow river and internal pollution treatment have kept increasing since then. Six major projects, including pollution interception around the lake, agricultural and rural non-point source treatment and ecological restoration construction, have been fully implemented, and the pollution loading into Lake Dianchi has been significantly reduced [61]. During the "13th Five-Year Plan period", the water quality of Lake Dianchi improved as a whole, and cyanobacteria blooms have continued to improve. However, due to temporal and spatial instability, non-point source pollution, internal source release and soil erosion have replaced point source pollution and become the main loading sources [43]. Therefore, river loading has remained at a low level after 2016, but the improvement effect of lake water quality has shown partial hysteresis.

4.3. Influence of Internal Pollution on Waihai Water Quality and Control Measures

Pearson's test results showed that there are other factors besides river loading affecting water quality at Waihai. According to Table 6, internal pollution was found to account for 7.6% and 8.9% of the total TN and TP loadings, respectively, suggesting that the role of internal pollution on the water quality of Lake Dianchi should not be ignored. Sediment nutrients can enter shallow lakes through not only molecular diffusion or concentration gradient diffusion (static diffusion) but also sediment resuspension and changes in conditions at the water–sediment interface; turbulent diffusion causes much higher internal release than static diffusion [73]. Zhu et al. concluded that a wind speed of above 8 m/s may cause a large amount of suspension of sediments in Lake Taihu, and the concentration of dissolved TP may increase up to 100% during strong winds [74]. Luo et al. used field investigations combined with data and mathematical interpolation methods to calculate that, when the wind speed reached 20 m/s, it could result in the suspension of about 2.75×10^8 m^3 of sediment in the upper 30 cm of Lake Taihu [75]. Zhang simulated water body changes in the middle of Lake Chaohu during the sediment resuspension period through laboratory experiments and concluded that the different intensities and durations of external disturbance directly affected the suspended state of sediment particles [76]. Lake Dianchi is a shallow lake with a low water-exchange rate, and a large number of pollutants are deposited at the bottom. When external loading is controlled, sediments in the lake will continue to affect the water quality [77]. In 2012, 6800 t of TN from the basin's non-point source was loaded into Lake Dianchi, and sediment n comprised nearly 67.5% of non-point pollution loading, showing that sediments of internal pollution are very serious [78], as well as leading to external loading reduction benefits that can only be offset to some extent with truncated external loading measures to reduce the trophic level of Lake Dianchi in a short time. The bottom mud of Lake Dianchi contains a variety of humus and organic matter, so Kunming enacted the measures of "environmental protection dredging" and complete reduction through harmless resource treatment of the bottom mud. By 2018, Kunming had cleared 15.17×10^6 m^3 of sediment and used the dredged sediment for ecological basement restoration and ecological forest construction in the low-land area around Lake Dianchi [60,79].

Over the years, local government and civil society have invested huge amounts of resources into river management and ecological restoration, but the restoration of Lake Dianchi has always been a long-term and systematic process. In this study, the amount of inflow volume entering Waihai was analyzed through model inversion. There are many tiny ditches around Lake Dianchi that are not included in the monitoring range, and the accumulation of pollutants in these ditches represents external loadings that cannot be ignored. Therefore, in the next stage of research, a circumnavigation survey of Lake Dianchi will be attempted, small intakes, including ditches, will be counted and water quality monitoring will be regularly carried out to obtain more accurate data regarding inflow volume and external loading.

5. Conclusions

1. The DYRESM can effectively capture extreme changes in water levels with an *RMSE* value of 0.0072 m between simulated and measured water levels and an *NSE* as high as 0.99.
2. During the period of 2007–2019, the multi-year average annual water inflow to Waihai was about 6.69×10^8 m^3, and there is a good fit between water inflow and precipitation in Kunming on an annual scale ($r = 0.74$), with a higher fitting coefficient between intra-annual inflow and precipitation ($r = 0.98$).
3. The external loading by rivers has decreased year by year, although river loading remained at a high level from 2007 to 2010. In 2011, the TN loading dropped to 2616 t and the TP loading dropped to 107 t due to a drought in the basin, and the river loading in subsequent years basically remained at a low level.
4. River loading was found to have clear intra-annual variation characteristics, and the contributions of TN and TP river loadings in the rainy season were 63% and 67% of the annual amount, respectively, indicating that river management should focus more on loading reduction in the rainy season.
5. Panlong River, Daqing River, Jinjia River, Xinbaoxiang River, Cailian River and Hai River are the focuses of treatment, and the sum of the loading of these rivers was found to account for 81.3% (TN) and 80.3% (TP) of the total river input.
6. Pearson's analysis results showed that there was no significant correlation between annual external loading and Waihai water quality, indicating the existence of other factors that influence water quality besides source input.
7. The contribution rates of internal pollution to the total amount of TN and TP were found to be 7.6% and 8.9%, respectively, indicating that the internal control of Lake Dianchi should not be ignored.

Author Contributions: Conceptualization, R.Z., L.L. and H.L.; methodology, L.L.; software, L.L.; validation, J.L., F.G., G.W., L.C. and J.Z.; investigation, M.P., F.H., C.L., D.M. and H.L.; resources, L.L.; data curation, L.L., M.P., F.H., C.L. and H.L.; writing-original draft preparation, R.Z., L.L., J.L., F.G., G.W., L.C., J.Z. and T.S.; writing-review and editing, R.Z., J.L., F.G., G.W., L.C., J.Z. and T.S.; supervision, L.L., M.P. and H.L.; project administration, L.L., M.P., F.H. and H.L.; funding acquisition, L.L. All authors have read and agreed to the published version of the manuscript.

Funding: This research was funded by Yunnan Provincial Department of Science and Technology (202001BB050078), Yunnan University (C176220100043), National Natural Science Foundation of China (41671205) and Dianchi Administration Bureau of Kunming (K207002003021).

Institutional Review Board Statement: Not applicable.

Informed Consent Statement: Not applicable.

Data Availability Statement: The data that support the findings of this study are available from the corresponding author upon reasonable request.

Acknowledgments: The authors would like to thank the English editors from MDPI for their language editing on the manuscript.

Conflicts of Interest: The authors declare no conflict of interest.

References

1. Song, F.; Hu, X.; Jin, X.; Yu, H. Analysis of lake management strategies of different types of lakes abroad and enlightenments for China. *J. Environ. Eng. Technol.* **2013**, *3*, 156–162. (In Chinese)
2. Yang, G.; Ma, R.; Zhang, L.; Jiang, J.; Yao, S.; Zhang, M.; Zeng, H. Lake status, major problems and protection strategy in China. *J. Lake Sci.* **2010**, *22*, 799–810. (In Chinese)
3. Park, Y.; Cho, K.; Park, J.; Cha, S.; Kim, J. Development of early-warning protocol for predicting chlorophyll *a* concentration using machine learning models in freshwater and estuarine reservoirs, Korea. *Sci. Total Environ.* **2015**, *502*, 31–41. [CrossRef] [PubMed]
4. Sinha, E.; Michalak, A.; Balaji, V. Eutrophication will increase during the 21st century as a result of precipitation changes. *Science* **2017**, *357*, 405–408. [CrossRef] [PubMed]
5. Zhao, Y.; Deng, X.; Zhan, J.; Xi, B.; Lu, Q. Progress on preventing and controlling strategies of lake eutrophication in China. *Environ. Sci. Technol.* **2010**, *33*, 92–98. (In Chinese)
6. Kong, F.; Gao, G. Hypothesis on cyanobacteria bloom-forming mechanism in large shallow eutrophic lakes. *Acta Ecol. Sin.* **2005**, *3*, 589–595. (In Chinese)
7. Rockwell, D.; Warren, G.; Bertram, P.; Salisbury, D.; Burns, N. The U.S. EPA Lake Erie Indicators Monitoring Program 1983–2002: Trends in phosphorus, silica, and chlorophyll *a* in the central basin. *J. Great Lakes Res.* **2005**, *31*, 23–34. [CrossRef]
8. Serrano, L.; Reina, M.; Quintana, X.; Romo, S.; Ptzig, M. A new tool for the assessment of severe anthropogenic eutrophication in small shallow water bodies. *Ecol. Indic.* **2017**, *76*, 324–334. [CrossRef]
9. Gibson, G.; Carlson, R.; Simpson, J.; Smeltzer, E.; Kennedy, R. *Nutrient Criteria Technical Guidance Manual Lakes and Reservoirs*; United States Environmental Protection Agency: Washington, DC, USA, 2000.
10. Zhou, Y.; Wang, L.; Zhou, Y.; Mao, X. Eutrophication control strategies for highly anthropogenic influenced coastal waters. *Sci. Total Environ.* **2020**, *705*, 135760.1–135760.11. [CrossRef]
11. Ministry of Ecological Environment, People's Republic of China. China Environmental Status Bulletin. 2020. Available online: https://www.mee.gov.cn/hjzl/sthjzk/zghjzkgb/ (accessed on 26 May 2021).
12. Ni, Z.; Wang, S.; Jin, X.; Jiao, L.; Li, Y. Study on the evolution and characteristics of eutrophication in the typical lakes on Yunnan-Guizhou Plateau. *Acta Sci. Circumstantiae* **2011**, *31*, 2681–2689. (In Chinese)
13. Zhu, Y.; Han, L.; Liu, L.; Qiu, Z.; Wang, Y.; Liu, Y.; He, Y.; Wang, H. Water quality assessment of Dianchi Lake in 2018. *Environ. Sci. Surv.* **2020**, *39*, 75–85. (In Chinese)
14. Gui, H.; Yan, Z. Reflection on eutrophication and non-point source control in Dianchi Lake. *Res. Environ. Sci.* **1999**, *12*, 43–44+59. (In Chinese)
15. Ma, J.; Li, H. Preliminary discussion on eutrophication status of lakes, reservoirs and rivers in China and overseas. *Resour. Environ. Yangtze Basin.* **2002**, *11*, 575–578. (In Chinese)
16. Dong, Y. Study on effects of rainfall and land use on watershed non-point source pollution: Case of Dianchi lake basin. *Yangtze River* **2018**, *49*, 24–33. (In Chinese)
17. Gao, K.; Zhu, Y.; Sun, F.; Chen, Y.; Liao, H.; Ma, H.; Hu, X. A study on the collaborative control of water quality of nitrogen and phosphorus between typical lakes and their inflow rivers in China. *J. Lake Sci.* **2021**, *33*, 1400–1414. (In Chinese)
18. Liu, R.; Zhu, L.; Lei, K.; Fu, G.; Deng, Y.; Zheng, Y.; Li, Z.; Li, F. Comprehensive treatment effect analysis of river inflowing into Dianchi Lake during '11th Five Year Plan' period. *Environ. Pollut. Control.* **2012**, *34*, 95–100. (In Chinese)
19. Jin, Z. Study and Application of Diagnosis and Treatment Techniques for Multi-Type River in Dianchi Lake Basin. Master's Thesis, Shanghai Jiao Tong University, Shanghai, China, 2020.
20. Li, Y.; Xu, X.; He, J.; Zheng, Y.; Zhang, K.; Chen, Y.; Li, Z. Point source pollution control and problem in Lake Dianchi basin. *J. Lake Sci.* **2010**, *22*, 633–639. (In Chinese)
21. Xiong, L.; Guo, S.; Fu, X.; Wang, M. Two-parameter monthly water balance model and its application. *Adv. Water Sci.* **1996**, *S1*, 80–86. (In Chinese) [CrossRef]
22. Zhang, T.; Wu, J.; Lin, J.; Wu, M.; Zhang, H. Analysis of water level change of Bosten Lake based on water balance. *J. China Hydrol.* **2015**, *35*, 78–83. (In Chinese)
23. Qin, Z.; Lei, K.; Huang, G.; Sun, M.; Dai, D.; Cheng, Q. Study on water balance process of Guanting Reservoir under variable conditions. *J. Environ. Eng. Technol.* **2021**, *11*, 56–64. (In Chinese)
24. Zan, C.; Huang, Y.; Li, J.; Liu, T.; Bao, A.; Xing, W.; Liu, Z. Analysis of water balance in Aral Sea and the influencing factors from 1990 to 2019. *J. Lake Sci.* **2021**, *33*, 1265–1275. (In Chinese)
25. Chen, X.; Huang, J.; Wang, S.; Zhou, G.; Gao, H.; Liu, M.; Yuan, Y.; Zheng, L.; Li, Q.; Qi, H. A new rainfall-runoff model using improved LSTM with attentive long and short lag-time. *Water* **2022**, *14*, 697. [CrossRef]
26. Schoppa, L.; Disse, M.; Bachmair, S. Evaluating the performance of random forest for large-scale flood discharge simulation. *J. Hydrol.* **2020**, *590*, 125531. [CrossRef]
27. Malik, M.; Dar, A.; Jain, M. Modelling streamflow using the SWAT model and multi-site calibration utilizing SUFI-2 of SWAT-CUP model for high altitude catchments, NW Himalaya's. *Model. Earth Syst. Environ.* **2022**, *8*, 1203–1213. [CrossRef]
28. Wang, J.; Liu, D.; Tian, S.; Ma, J.; Wang, L. Coupling reconstruction of atmospheric hydrological profile and dry-up risk prediction in a typical lake basin in arid area of China. *Sci. Rep.* **2022**, *12*, 6535. [CrossRef] [PubMed]
29. Guo, Y.; Li, H.; Wang, S.; Bai, Y.; Ren, L.; Ding, I. Contribution analysis of external source pollution load and environmental capacity estimation of reservoirs in Central Yunnan Plateau. *Environ. Sci.* **2022**, 1–18. (In Chinese) [CrossRef]

30. Imberger, J.; Patterson, J. A dynamic reservoir simulation model: DYRESM 5. In *Transport Models/Inland & Coastal Waters*; Academic Press: New York, NY, USA, 1981; pp. 310–361. [CrossRef]

31. Andrea, F.; Rogora, M.; Sibilla, S.; Dresti, C. Relevance of inflows on the thermodynamic structure and on the modeling of a deep subalpine lake (Lake Maggiore, Northern Italy/southern Switzerland). *Limnologica* **2017**, *63*, 42–56.

32. Fadel, A.; Lemaire, B.; Vinçonleite, B.; Atoui, A.; Slim, K.; Tassin, B. On the successful use of a simplified model to simulate the succession of toxic cyanobacteria in a hypereutrophic reservoir with a highly fluctuating water level. *Environ. Sci. Pollut. Res. Int.* **2017**, *24*, 20934–20948. [CrossRef]

33. Li, J.; Li, H.; Luo, L.; Gong, F.; Zhang, R.; Liu, F.; Wu, S.; Luo, B. Water level retrieval for the past and prediction for the next 30 years at Lake Fuxian. *J. Lake Sci.* **2022**, *34*, 1–15. (In Chinese)

34. Liu, X.; Gao, Y.; Du, G.; Wang, W.; Wei, X. Experimental study on effects of iron ion on algae growth of in Dianchi Lake. *Environ. Pollut. Control.* **2006**, *28*, 324–326. (In Chinese)

35. Hu, M.; Zhu, T.; Jiang, Q.; Zou, R.; Wu, Z.; Zhang, X.; Ye, R.; Liu, Y. Simulation study on nitrogen and phosphorus reycling response of changing dissolved oxygen concentration in Lake Dianchi. *Acta Sci. Nat. Univ. Pekin.* **2021**, *57*, 481–488. (In Chinese)

36. Pan, M.; Gao, L. The influence of socio-economic development on water quality in the Dianchi Lake. *Strateg. Study CAE* **2010**, *12*, 117–122. (In Chinese)

37. He, K. Comprehensive evaluation and trend analysis for water quality of Dianchi Lake. *Yangtze River* **2012**, *43*, 37–41. (In Chinese)

38. Feng, Z. Study on pH Value Characteristics and its Change Mechanism of the Main Rivers Flowing into Dianchi Lake. Master's Thesis, Yunnan Normal University, Kunming, China, 2018.

39. Chen, R. Dynamic Characteristics and Genetic Analysis of Groundwater Chemical Field in the North of Kunming Basin. Master's Thesis, Kunming University of Science and Technology, Kunming, China, 2008.

40. Yuan, H.; Hou, L.; Liang, Q.; Li, J.; Ren, J. Correlation between microplastics pollution and eutrophication in the near shore waters of Dianchi Lake. *Environ. Sci.* **2021**, *42*, 3166–3175. (In Chinese)

41. Wang, H.; Chen, Y. Change trend of eutrophication of Dianchi Lake and reason analysis in recent 20 years. *Environ. Sci. Surv.* **2009**, *28*, 57–60. (In Chinese)

42. Chen, G.; Sang, X.; Gu, S.; Yang, X.; Zhou, Z.; Li, Y. Joint disposals of multi-source water resources for rehabilitating healthy water cycle in Lake Dianchi Basin. *J. Lake Sci.* **2018**, *30*, 57–69. (In Chinese)

43. Yang, F.; Xu, Q.; Song, Y.; Zhou, X.; Liu, X.; Yan, C.; Huang, L. Evolution trend, treatment process and effect of water ecological environment in Dianchi Lake Basin. *J. Environ. Eng. Technol.* **2022**, *12*, 633–643. (In Chinese)

44. David, F.; David, P.; Conrad, A. Modelling the relative importance of internal and external nutrient loads on water column nutrient concentrations and phytoplankton biomass in a shallow polymictic lake. *Ecol. Model.* **2008**, *211*, 411–423.

45. Hetherington, A.; Schneider, R.; Rudstam, L.; Gal, G.; Degaetano, A.; Walter, M. Modeling climate change impacts on the thermal dynamics of polymictic Oneida Lake, New York, United States. *Ecol. Model.* **2015**, *300*, 1–11. [CrossRef]

46. Stull, R. *An Introduction to Boundary Layer Meteorology*; Springer: Berlin, Germany, 1988.

47. Patten, B.; Egloff, D.; Richardson, T. Systems Analysis and Simulation in Ecology. In *Total Ecosystem Model for a Cove in Lake Texoma*; Academic Press: Cambridge, MA, USA, 1975; pp. 205–421.

48. Luo, L.; Hamilton, D.; Han, B. Estimation of total cloud cover from solar radiation observations at Lake Rotorua, New Zealand. *Sol. Energy.* **2010**, *84*, 501–506. [CrossRef]

49. Spigel, R.; Imberger, J.; Rayner, K. Modeling the diurnal mixed layer. *Limnol. Oceanogr.* **1986**, *31*, 533–556. [CrossRef]

50. Saddek, T.; Xavier, C. Application of the DYRESM-CAEDYM model to the Sau Reservoir situated in Catalonia, Spain. *Desalination Water Treat.* **2016**, *57*, 12453–12466.

51. Bueche, T.; Vetter, M. Simulating water temperatures and stratification of a pre-alpine lake with a hydrodynamic model: Calibration and sensitivity analysis of climatic input parameters. *Hydrol. Process.* **2014**, *28*, 1450–1464. [CrossRef]

52. Imerito, A. Dynamic Reservoir Simulation Model DYRESM v4: V4.0 Science Manual. Available online: http://www.cwr.uwa.edu.au/software1/CWRDownloads/modelDocs/DYRESM_Science.pdf (accessed on 2 February 2013).

53. Song, J.; Yang, H.; Li, C. A further study of causes of the severe drought in Yunnan Province during the 2009/2010 winter. *Chin. J. Atmos. Sci.* **2011**, *35*, 1009–1019. (In Chinese)

54. Jin, J.; Shu, Z.; Chen, M.; Wang, G.; Sun, Z.; He, R. Meteo-hydrological coupled runoff forecasting based on numerical weather prediction product. *Adv. Water Sci.* **2019**, *30*, 316–325. (In Chinese)

55. Yang, F. Analysis of COD characteristics and accumulation in Dianchi Lake. Master's Thesis, Chinese Research Academy of Environmental Sciences, Beijing, China, 2017.

56. Chen, J. Analysis of water quality of main into-lake rivers in Dianchi Lake. *J. Yunnan Agric. Univ.* **2005**, *20*, 569–572. (In Chinese)

57. Gao, W.; Zhou, F.; Guo, H.; Zheng, Y.; Yang, C.; Zhu, X.; Li, N.; Liu, W.; Sheng, H.; Chen, Q.; et al. High-resolution nitrogen and phosphorus emission inventories of Lake Dianchi Watershed. *Acta Sci. Circumstantiae* **2013**, *33*, 240–250. (In Chinese)

58. Li, Z.; Zheng, Y.; Zhang, D.; Ni, J. Impacts of 20-year socio-economic development on aquatic environment of Lake Dianchi Basin. *J. Lake Sci.* **2012**, *24*, 875–882. (In Chinese)

59. Dong, W. The investigation and study on the management of non-point source pollution in Dianchi Watershed. Master's Thesis, Kunming University of Science and Technology, Kunming, China, 2013.

60. Zhang, Z. Analysis and thinking of control of the Dianchi Lake water pollution. *Environ. Eng.* **2014**, *32*, 26–29+35. (In Chinese)

61. He, J.; Xu, X.; Yang, Y.; Wu, X.; Wang, L.; Li, S.; Zhou, H. Problems and effects of comprehensive management of water environment in Lake Dianchi. *J. Lake Sci.* **2015**, *27*, 195–199. (In Chinese)
62. Xu, X.; Wu, X.; He, J.; Wang, L.; Zhang, Y.; Yan, Y.; Chen, Y.; Ye, H. Research on the pollution characteristics of Dianchi watershed(1988-2014) and identification of countermeasures. *J. Lake Sci.* **2016**, *28*, 476–484. (In Chinese)
63. Wu, X.; Zhang, Y.; He, J.; Zhou, H.; Deng, W. Characteristics of urban residents domestic sewage in Dianchi Lake Watershed. *China Water Wastewater* **2021**, *37*, 89–95. (In Chinese)
64. Li, L.; Wang, S.; Wang, H.; Zhang, R.; Jiao, L.; Ding, S.; Yu, Y. Temporal and spatial variations of phosphorus loading and the forms, compositions and contributions in inlet river of Lake Dianchi. *J. Lake Sci.* **2016**, *28*, 951–960. (In Chinese)
65. Zhang, J.; Yang, F.; Zeng, W.; Jiang, H.; Wang, J. Performance evaluation of the financial investment policy for water pollution control in Dianchi Lake Basin. *Acta Sci. Circumstantiae* **2015**, *35*, 596–601. (In Chinese)
66. Ren, J.; Jia, H.; Jiao, L.; Wang, Y.; Yang, S.; Wu, Q.; Gao, Q.; Cui, Z.; Hao, Z. Characteristics of nitrogen and phosphorus formation in atmospheric deposition in Dianchi Lake and their contributions to lake loading. *Environ. Sci.* **2019**, *40*, 582–589. (In Chinese)
67. Wang, J.; He, L.; Yang, C.; Dao, G.; Du, J.; Han, Y.; Wu, G.; Wu, Q.; Hu, H. Comparison of algal bloom related meteorological and water quality factors and algal bloom conditions among Lakes Taihu, Chaohu, and Dianchi (1981–2015). *J. Lake Sci.* **2018**, *30*, 897–906. (In Chinese)
68. He, X.; Li, Q.; Duan, Q.; Xiong, Y.; Liu, R.; Li, Y.; Guo, Y.; Wang, L.; Zhao, X. Variation analyzing on fertilization and soil nutrients in Yunnan Province. *Soil Fertil. Sci. China* **2011**, *3*, 21–26. (In Chinese)
69. Bu, Q.; Jin, X.; Wang, S. Study of the potential exchangeable phosphorus in sediments of shallow lakes in the middle and lower Yangtze River. *Geogr. Res.* **2007**, *26*, 117–124. (In Chinese)
70. Guo, H.; Wang, X.; Yi, X. Study on eutrophication control strategy based on the succession of water ecosystem in the Dianchi Lake. *Geogr. Res.* **2013**, *32*, 998–1006. (In Chinese)
71. Dong, H.; Liu, P.; Dong, Y.; Zhou, Q.; Zeng, G. Suggestions on reusing wastewater to ensure water resource sustainability. *Environ. Sci. Surv.* **2017**, *36*, 24–27. (In Chinese) [CrossRef]
72. Li, Y.; Zhang, N. Assessment of groundwater overdraft zones in Kunming Basin. *J. Yangtze River Sci. Res. Inst.* **2017**, *34*, 35–38+44. (In Chinese)
73. Luo, L.; Qin, B.; Zhu, G.; Sun, X.; Hong, D.; Gao, Y.; Xie, R. Nutrient fluxes induced disturbance in Meiliang Bay of Lake Taihu. *Sci. China Ser. D* **2005**, *35*, 166–172. (In Chinese) [CrossRef]
74. Zhu, G.; Qin, B.; Gao, G. Direct evidence of violent release of endogenous phosphorus from large shallow lakes caused by wind and waves disturbance. *Chin. Sci. Bull.* **2005**, *50*, 66–71. (In Chinese) [CrossRef]
75. Luo, L.; Qin, B.; Zhu, G. Calculation of total and resuspendable sediment volume in Lake Taihu. *Oceanol. Limnol. Sin.* **2004**, *35*, 491–496. (In Chinese)
76. Zhang, Y. Release of heavy metals from the water-sediment interface of Chaohu Lake under resuspension disturbance. Master's Thesis, Hefei University of Technology, Hefei, China, 2021.
77. Chen, Y.; Tang, L.; Zhang, D.; Li, J.; Zhou, J.; Guan, X. Spatial and temporal dynamic variation of nitrogen in sediment of Dianchi Lake. *Soils* **2007**, *39*, 879–883. (In Chinese)
78. Wang, M.; Yan, H.; Jiao, L.; Wang, S.; Liu, W.; Luo, J.; Luo, Z. Characteristics of internal nitrogen loading and influencing factors in Dianchi Lake sediment. *China Environ. Sci.* **2015**, *35*, 218–226. (In Chinese)
79. Ling, S. Study on the improvement of water quality and water environment of Dianchi Lake by Central Yunnan Water Diversion Project. Master's Thesis, Kunming University of Science and Technology, Kunming, China, 2018.

 *water*

Article

Modeling the Effects of Climate Change and Land Use/Land Cover Change on Sediment Yield in a Large Reservoir Basin in the East Asian Monsoonal Region

Huiyun Li [1], Chuanguan Yu [2], Boqiang Qin [1,*], Yuan Li [3], Junliang Jin [4], Liancong Luo [5,*], Zhixu Wu [2], Kun Shi [1] and Guangwei Zhu [1]

[1] State Key Laboratory of Lake Science and Environment, Nanjing Institute of Geography and Limnology, Chinese Academy of Sciences, Nanjing 210008, China; hyli@niglas.ac.cn (H.L.); kshi@niglas.ac.cn (K.S.); gwzhu@niglas.ac.cn (G.Z.)

[2] Hangzhou Bureau of Ecology and Environment Chun'an Branch, Hangzhou 311700, China; ycggates@126.com (C.Y.); caepb@126.com (Z.W.)

[3] School of Tourism and Urban & Rural Planning, Zhejiang Gongshang University, Hangzhou 310018, China; liyuan@zjgsu.edu.cn

[4] Yangtze Institute for Conservation and Development, Nanjing Hydraulic Research Institute, Nanjing 210098, China; jljin@nhri.cn

[5] Institute for Ecological Research and Pollution Control of Plateau Lakes, School of Ecology and Environ-Mental Science, Yunnan University, Kunming 650504, China

* Correspondence: qinbq@niglas.ac.cn (B.Q.); billluo@ynu.edu.cn (L.L.)

Citation: Li, H.; Yu, C.; Qin, B.; Li, Y.; Jin, J.; Luo, L.; Wu, Z.; Shi, K.; Zhu, G. Modeling the Effects of Climate Change and Land Use/Land Cover Change on Sediment Yield in a Large Reservoir Basin in the East Asian Monsoonal Region. *Water* **2022**, *14*, 2346. https://doi.org/10.3390/w14152346

Academic Editors: Fei Zhang, Ngai Weng Chan, Xinguo LI and Xiaoping Wang

Received: 30 May 2022
Accepted: 25 July 2022
Published: 29 July 2022

Publisher's Note: MDPI stays neutral with regard to jurisdictional claims in published maps and institutional affiliations.

Abstract: This research addresses the separate and combined impacts of changes in climate and land use/land cover on the hydrological processes and sediment yield in the Xin'anjiang Reservoir Basin (XRB) in the southeast of China by using the soil and water assessment tool (SWAT) hydrological model in combination with the downscaled general circulation model (GCM) projection outputs. The SWAT model was run under a variety of prescribed scenarios including three climate changes, two land use changes, and three combined changes for the future period (2068–2100). The uncertainty and attribution of the sediment yield variations to the climate and land use/land cover changes at the monthly and annual scale were analyzed. The responses of the sediment yield to changes in climate and land use/land cover were considered. The results showed that all scenarios of climate changes, land use/land cover alterations, and combined changes projected an increase in sediment yield in the basin. Under three representative concentration pathways (RCP), climate change significantly increased the annual sediment yield (by 41.03–54.88%), and deforestation may also increase the annual sediment yield (by 1.1–1.2%) in the future. The comprehensive influence of changes in climate and land use/land cover on sediment yield was 97.33–98.05% (attributed to climate change) and 1.95–2.67% (attributed to land use/land cover change) at the annual scale, respectively. This means that during the 2068–2100 period, climate change will exert a much larger influence on the sediment yield than land use/land cover alteration in XRB if the future land use/land cover remains unchanged after 2015. Moreover, climate change impacts alone on the spatial distribution of sediment yield alterations are projected consistently with those of changes in the precipitation and water yield. At the intra-annual scale, the mean monthly transported sediment exhibits a significant increase in March–May, but a slight decrease in June–August in the future. Therefore, the adaptation to climate change and land use/land cover change should be considered when planning and managing water environmental resources of the reservoirs and catchments.

Keywords: climate change; land use/land cover change; sediment response; multiple scenarios; modeling

---

1. Introduction

Catchment sediment yield is mainly controlled by soil properties, topography, climate condition, and land use/land cover types [1–4]. In contrast, the soil properties and topography are relatively stable, while climate and land use/land cover are variable over a specific time period [5,6]. Climate change, mainly in the form of temperature and precipitation, has a direct impact on runoff and an indirect impact on sediment by changing the process of the water cycle in the basin, and further influences the phytoplankton community [7–9]. Land use/land cover changes caused by anthropogenic activities may re-distribute the rainfall-runoff by changing the processes of infiltration, evapotranspiration, and groundwater recharge, which has a profound impact on the water and sediment production mechanism [10,11].

The impacts of climate change on streamflow and sediment yield have been investigated in a number of studies [12–18]. A previous study indicated that the runoff increased by 1.3% and the sediment yield increased by 2% for every 1% increase in rainfall in eight large Chinese catchments [12]. Similarly, a preliminary study of a watershed in Spain showed that higher precipitation is usually associated with more runoff and soil loss [13]. This is not only because precipitation increases soil moisture, but also because it saturates soil moisture or produces soil crusts [14]. In contrast, Zhao et al. showed that the reduction in precipitation was one of the main factors, leading to the sharp reduction in the discharge and sediment yield in the middle reaches of the Yellow River [15]. However, the impacts of precipitation change on soil erosion are complicated and are not always negative. Increasing rainfall may increase the plant biomass and vegetation canopy, thus reducing the runoff and erosion [16]. In addition to precipitation, temperature is also one of the important meteorological factors affecting the sediment of the basin [17,18]. For example, Syvitski divided the watersheds into climatic zones according to different temperatures, and found that the average temperature of the watershed has an important impact on sediment transport [18].

On the other hand, the joint effects of climate variability and vegetation change on hydrological process have been a key research point. Such synergistic influences on hydrological processes and sediment yields are complex [19]. Some studies have found that sediment alteration was dominantly influenced by land use/land cover changes, while some showed that climate variability was a more important impact factor [20]. It is essential to accurately distinguish and quantify the effects of climate variability/climate change on streamflow and sediment for catchment and reservoir management in the future under different conditions [21–24]. Compared to the influence on streamflow, few works have concerned the sediment spatial and temporal changes in response to combining the variations in the land use/land cover with climate change for an uncertain future. Therefore, a thorough study on the impacts of multiple climatic conditions and land use/land cover scenarios on sediment is needed [25].

An IPCC Special Report stated that a global warming of 1.5 °C above pre-industrial levels has significantly affected the hydrological process including the quality and quantity of water resources in many regions [26,27]. Until now, numerous studies on assessing the response of hydrological circles to climate-driven force have widely applied the general circulation model (GCM) projections of the coupled model inter-comparison project phase 5 (CMIP5) [4,28]. A tentative conclusion is that RCP2.6, RCP4.5, RCP6.0, and RCP 8.5 are responsible for a 16.3%, 14.3%, 36.7%, and 71.4% increase in future streamflow, and a 16.5%, 32.4%, 81.8%, and 170% increase in future sediment yield, respectively, in northeastern China [4]. An increase in monthly streamflow (maximum increases by 52–170% under different RCP scenarios) was reported, along with a monthly average decrease in sediment concentrations of 10% projected in southwest Iran in the future [28]. Although GCM outputs have been extensively employed to study the impacts of climate change on the hydrological process in many locations, it is problematic to use GCM outputs directly in hydrological models at regional and local scales because of the low resolution of GCM projections [29]. Therefore, downscaling methods are often applied to obtain regional scale analysis of meteorological variables from coarse-scale GCM outcomes to allow the conclusions on streamflow and sediment regime changes to be more reliable [30].

The Xin'anjiang Reservoir, which is the largest reservoir in the Yangtze River Delta in China, plays quite an important role in the local water supply, fishery, water transportation, and crop irrigation [31]. The Xin'anjiang Reservoir is famous for its excellent water quality; however, the pressure of water environment protection in the reservoir is increasing year by year [32]. The Jiekou section, located in the estuary area of the Xin'anjiang Reservoir, in particular, is facing the problem of a decrease in water transparency and the risk of algal blooms [33]. This might be related to the climate variability and land use/land cover change in the basin. Previous studies have noted that the annual streamflow through the Jiekou section, accounting for over 60% of the total inflows of the Xin'anjiang Reservoir, showed an obvious increasing trend in the last few decades caused by rainstorms [34,35]. However, few studies have attempted to identify how climate variability and land use/land cover change affect sediment yield. In this study, we focused on identifying and quantifying the effects of climate change and land use/land cover change on the sediment yield using a hydrological modeling approach. With the help of our research results, a deeper understanding of sediment response to climate-driven forcing and land use/land cover changes in XRB would be beneficial for water quality protection and bloom prevention of the reservoir in the East Asian monsoonal region.

2. Data and Methods

2.1. Study Area

The Xin'anjiang River drains into the Xin'anjiang Reservoir, Chun'an, Zhejiang Province, southeast China, situated within a watershed that spans an area of roughly 10,442 km^2 (Figure 1) [36]. The reservoir has a surface area of 573 km^2 and a water storage capacity of 178.4×10^8 m^3 when the normal water storage level is 10^8 m asl [37]. The longest path of the river is over 370 km, and two river gauging stations are located at Tunxi and Yuliang, respectively. The basin is dominated by a typical subtropical humid monsoon climate and enters the East Asian rainy season, also known as the plum rain, in June and July every year [38]. For the last 50 years, the mean annual precipitation has been about 1621 mm, the mean annual runoff is about 1018 mm, and the mean annual air temperature has ranged from 16.7 °C to 18.9 °C. Approximately 42% of the annual precipitation is contributed by monsoons (June–September), and the maximum humidity is recorded as 100% in June and July.

Jiekou is the main entrance for the streamflow and sediment of the Xin'anjiang River to Xin'anjiang Reservoir by controlling around 60% of the area of the whole basin [39]. The elevation of the basin varies from -1 m to 1764 m from the mean sea level. The terrain is complex and diverse with mainly a geomorphic type of mountains. The zonal soil types of the basin are mainly red soil, yellow soil, and yellow brown soil, which are distributed vertically according to the altitude. The area is covered with dense forests, which is the most widely distributed land-use type. The cultivated land is concentrated at the periphery of urban land [40].

2.2. Data Description

2.2.1. Hydrometeorological Data

Daily meteorological data recorded including air temperature (°C), precipitation (mm), relative humidity (%), solar radiation (MJ/m^2/day), and wind speed (m/s) from 1973 to 2018 at two meteorological stations (Figure 1) were downloaded from the website of the National Meteorological Information Center (China Meteorological Administration, CMA) (http://data.cma.cn/en (accessed on 1 January 2019)) [41]. The observed daily streamflow data for the period of 2001–2014 at two hydrological stations (Figure 1) were collected from the Hydrological Data Yearbook published by the Ministry of Water Resources of the People's Republic of China (MWR) [42]. The mean sediment transport rate investigated from 2006 to 2014 was obtained from the same data source. The time series above was checked for outliers and errors in order to be used in hydrological modeling.

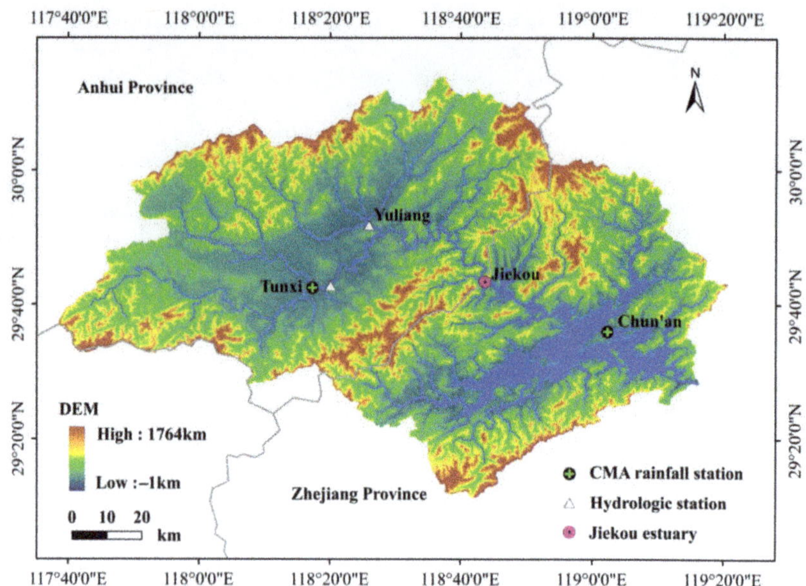

Figure 1. The location of the Xin'anjiang Reservoir Basin, China.

2.2.2. Geospatial Data

The basic geospatial datasets required to construct the model include a digital elevation model (DEM), a soil classification map, and land use information. The DEM map with a 90 m spatial resolution used for watershed delineation and sub-basin discretization was downloaded by the Geospatial Data Cloud of China. The 1 km resolution soil map was originally derived from the Harmonized World Soil Database (HWSD), which is produced by the Food and Agriculture Organization of the United Nations [43]. The soil data over China were derived from the results of the Second National Land Survey organized by China's State Council from 2007 to 2009. This was produced by the Institute of Soil Science, Chinese Academy of Sciences.

To estimate the effect of land use/land cover change, two land use/land cover maps with a spatial resolution of 30 m for XRB were interpreted from Landsat imagery in 1987 and 2015, respectively. The cloud was masked based on the pixel_qa band of the Landsat surface reflectance data (https://developpers.google.com/earth-engine/datasets/catalog/LANDSAT_LC08_C01_T1_SR (accessed on 1 May 2020)) after the images were obtained. A median imagery was output by calculating the median value at each pixel of all images in one collection (https://developers.google.com/earth-engine/reducers_image_collection (accessed on 1 May 2020)). For each median imagery, the land use/land cover information was extracted with a support vector machine classification algorithm in the ENVI (version 5.3). The land use/land cover here is classified into five classes for the SWAT model, namely forest, water body, cultivated land, urban land, and bare land [44].

2.2.3. RCP Data

Representative concentration pathways (RCPs) including a stringent mitigation scenario (RCP2.6), an intermediate scenario (RCP4.5), and one scenario with very high GHG emissions (RCP8.5) [45] were used to estimate the impacts of climate change. Eighteen datasets obtained from a coupled model inter-comparison project phase 5 (CMIP5) GCM for XRB were downloaded from the website of the World Climate Research Program (https://esgf-node.llnl.gov/search/cmip5/ (accessed on 1 February 2020)). The Taylor diagram method was adopted to assess the performance of datasets from CMIP5 GCMs in simulating the historical meteorological elements [46]. Four assessment criteria—the correlation coefficient (r), root mean square error ($RMSE$), standard deviation of observed values (σ_O), and standard deviation of simulated values (σ_S)—were used to identify the most applicable dataset. More detailed information about the Taylor diagram method can be found in Taylor [46].

The selected CMIP5 datasets comprise three meteorological elements (daily air temperature and precipitation) for a historical period (1901–2005) and a projection period (2006–2100, RCP2.6, RCP4.5, and RCP8.5 scenarios). The original resolution data were downscaled into $0.5° \times 0.5°$ by the China Meteorological Data Service Center (CMDC) using a statistical downscaling method.

2.3. Methodology

An integrated framework was designed to evaluate the effect of climate change and land use/land cover change on the streamflow and sediment yield using XRB as a case study. To set up the structure of this approach, we (1) assessed the accuracy and availability of the downscaled GCM data, and the interpreted land use/land cover map from remote sensing imagery; (2) designed individual and combined climate and land use/land cover change scenarios; (3) modeled streamflow and sediment yield response under uncertainty; and (4) evaluated the streamflow and sediment variation under climate change and land use/land cover change. The simulation baseline is in the period of 1973–2005 and the future is in the period of 2068–2100.

2.3.1. Climate Change and Land Use/Land Cover Change Scenarios

Three RCP scenarios were selected in this study to assess how different emissions impact streamflow and sediment yield, namely, RCP2.6, RCP4.5, and RCP8.5. These three scenarios represent the total radiative force in 2100 relative to pre-industrial values, which are +2.6, +4.5 and +8.5 W/m^2, respectively. The calibrated SWAT model was used to simulate the following eight scenarios: SN_B, $SN_{2.6}$, $SN_{4.5}$, $SN_{8.5}$, SN_B^{LC}, $SN_{2.6}^{LC}$, $SN_{4.5}^{LC}$, and $SN_{8.5}^{LC}$, respectively, as listed in Table 1. The SN_B^{LC}, $SN_{2.6}^{LC}$, $SN_{4.5}^{LC}$, and $SN_{8.5}^{LC}$ were essentially the SN_B, $SN_{2.6}$, $SN_{4.5}$, and $SN_{8.5}$ scenarios with the addition of the land use/land cover change. We present the land utilization condition in 1987 as the baseline of the land use/cover, based on assuming that there were no significant changes in the land use/land cover during the baseline period (1973–2005). Similarly, we used the land use/land cover map in 2015 as the representative land use/land cover in the future period (2068–2100). We assumed that there would be no significant changes in the land use/land cover between the future period (2068–2100) and that in 2015. More details on our scenarios can be found in Table 1.

Table 1. The scenario analysis for characterizing the effects of climate change and land use/land cover change on the streamflow and sediment.

Scenario	Simulation Time	Land Use/Cover	Climate	Description
SN_B	1973–2005	LULC1987	History	Baseline
$SN_{2.6}$	2068–2100	LULC1987	RCP2.6	With a stringent mitigation scenario and no land use/land cover change
$SN_{4.5}$	2068–2100	LULC1987	RCP4.5	With an intermediate scenario and no land use/land cover change
$SN_{8.5}$	2068–2100	LULC1987	RCP8.5	With a very high greenhouse gas emission scenario and no land use/land cover change
SN_B^{LC}	1973–2005	LULC2015	History	With a land use/land cover change and no climate change
$SN_{2.6}^{LC}$	2068–2100	LULC2015	RCP2.6	With land use/land cover change and a stringent mitigation scenario
$SN_{4.5}^{LC}$	2068–2100	LULC2015	RCP4.5	With land use/land cover change and intermediate scenario
$SN_{8.5}^{LC}$	2068–2100	LULC2015	RCP8.5	With land use/land cover change and a very high greenhouse gas emission scenario

2.3.2. SWAT Hydrological Model

The SWAT (Soil & Water Assessment Tool) model developed by the USDA (the United States Department of Agriculture) is a semi-distributed, process-based, continuous, daily time-step hydrological model. It has been widely applied to represent the main hydrological processes within small and large basins [11,25,47]. In this study, ArcSWAT (an ArcGIS-ArcView extension and interface for SWAT) running on the ArcGIS (version 10.2) platform as an interface was used to assess the streamflow and sediment yield. Several sub-basins and multiple HRUs (Hydrologic Respond Units) are divided according to the land use types, soil classes, and slopes. Erosion caused by rainfall and runoff is calculated with the Modified Universal Soil Loss Equation (MULSE) in the SWAT model [48]. Parameter sensitivity analysis, calibration, and validation are carried out by SWAT-CUP (SWAT Calibration and Uncertainty Programs), which is an automatic sensitivity analysis tool in the SWAT model [49–51]. Sensitivity analysis is the procedure used to identify the most influential parameters for calibration using SUFI-2 (the global sensitivity analysis of the sequential uncertainty fitting) algorithm. In order to evaluate the performance of the SWAT model in streamflow and sediment yields simulations, the Nash–Sutcliffe coefficient of efficiency (NSE) and coefficient of determination (r^2) between the observed and estimated values were calculated by [52]:

$$NSE = 1 - \frac{\sum_{i=1}^{n}(O_i - S_i)^2}{\sum_{i=1}^{n}(O_i - \overline{O})^2} \tag{1}$$

$$r^2 = \frac{\left[\sum_{i=1}^{n}(O_i - \overline{O})(S_i - \overline{S})\right]^2}{\sum_{i=1}^{n}(O_i - \overline{O})^2 \sum_{i=1}^{n}(S_i - \overline{S})^2} \tag{2}$$

where O_i and S_i are the observed and simulated hydrological parameters and $\overline{O}$ and $\overline{S}$ are the mean of observed and simulated values, respectively. The criterion considers the model performance to be: very good if $0.75 \leq NSE < 1.00$ and $r^2 = 1.00$; good if $0.65 < NSE \leq 0.75$ and $0.80 \leq r^2 < 1.00$; satisfactory if $0.40 < NSE \leq 0.65$ and $0.50 \leq r^2 < 0.80$; unsatisfactory if $NSE \leq 0.40$ and $r^2 < 0.50$ [53–55].

2.3.3. Sediment Response to Changes of Climate and Land Use/Land Cover

For a given catchment, the total change in the mean annual sediment between independent periods with different climatic RCP scenarios and land use/land cover characteristics can be estimated as:

$$\Delta D_{RCPj}^{LC} = \Delta D_{RCPj} + \Delta D^{LCj}, \; j = 2.6, \, 4.5 \text{ and } 8.5 \tag{3}$$

where ΔD_{RCPj}^{LC} indicates the total change in the mean annual sediment between the future and baseline and ΔD_{RCPj} is the change in the mean annual sediment because of the climate change (different RCP scenarios, j = 2.6, 4.5 and 8.5, respectively) between the two periods. We assumed that there were almost no other regulations or diversions except for land use/land cover change in the catchment. ΔD^{LCj} indicates the change in the mean annual sediment as a result of change in the land use/land cover change between the two periods.

To separate the sediment yield impacts caused by climate variability and land use/land cover change, an effective method used to quantify ΔD_{RCPj} and ΔD^{LC} can be seen in the following expressions [56]:

$$\Delta D_{RCPj} = \frac{(D_{RCPj} - D_B) + \left(D_{RCPj}^{LC} - D^{LC}\right)}{N}, \; N = 2, \, j = 2.6, \, 4.5 \text{ and } 8.5 \tag{4}$$

$$\Delta D^{LCj} = \frac{(D^{LC} - D_B) + \left(D_{RCPj}^{LC} - D_{RCPj}\right)}{N}, \; N = 2, \, j = 2.6, \, 4.5 \text{ and } 8.5 \tag{5}$$

where $D_{RCPj}(j = 2.6, 4.5, \text{ and } 8.5)$ are the mean annual sediment under the RCP2.6, RCP4.5, and RCP8.5 scenarios, respectively, with the historical land use/land cover condition. D_B is the mean annual sediment of the baseline. D_{RCPj}^{LC} is the mean annual sediment under different RCP scenarios after catchment land use/land cover change. D^{LC} indicates the mean annual sediment caused by land use/land cover change without climate variability. The climate condition when simulating D^{LC} was as the same as simulating D_B.

As a result of the above, the difference in sediment between RCP scenarios (D_{RCPj}) and the baseline (D_B) can be considered as the impacts of RCP scenarios on sediment change (1987 land use/land cover condition). Similarly, the difference in sediment between D_{RCPj}^{LC} and D^{LC} can be considered as the impacts of RCP scenarios on sediment change (2015 land use/land cover condition). On the other hand, the effects of land use/land cover change on sediment can be determined by applying the difference between D^{LC} and D_B or between D_{RCPj}^{LC} and D_{RCPj}. The difference between sediment in different RCP scenarios after land use/land cover change (2015 land use/land cover condition) and the baseline represents the combined effects of climate variability and land use/land cover change. The combined effects can also be described as:

$$\Delta D_{RCPj}^{LC} = D_{RCPj}^{LC} - D_B, \ j = 2.6, \ 4.5 \text{ and } 8.5 \tag{6}$$

Therefore, the percentage contributions of different RCP scenarios (α_{RCPj}) and land use/land cover change (α^{LC}) to the variations in sediment can be expressed by:

$$\alpha_{RCPj} = \frac{\Delta D_{RCPj}}{\Delta D_{RCPj}^{LC}} \times 100\%, \ j = 2.6, \ 4.5 \text{ and } 8.5 \tag{7}$$

$$\alpha^{LCj} = \frac{\Delta D^{LCj}}{\Delta D_{RCPj}^{LC}} \times 100\%, \ j = 2.6, \ 4.5 \text{ and } 8.5 \tag{8}$$

3. Results and Discussion

3.1. Climate Change Analysis under Varying Scenarios

Monthly meteorological data from CMA were used to assess the performance of GCM outputs in climate in XRB. The arithmetic average value of the records of Tunxi Station and Chun'an Station represented the average value of the basin. Eighteen meteorological datasets including three elements (maximum temperature, minimum temperature, and precipitation) from downscaled CMIP5 GCMs were used to plot a Taylor diagram against the CMA data (see Figure 2). For the monthly minimum temperature, r values between the CMA data and eighteen GCM outputs were 0.94–0.97, and all $RMSE$ values were less than 2.5 °C. Meanwhile, the σ_S of all GCM monthly minimum temperatures and σ_O were very close. Hence, all eighteen GCMs were suitable to simulate the historical data of the monthly minimum temperature (1973–2005). Eleven datasets outperformed the other GCM outputs for the monthly maximum temperature with higher r and lower $RMSE$ (Figure 2). It can be seen from the Taylor diagram that the simulation results of all eighteen GCMs on monthly precipitation were not as good as those on monthly temperature. The highest r for precipitation was around 0.38, and the lowest $RMSE$ was around 8.0 mm. Taken together, a certain dataset, namely, CSIRO-Mk3-6-0, was selected to evaluate the effect of climate change between the future and historical periods due to its best performance in climate simulation in the basin.

Figure 3a,c,e shows the time series of the downscaled CSIRO-Mk3-6-0 annual maximum and minimum temperature averaged over XRB in the baseline period (1973–2005) and 2006–2100. The mean annual maximum and minimum temperatures of the basin at the baseline were 20.79 °C and 11.43 °C, respectively, while those in the last 33 years of the 21st century (the simulation period, 2068–2100) will be increased dramatically by 1.91–5.11 °C and 1.75–4.46 °C relative to that at the baseline. The mean monthly maximum temperature and minimum temperature from 2068 to 2100 under RCP2.6, RCP4.5, and RCP8.5 will be increased by 1.36–7.14 °C and 0.72–7.18 °C relative to that at the baseline. XRB has four distinctive seasons, with the highest increases in seasonal maximum and the minimum temperature of 2.46–5.65 °C and 2.59–5.97 °C in fall (September–November) under different RCPs, respectively.

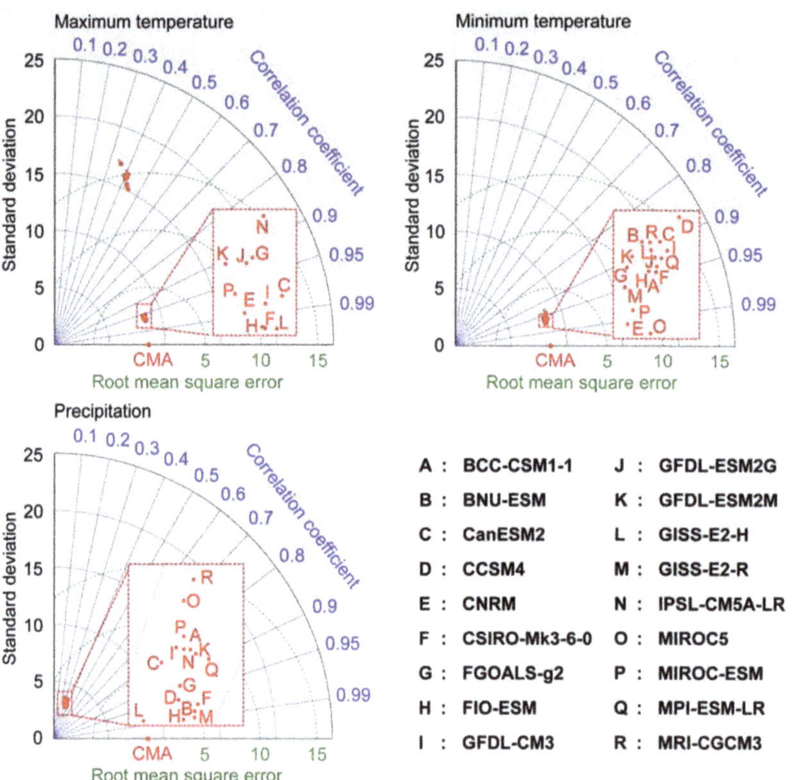

Figure 2. The Taylor diagram of the monthly maximum temperature, minimum temperature, and precipitation simulated by the 18 GCM models.

The basin has a subtropical monsoon climate, and the precipitation is significantly affected by monsoon circulation. Figure 3e shows that the mean annual precipitation in 1973–2005 was 1662.48 mm, while that in 2068–2100 increased significantly by 97.69–285.72 mm relative to that in 1973–2005 under RCP2.6, RCP4.5, and RCP8.5. The temporal distribution of precipitation in XRB is nonuniform. The precipitation from spring (March–May) and summer (June–August) accounted for 36.78% and 35.36% of the total precipitation in a year, respectively. The precipitation was low in the fall and winter seasons from September to February of the next year, which accounted for 27.86% of the total precipitation in a year. In 2068–2100, no significant changes in precipitation were observed in the spring, fall, and winter seasons. There was abundant precipitation in summer (June–August), with an increase of 4.79–9.35% under three RCP scenarios relative to the same period in 1973–2005. The most obvious increase in precipitation occurred in June and accounted for 19.45–35.99% of the same month in 1973–2005 under different RCPs.

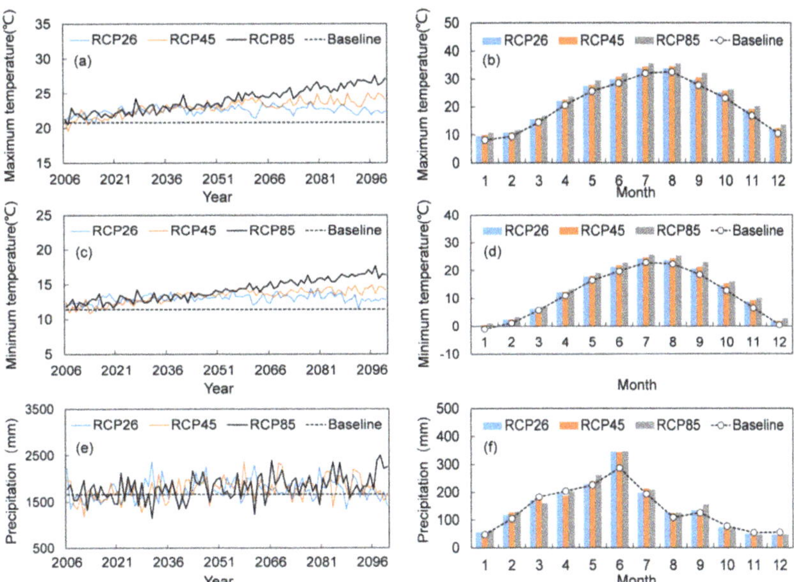

Figure 3. The time series of annual maximum temperature (**a**), annual minimum temperature (**c**), annual precipitation (**e**), monthly maximum temperature (**b**), monthly minimum temperature (**d**), and monthly precipitation (**f**) averaged over XRB projected by downscaled CMIP5 GCM (CSIRO-Mk3-6-0) in 2006–2100 under RCP2.6, RCP4.5 and RCP8.5, respectively. The solid lines and histograms indicate CSIRO-Mk3-6-0 outputs of different RCPs, the dotted line is the mean values of annual meteorological elements of the basin in baseline (1973–2005). The dashed line with circles indicates the mean values of monthly meteorological elements.

3.2. Land Use/Land Cover Change Analysis under Varying Scenarios

The land use/land cover classification map of 1987 and 2015 were interpreted from Landsat imagery (see Figure 4). Take the patterns of land use in 1987 as the representative underlying surface type in the baseline period, while the patterns of land use in 2015 represent the underlying surface type condition in the future. It is assumed that there will be no significant changes in land use/land cover from 2015 to the end of the 21st century. By comparing the land use/land cover classification maps of the XRB, it was found that the spatial distribution of land use/land cover in the two periods were different, especially the type of urban area. The urban area increased significantly due to deforestation and the conversion of cultivated land. The urban area in 2015 and in the future will be increased by 547% relative to that at the baseline period. The forest and cultivated land areas in 2015 and the future will be decreased by 2.94% and 13.35% relative to that at the baseline period, respectively.

3.3. Results of Sensitivity Analysis and Model Performance Assessment

Table 2 lists the results of the global sensitivity analysis by using SWAT-CUP, based on the sensitivity ranking of the parameters. For the simulated streamflow, CN2, CH_K2, SOL_Z, SURLAG, ESCO, GW_DELAY, GWQMN, SOL_K, CANMX, SOL_AWC, ALPHA_BF, and CH_N2 were the first 12 high sensitivity parameters, while USLE_P, SLSUBBSN, BIOMIX, SPEXP, and SPCON were the top five high-sensitivity parameters for the simulated sediments. In the streamflow parameters, the SCS runoff curve number 'CN2' ranked first, much higher than the others. For a given catchment, CN2 controls the main runoff confluence process and represents the confluence capacity of different underlying surfaces. In the sediment parameters, the USLE (Universal Soil Loss Equation) equation support practice factor 'USLE_P' is the most sensitive, which indicates the ratio of soil loss under soil and water conservation measures to soil loss under corresponding slope conditions. Table 2 shows that parameters representing the surface runoff, soil properties, groundwater return flow, ground water, and land cover management are sensitive. Consequently, it is important for the streamflow and sediment simulation to accurately estimate these parameters.

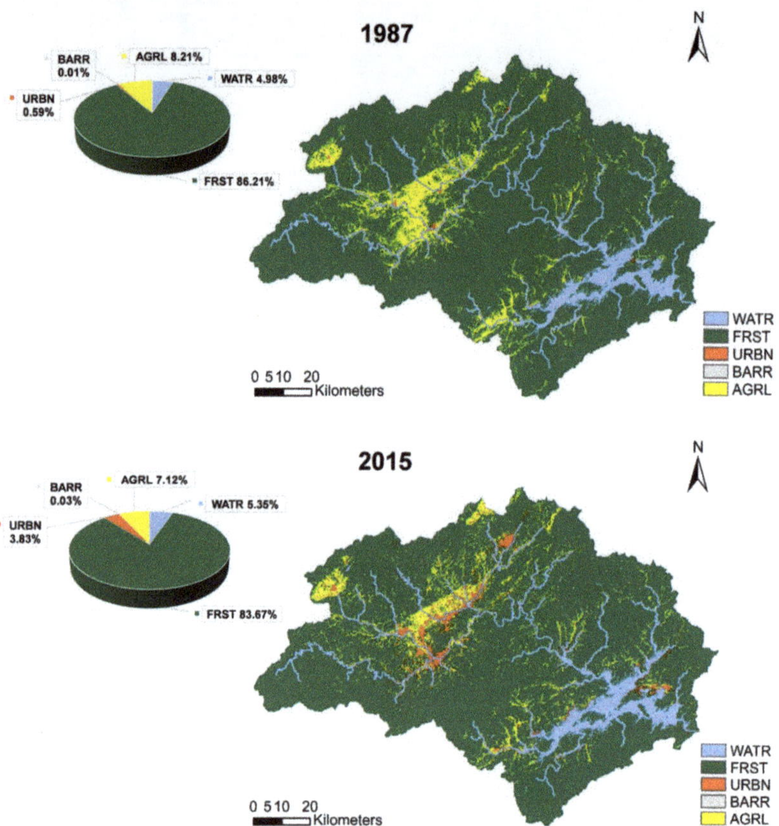

Figure 4. The land use/land cover classification maps in 1987 and 2015 under the SN and SN^{LC} scenarios and the proportion of area of each land use/land cover type. The land use/land cover in 1987 and 2015 represent the land use/land cover in the simulation baseline period (1973–2005) and the future (2068–2100), respectively. WATR indicates water body, FRST indicates forest, URBN indicates urban land, BARR indicates bare land, AGRL indicates cultivated land.

The SWAT model was calibrated and validated on a monthly scale in 2001–2010 and 2011–2014 for the streamflow for two stations (Tunxi and Yuliang station), respectively. The results are shown in Figure 5 and Table 3. For Tunxi Station, the observed and simulated streamflow fit well with the values of $NSE = 0.83$ and $r^2 = 0.85$ for the calibration period and $NSE = 0.89$ and $r^2 = 0.90$ for the validation period. For Yuliang Station, the observed and simulated streamflow were in satisfactory agreement with values of $NSE = 0.64$ and $r^2 = 0.69$ for the calibration period, and $NSE = 0.73$ and $r^2 = 0.88$ for the validation period. Based on the whole period of sediment monitoring data (2006–2014), the model was calibrated and validated on a monthly scale in 2006–2012 and 2013–2014 for the sediment for the same stations, respectively (see Figure 6 and Table 3). NSE and r^2 between the observed and simulated sediment transport rate were larger than 0.60 and 0.62 for the calibration period for two stations. The results of the model validation at a monthly time step were good ($NSE = 0.74$ and $r^2 = 0.81$) for Tunxi Station and satisfactory ($NSE = 0.47$ and $r^2 = 0.55$) for Yuliang Station. The hydrological model captured the low and some of the peak values of the flow and sediment very well. Overall, the SWAT model was mainly satisfied with the observed data in the XRB. That is to say, it is acceptable to use the calibrated parameters incorporated with the SWAT database for further simulations.

Table 2. The results of the sensitivity analysis and calibration for the SWAT model.

Parameter	Definition	Sensitivity Analysis		Calibration		
		t-Statistics	*p*-Value	Min	Max	Optimal
Streamflow						
CN2 *	SCS runoff curve number for moisture condition II	−35.47	0.00	−0.5	0.5	0.047
CH_K2	Effective hydraulic conductivity in main channel alluvium (mm/h)	−2.64	0.01	−0.01	500	378.873
SOL_Z *	Depth to bottom of first soil layer (mm)	2.57	0.01	−0.5	0.5	0.148
SURLAG	Surface runoff lag time (days)	−0.96	0.34	0.05	24	17.324
ESCO	Soil evaporation compensation factor	0.64	0.53	0	1	0.347
GW_DELAY	Groundwater delay (days)	0.58	0.56	30	450	62.025
GWQMN	Threshold depth of water in the shallow aquifer for return flow to occur (mm H_2O)	−0.50	0.62	0	5000	46.250
SOL_K *	Saturated hydraulic conductivity of first soil layer (mm/h)	0.39	0.70	−0.8	0.8	0.638
CANMX	Maximum canopy storage (mm H_2O)	0.29	0.77	0	100	90.425
SOL_AWC *	Available water capacity of first soil layer (mm/mm)	0.19	0.85	−0.5	0.5	−0.251
ALPHA_BF	Baseflow alpha factor (days)	−0.14	0.89	0	1	0.768
CH_N2	Manning's "n" value for the main channel	−0.01	0.99	−0.01	0.3	0.295
Sediment						
USLE_P	USLE equation support practice factor	−39.88	0.00	0	1	0.020
SLSUBBSN *	Average slope length (m)	−12.38	0.00	−0.9	0.9	−0.498
BIOMIX	Biological mixing efficiency	−6.52	0.00	0	1	0.051
SPEXP	Exponent parameter for calculating sediment re-entrained in channel sediment routing	1.07	0.29	1	1.5	1.429
SPCON	Linear parameter for calculating the maximum amount of sediment that can be re-entrained during channel sediment routing	0.77	0.44	0.0001	0.01	0.006

Note: * The asterisk means the existing parameter value is multiplied by (1+ a given value).

This study used the SUFI-2 approach to analyze the sediment uncertainty, mainly resulting from the uncertainties in the CMIP5 GCM projections and land use/land cover information. In SUFI-2, the parameter uncertainty, described by a multivariate uniform distribution in a parameter hypercube, accounted for all sources of uncertainties in the hydrological model. The propagation of parameter uncertainty led to the output uncertainty, which was quantified by the 95% prediction uncertainty (95PPU) band. Latin hypercube sampling was used to calculate the 95PPU at the 2.5% and 97.5% levels of the cumulative distribution function of the output variables [48]. Two indices, the *p*-factor (the percent of observations bracketed by the 95PPU) and *r*-factor (the relative width of 95% probability band), were calculated to evaluate the goodness of calibration uncertainty on the basis of the *p*-factor approaching 100% and the *r*-factor approaching 1. For streamflow, it is considered to be satisfactory if the *p*-factor >70% while having an *r*-factor of around 1 [47,48]. For the sediment, a smaller *p*-factor and a larger *r*-factor could be acceptable (SWAT-CUP user-manual). In this study, the 95PPU of streamflow brackets was 88% of the observations for Tunxi Station and 76% of the observations for Yuliang Station, while the *r*-factor equaled 1.01 and 1.25, respectively. The uncertainty analysis results of the sediment showed that the 95PPU bracketed 51% and 39% of the observations for Tunxi and Yuliang Stations, respectively. Meanwhile, the *r*-factor equaled 0.62 for Tunxi and 0.51 for Yuliang, which are very close to a suggested value of 1.

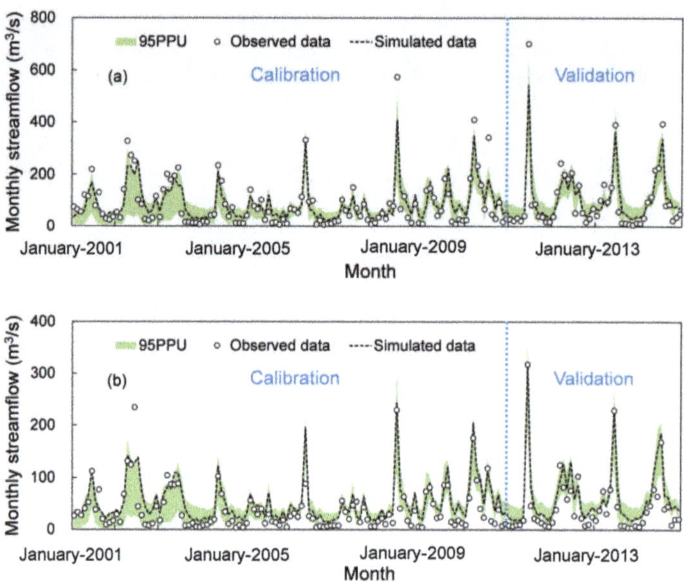

Figure 5. A comparison between the observed and modeled monthly streamflow in XRB during the calibration (2001–2010) and validation (2011–2014) periods. (**a**) The comparison result for Tunxi Station; (**b**) the comparison result for Yuliang Station.

Table 3. The SWAT performance during the calibration and validation periods.

Variables	Stations	Periods	Monthly Average		NSE	r^2	Classes
			Observed	Simulated			
Streamflow (m^3/s)	Tunxi	Calibration	78.3	66.6	0.83	0.85	Very good/Good
		Validation	108.9	114.7	0.89	0.90	Very good/Good
	Yuliang	Calibration	35.7	38.8	0.64	0.69	Satisfactory/Satisfactory
		Validation	46.2	67.7	0.73	0.88	Good/Good
Sediment (thousand tons)	Tunxi	Calibration	38.9	43.3	0.70	0.71	Good/Satisfactory
		Validation	38.1	54.0	0.74	0.81	Good/Good
	Yuliang	Calibration	16.0	24.1	0.60	0.62	Satisfactory/Satisfactory
		Validation	28.0	29.8	0.47	0.55	Satisfactory/Satisfactory

3.4. Separating Impacts of Climate Variability and Land Use/Land Cover Change on Sediment

In the section of the estuary into the Xin'anjiang Reservoir, Jiekou, the mean annual transported sediment was 48.93×10^4 tons/yr at the baseline (1973–2005). In the future period (2068–2100), the mean annual transported sediment will be 69.53–76.31×10^4 tons/yr, with a variation of 42.10–55.97% relative to that in the baseline period (combining effects of climate change and land use/land cover change). We quantified the contribution of climate change and land use/land cover change impacting the transported sediment at the mean annual scale by using the framework described in Section 2.3.3. The results showed that the joint climate and land use/land cover changes caused an increase in the mean annual transported sediment of 20.60–27.39×10^4 tons/yr (see Table 4). The mean annual transported sediment is expected to increase under both the individual and combined climate and land use/land cover change impacts. Changes in the mean annual transported sediment will be mainly driven by climate change if the land use/land cover conditions in the future are kept as in 2015. In this case, the land use/land cover change might weaken the influence on sediment attributed to climate change. Specifically, the increases in the annual transported sediment for the predication period (2068–2100) due to climate variability are 20.07–26.85 $\times 10^4$ tons/yr, which represent a contribution of 97.45–98.05%, while the land use/land cover change will lead to an increase in the annual transported sediment by 0.53–0.60 $\times 10^4$ tons/yr, with a contribution ranging from 1.95% to 2.67%. RCP8.5 showed smaller effects in increasing the influence on the sediment attributed to climate change than RCP2.6 and RCP4.5. However, the results may be very different if the future land use/land cover condition changes significantly compared with 2015.

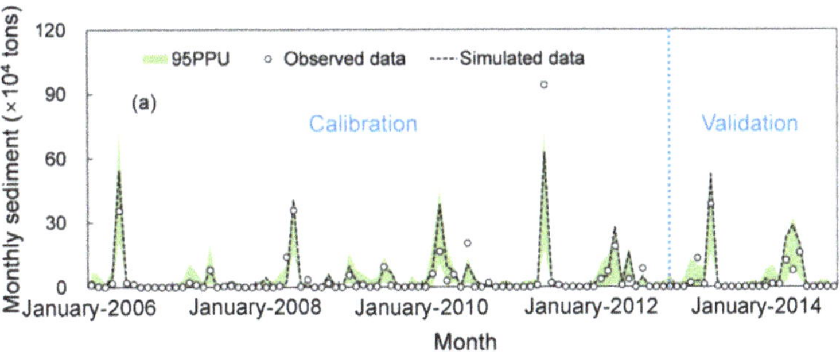

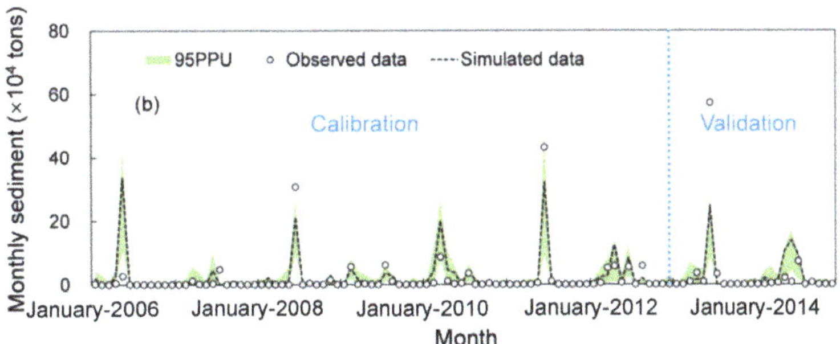

Figure 6. The comparison between the observed and modeled monthly transported sediment in XRB during the calibration (2006–2012) and validation (2013–2014) periods. (**a**) The comparison result for Tunxi Station; (**b**) the comparison result for Yuliang Station.

Table 4. The results of separating the impacts of climate change and land use/land cover change on the sediment in the XRB.

Scenarios	$\Delta D_{RCPj}(10^4 \text{ tons})$	$\Delta D^{LCj}(10^4 \text{ tons})$	$\Delta D^{LC}_{RCPj}(10^4 \text{ tons})$	$\alpha_{RCPj}(\%)$	$\alpha^{LCj}(\%)$
RCP2.6	26.85	0.53	27.39	98.05	1.95
RCP4.5	20.07	0.53	20.60	97.45	2.55
RCP8.5	21.87	0.60	22.47	97.33	2.67

The spatial distribution of the sediment yield (sediment from the sub-basin that is transported into the reach during the time step) in the baseline period (with the land use/land cover map in 1987) and the relative variation ratio of the sediment yield under RCP2.6, RCP4.5, and RCP8.5 are shown in Figure 7. The relative variation ratio is described as a percentage of sediment yield in the baseline period. This means that we used the difference between the modeling value of the sediment yield in different RCPs and at the baseline as a numerator and the mean value of the sediment yield in the baseline period as a denominator. In the baseline period, the sediment yields from each sub-basin were 0.02–2.07 tons/ha/yr, with an average value for the whole XRB of 0.75 tons/ha/yr. In future scenarios (RCP2.6, RCP4.5 and RCP8.5), sediment yield had a strong response to climate change. Compared to the baseline period, our modeling analysis predicted dramatic increases in the sediment yield for each sub-basin under all three RCPs, especially under

RCP2.6 (with an increase of 19.20–85.70%) and RCP8.5 (with an increase of 34.18–68.05%). The increases in future sediment yield under the scenario of RCP2.6 were mainly concentrated downstream of the basin and in the area around Xin'anjiang Reservoir, while those under the scenario of RCP8.5 were mainly concentrated in the upstream of the basin. This is mainly because the precipitation significantly increases from April to July, which is the cultivation season of the main crops in the XRB, under the scenario of RCP8.5. The area of cultivated land in the upstream sub-basins of Jiekou is relatively larger than that in the downstream sub-basins. Consequently, frequent farming activities lead to an increase in the sediment loss with rainfall runoff, increasing the sediment input of the reservoir. It can be found that the spatial distribution of sediment yield change is consistent with that of water yield change, which is the net amount of water that leaves the sub-basin and contributes to streamflow in the reach.

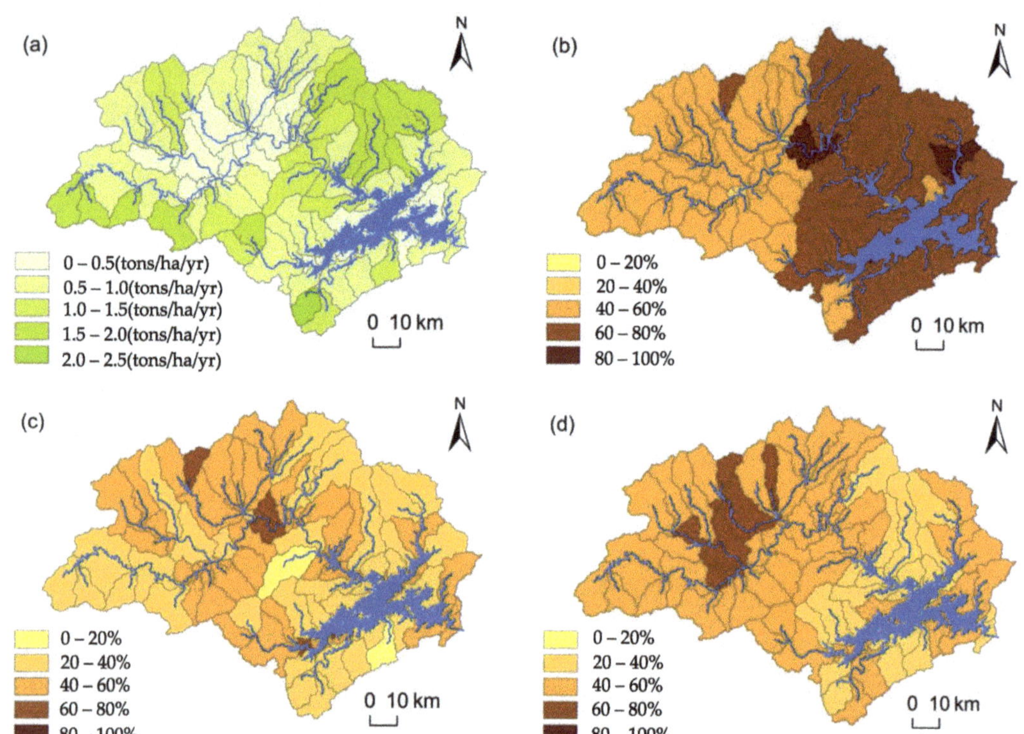

Figure 7. The spatial characteristics of annual sediment yield in XRB. (**a**) The mean annual sediment yield of every sub-basin in the baseline period. (**b**) The relative changes in the sediment yield between RCP2.6 and baseline. (**c**) The relative changes in the sediment yield between RCP4.5 and the baseline. (**d**) The relative changes in the sediment yield between RCP8.5 and the baseline.

Figure 8 shows the inter- and intra-annual variability in sediment transported with water out of reach in Jiekou under the baseline period, RCP2.6, RCP4.5, and RCP8.5. It can be seen that the inter-annual variation in the transported sediment in Jiekou is significant. Compared to the baseline, the mean annual transported sediment under RCP2.6, RCP4.5, and RCP8.5 increased dramatically by 40.91–54.75% when there was no land use/land cover change from 1987 to the future. The largest increase in the mean annual transported sediment is under scenario RCP2.6, followed by scenario RCP8.5. Through correlation analysis, the annual transported sediment was positively correlated with rainfall (r is 0.72 in the baseline period; r is 0.58–0.73 under different RCPs) and runoff (r is 0.77 in baseline period; r is 0.65–0.91 under different RCPs), respectively, indicating that rainfall and runoff have a great impact on sediment output. Meanwhile, the nonparametric Mann–Kendall test [57,58], commonly used to assess the significance of trends in hydrology and climatology, was used to detect trends in the time series of the annual transported sediment. The annual transported sediment (2068–2100) exhibited a positive trend for the Jiekou site at the $\alpha = 0.1$ level of significance under RCP2.6 and RCP4.5, respectively, but no significant trend under RCP8.5 ($\alpha = 0.1$).

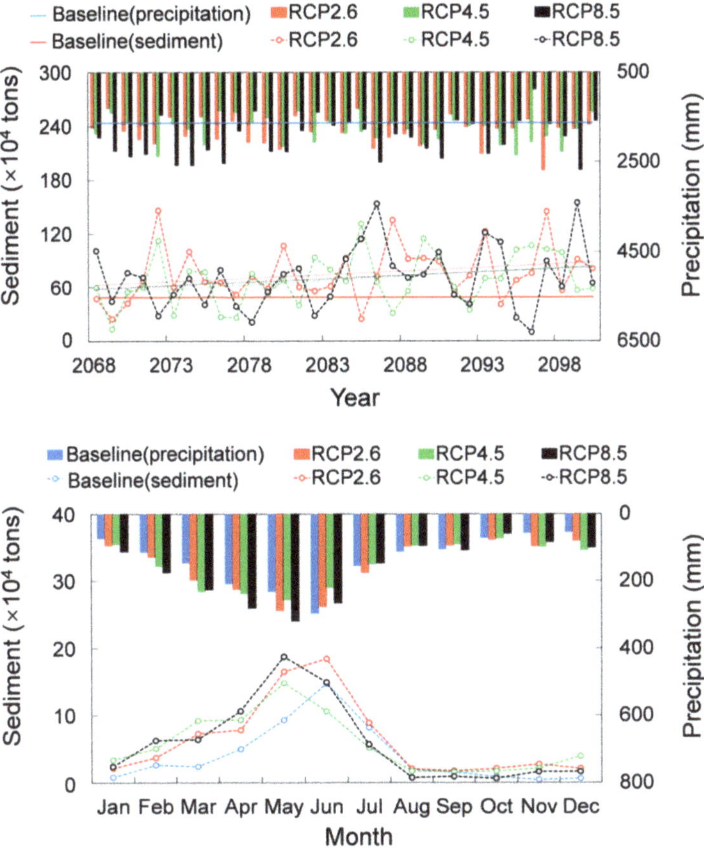

Figure 8. The inter- and intra-annual (monthly) variability in the sediment transported with water out of reach at Jiekou (estuary of Xin'anjiang River) under different climate scenarios.

A non-uniform distribution of the mean monthly transported sediment in Jiekou shows that the transported sediment in spring and summer accounted for 85.36% of the total sediment output in 1973–2005 and 73.95–80.88% in 2068–2100. Under scenario RCP2.6, the mean monthly transported sediment exhibited a significant increase in March–June, but no obvious change in June–February of the following year. Under scenario RCP4.5, the mean monthly transported sediment increased significantly in January–May but decreased in June–August. Under scenario RCP8.5, there was a significant increase in February–May, but a slight decrease in June–October. The intra-annual (monthly) distribution of the transported sediment is consistent with that of rainfall. It is indicated that the transported sediment is not only related to rainfall intensity, but also to the time distribution of rainfall.

3.5. Implication for Water Quality Management of Reservoir/Lake

Influenced by the temperate monsoon climate, more than 60% of the annual precipitation was recorded in April–August in XRB. According to the CMIP5 outputs, the monthly precipitation increased significantly in June and the frequency of heavy rainfall events increased in the flooding season under the three RCP scenarios. Correspondingly, the reservoir's inflow volume increased sharply after heavy rainfall. Compared with small and medium rainfall, it is easier for heavy rainfall or rainstorms to cause massive soil erosion. Seventeen HJ-1 A/B images during 10 heavy rainfall events from 2009 to 2014 were used to illustrate the relation of total suspended matter (TSM) concentration in the estuary of the Xin'anjiang River to the amount of precipitation of the basin. A significant positive correlation could be found between the TSM concentrations and rainfall amount ($p < 0.005$) [41].

In the Xin'anjiang Reservoir, significant turbid density flow always follows heavy rainfall events and rainstorms, which affects the reservoir water quality, especially in Jiekou estuary. It was investigated that the first small peak flow in March 2018 in Jiekou caused great changes in the water transparency and nutrient concentration, which indicates that the first peak inflow discharge of each year has a great impact on the water quality of the reservoir [33]. The particulate matter contributes to most nutrient inputs, which means that heavy rainfall events could lead to very high nutrient input into the reservoir/lake due to massive erosion from the upstream catchment and the area surrounding the reservoir/lake [59]. A large number of external nutrients carried by heavy rainfall (or rainstorms) and floods as well as the sediment resuspension caused by flood scouring increase the nutrient concentration of the water body in the reservoir/lake [60]. The degrees of eutrophication are aggravated correspondingly and suitable conditions for algae growth are provided. Therefore, understanding the effects of rainfall increases in the flooding season (especially the frequency of heavy rainfall events or rainstorm increases) on sediment yield in the basin could help water managers to strengthen the management of heavy rainfall runoff. It is also advantageous to the protection of water environment for reservoirs/lakes.

4. Conclusions

This study demonstrated that the sediment load and streamflow of XRB would significantly increase in the future under the integrated impacts of climate change and land use/land cover change. Sediment generated from the sub-basins above the Jiekou section (transported into the reservoir) will increase by 42.10–55.97% in 2068–2100, relative to that in the baseline period. Rainfall and temperature are the major climatic affecting factors in these increases, and the land use/land cover change can be attributed to the deforestation and urbanization during the simulating period. We found that more than 90% of these increases in sediment will be caused by climate change if the land use/land cover situation in the future are not obviously changed. While climate change combined with land use/land cover change in all three RCPs projected an increase in sediment, there were disagreements on the spatiotemporal distribution of sediment yield under multiple scenarios. In terms of space, the increases in the future sediment yield are mainly concentrated in the downstream of the basin under RCP2.6 but in the upstream of the basin under RCP8.5. In terms of time, more precipitation and floods in the wet season may occur in the future. Consequently, this will increase the sediment yield by 22.07–46.12% in the wet season (March–August) with respect to the baseline scenario. Therefore, it is important to emphasize increasing adaptation to climate change and land use/land cover change when designing and managing water environmental resources of the reservoirs and catchments.

In summary, climate change and land use/land cover can exert a great influence on the sediment yield in this humid and monsoonal climate region with separated or combined effects. The projections of future changes in sediment yield suggest the great challenge that lakes or reservoirs will face, because increasing sediment yield is associated with the high input of nutrients, especially phosphorus, which is a critical element for phytoplankton proliferation and algal bloom occurrence. Our findings can greatly benefit managers/decision-makers in improving their understanding of these effects on rainfall–runoff processes and soil erosion as well as nutrient delivery in the catchment. Moreover, it can help them to design and adopt reasonable measures for watershed management and local governments regarding environmental conditions including climate change and land use/land cover change.

Author Contributions: Conceptualization, B.Q. and L.L.; Methodology, H.L.; Software, H.L.; Validation, H.L., Y.L., and J.J.; Formal analysis, H.L. and Y.L.; Investigation, C.Y. and Z.W.; Resources, Y.L. and C.Y.; Data curation, J.J. and C.Y.; Writing—original draft preparation, H.L.; Writing—review and editing, B.Q. and L.L.; Visualization, K.S. and G.Z.; Supervision, B.Q. All authors have read and agreed to the published version of the manuscript.

Funding: This research was funded by the National Natural Science Foundation of China (42171034, 41671205 and 41830757) and Yunnan University (C176220100043).

Institutional Review Board Statement: Not applicable.

Informed Consent Statement: Not applicable.

Data Availability Statement: The relevant data was available on the request of corresponding authors.

Conflicts of Interest: The authors declare no conflict of interest.

References

1. Van, O.K.; Govers, G.; Desmet, P. Evaluating the effects of changes in landscape structure on soil erosion by water and tillage. *Landsc. Ecol.* **2000**, *15*, 577–589.
2. Zhang, X.C.; Liu, W.Z. Simulating potential response of hydrology, soil erosion, and crop productivity to climate change in Changwu tableland region on the Loess Plateau of China. *Agric. For. Meteorol.* **2005**, *131*, 127–142. [CrossRef]
3. Khoi, D.N.; Suetsugi, T. Impact of climate and land-use changes on hydrological processes and sediment yield—A case study of the Be River catchment, Vietnam. *Hydrol. Sci. J.* **2014**, *59*, 1095–1108. [CrossRef]
4. Li, Z.Y.; Fang, H.Y. Modeling the impact of climate change on watershed discharge and sediment yield in the black soil region, northeastern China. *Geomorphology* **2017**, *293*, 255–271. [CrossRef]
5. Boix-Fayos, C.; Martínez-Mena, M.; Arnau-Rosalén, E.; Calvo-Cases, A.; Castillo, V.; Albaladejo, J. Measuring soil erosion by field plots: Understanding the sources of variation. *Earth Sci. Rev.* **2006**, *78*, 267–285. [CrossRef]
6. Montgomery, D.R.; Brandon, M.T. Topographic controls on erosion rates in tectonically active mountain ranges. *Earth Planet. Sci. Lett.* **2002**, *201*, 481–489. [CrossRef]
7. Wang, H.; Saito, Y.; Zhang, Y.; Bi, N.; Sun, X.; Yang, Z. Recent changes of sediment flux to the western Pacific Ocean from major rivers in East and Southeast Asia. *Earth Sci. Rev.* **2011**, *108*, 80–100. [CrossRef]
8. Langbein, W.B.; Schumm, S.A. Yield of sediment in relation to mean annual precipitation. *EOS* **1958**, *39*, 1076–1084. [CrossRef]
9. Stockwell, J.D.; Doubek, J.P.; Adrian, R.; Anneville, O.; Carey, C.C.; Carvalho, L.; Domis, L.N.D.S.; Dur, G.; Frassl, M.A.; Grossart, H.P.; et al. Storm impacts on phytoplankton community dynamics in lakes. *Glob. Chang. Biol.* **2020**, *26*, 2756–2784. [CrossRef]
10. Shi, Z.H.; Huang, X.D.; Ai, L.; Fang, N.F.; Wu, G.L. Quantitative analysis of factors controlling sediment yield in mountainous watersheds. *Geomorphology* **2014**, *226*, 193–201. [CrossRef]
11. Wang, M.; Hamilton, D.P.; McBridea, C.G.; Abellc, J.M.; Hicksa, B.J. Modelling hydrology and water quality in a mixed land use catchment and eutrophic lake: Effects of nutrient load reductions and climate change. *Environ. Modell. Softw.* **2018**, *109*, 114–133.
12. Lü, X.X.; Ran, L.S.; Liu, S.; Jiang, T.; Zhang, S.R.; Wang, J.J. Sediment loads response to climate change: A preliminary study of eight large Chinese rivers. *Int. J. Sediment Res.* **2013**, *28*, 1–14. [CrossRef]
13. Zabaleta, A.; Meaurio, M.; Ruiz, E.; Antigüedad, I. Simulation climate change impact on runoff and sediment yield in a small watershed in the Basque Country, northern Spain. *J. Environ. Qual.* **2014**, *43*, 235–245. [CrossRef]
14. Imeson, A.C.; Lavee, H. Soil erosion and climate change: The transect approach and the influence of scale. *Geomorphology* **1998**, *23*, 219–227. [CrossRef]
15. Zhao, G.J.; Mu, X.M.; Tian, P.; Wang, F.; Gao, P. The variation trend of streamflow and sediment flux in the middle reaches of Yellow River over the Past 60 years and the influencing Factors. *Res. Sci.* **2012**, *34*, 1070–1078. (In Chinese)
16. Nearing, M.A.; Jetten, V.; Baffaut, C.; Cerdan, O.; Couturier, A.; Hernandez, M.; Le Bissonnaise, Y.; Nicholsa, M.H.; Nunesf, J.P.; Renschlerg, C.S.; et al. Modeling response of soil erosion and runoff to changes in precipitation and cover. *Catena* **2005**, *61*, 131–154. [CrossRef]
17. Jansen, I.M.L.; Painter, R.B. Predicting sediment yield from climate and topography. *J. Hydrol.* **1974**, *21*, 371–380. [CrossRef]
18. Syvitski, J.P.M. The influence of climate on the flux of sediment to the coastal ocean. In Proceedings of the Oceans 2003. Celebrating the Past ... Teaming toward the Future (IEEE Cat. No.03CH37492), San Diego, CA, USA, 22–26 September 2003; Available online: https://ieeexplore.ieee.org/abstract/document/1283423 (accessed on 1 May 2020).

19. Phan, D.B.; Wu, C.C.; Hsieh, S.C. Impact of climate change on stream discharge and sediment yield in northern Viet Nam. *Water Resour.* **2011**, *38*, 827–836. [CrossRef]

20. Huang, X. The Effect and Mechanism of Vegetation Cover and Rainfall Change on Runoff and Sediment Yield in Du Watershed, China. Ph.D. Thesis, Huazhong Agricultural University, Wuhan, China, 2019.

21. Walling, D.E.; Fang, D. Recent trends in the suspended sediment loads of the world's rivers. *Glob. Planet Chang.* **2003**, *39*, 111–126. [CrossRef]

22. Martin, J.M.; Meybeck, M. Elemental mass-balance of material carried by major world rivers. *Mar. Chem.* **1979**, *7*, 173–206. [CrossRef]

23. Walling, D.E. *The Impact of Global Change on Erosion and Sediment Transport by Rivers: Current Progress and Future Challenges*; United Nations Educational, Scientific and Cultural Organization: Paris, France, 2009; pp. 1–26.

24. Montgomery, D.R. Soil erosion and agricultural sustainability. *Proc. Natl. Acad. Sci. USA* **2007**, *104*, 13268–13272. [CrossRef] [PubMed]

25. Sinha, R.K.; Eldho, T.I. Effects of historical and projected land use/cover change on runoff and sediment yield in the Netravati River basin, Western Ghats, India. *Environ. Earth Sci.* **2018**, *77*, 111. [CrossRef]

26. IPCC. Climate change 2014: Synthesis report. In *Contribution of Working Groups I, II and III to the Fifth Assessment Report of the Intergovernmental Panel on Climate Change*; Pachauri, R.K., Meyer, L.A., Eds.; Core Writing Team, IPCC: Geneva, Switzerland, 2014; p. 151.

27. IPCC. *Global Warming of 1.5 °C. An IPCC Special Report on the Impacts of Global Warming of 1.5 °C above Pre–Industrial Levels and Related Global Greenhouse Gas Emission Pathways, in the Context of Strengthening the Global Response to the Threat of Climate Change, Sustainable Development, and Efforts to Eradicate Poverty*; World Meteorological Organization: Geneva, Switzerland, 2018; p. 32.

28. Vaighan, A.A.; Talebbeydokhti, N.; Bavani, A.M. Assessing the impacts of climate and land use change on streamflow, water quality and suspended sediment in the Kor River Basin, Southwest of Iran. *Environ. Earth Sci.* **2017**, *76*, 543. [CrossRef]

29. Thrasher, B.; Maurer, E.P.; McKellar, C.; Duffy, P.B. Technical note: Bias correcting climate model simulated daily temperature extremes with quantile mapping. *Hydrol. Earth Syst. Sci.* **2012**, *16*, 3309–3314. [CrossRef]

30. Peng, H.; Jia, Y.W.; Qiu, Y.Q.; Niu, C.W.; Ding, X.Y. Assessing climate change impacts on the ecohydrology of the Jinghe River basin in the Loess Plateau, China. *Hydrol. Sci. J.* **2013**, *58*, 651–670. [CrossRef]

31. Zhou, Y.Q.; Jeppesen, E.; Zhang, Y.L.; Shi, K.; Liu, X.H.; Zhu, G.W. Dissolved organic matter fluorescence at wavelength 275/342 nm as a key indicator for detection of point-source contamination in a large Chinese drinking water lake. *Chemosphere* **2016**, *144*, 503–509. [CrossRef] [PubMed]

32. Zhang, Y.L.; Wu, Z.X.; Liu, M.L.; He, J.B.; Shi, K.; Zhou, Y.Q.; Wang, M.Z.; Liu, X.H. Dissolved oxygen stratification and response to thermal structure and long-term climate change in a large and deep subtropical reservoir (Lake Qiandaohu, China). *Water Res.* **2015**, *75*, 249–258. [CrossRef]

33. Da, W.Y.; Zhu, G.W.; Li, Y.X.; Wu, Z.X.; Zheng, W.T.; Lan, J.; Wang, Y.C.; Xu, H.; Zhu, M.Y. High frequent dynamics of water quality and phytoplankton community in inflowing river mouth of Xin'anjiang Reservoir, China. *Environ. Sci.* **2020**, *41*, 713–727.

34. Zheng, Y.N.; Wen, X.; Fang, G.H. Research on climate change and runoff response in Xin'an river basin. *J. Water Res. Water Eng.* **2015**, *26*, 106–110.

35. Pan, Y.; Luo, Y.; Wang, Y.; Zhang, Q.; Zhu, Z. Characteristics of evolution of precipitation and runoff in Xin'an River Basin. *Res. Soil Water Conserv.* **2018**, *25*, 121–125.

36. Li, H.Y.; Wang, Y.C.; Shan, L.; Luo, L.C.; Zhu, G.W.; Xu, H.; Shi, P.C. Effect of rainstorm runoff on total phosphorus load in Xin'anjiang. *Res. Environ. Sci.* **2022**, *35*, 887–895. (In Chinese)

37. Liu, M.; Zhang, Y.L.; Shi, K.; Zhu, G.W.; Wu, Z.X.; Liu, M.L.; Zhang, Y.B. Thermal stratification dynamics in a large and deep subtropical reservoir revealed by high-frequency buoy data. *Sci. Total Environ.* **2019**, *651*, 614–624. [CrossRef] [PubMed]

38. Zhang, Y.L.; Wu, Z.X.; Liu, M.L.; He, J.B.; Shi, K.; Wang, M.Z.; Yu, Z.M. Thermal structure and response to long-term climatic changes in Lake Qiandaohu, a deep subtropical reservoir in China. *Limnol. Oceanogr.* **2014**, *59*, 1193–1202. [CrossRef]

39. Zhu, G.W.; Cheng, X.L.; Wu, Z.X.; Shi, P.C.; Zhu, M.Y.; Xu, H.; Guo, C.X.; Zhao, X.C. Spatio-temporal variation of nutrient concentrations and environmental challenges of Qiandaohu Reservoir, China. *Res. Environ. Sci.* **2022**, *35*, 852–863. (In Chinese)

40. Chen, D.Q.; Li, H.P.; Zhang, W.S.; Pueppke, S.G.; Pang, J.P.; Diao, Y.Q. Spatiotemporal Dynamics of Nitrogen Transport in the Qiandao Lake Basin, a Large Hilly Monsoon Basin of Southeastern China. *Water* **2020**, *12*, 1075. [CrossRef]

41. Zhang, Y.B.; Shi, K.; Zhang, Y.L.; Moreno-Madriñán, M.J.; Zhu, G.W.; Zhou, Y.Q.; Li, Y. River plume monitoring in a deep valley reservoir using HJ-1 A/B images. *J. Hydrol.* **2020**, *587*, 125031. [CrossRef]

42. MWR. *Hydrological Data of River Basins in China*; Ministry of Water Resources of the People's Republic of China: Beijing, China, 2015. (In Chinese)

43. FAO; IIASA; ISRIC; ISSCAS; JRC. *Harmonized World Soil Database (Version 1.2)*; FAO: Rome, Italy; IIASA: Laxenburg, Austria, 2012.

44. Li, Y.; Zhang, Y.L.; Shi, K.; Zhu, G.W.; Zhou, Y.Q.; Zhang, Y.B.; Guo, Y.L. Monitoring spatiotemporal variations in nutrients in a large drinking water reservoir and their relationships with hydrological and meteorological conditions based on Landsat 8 imagery. *Sci. Total Environ.* **2017**, *599–600*, 1705–1717. [CrossRef]

45. IPCC. Summary for policymakers. In *Climate Change 2013: The Physical Science Basis. Contribution of Working Group I to the Fifth Assessment Report of the Intergovernmental Panel on Climate Change*; Stocker, T.F., Qin, D., Plattner, G.-K., Tignor, M., Allen, S.K., Boschung, J., Nauels, A., Xia, Y., Bex, V., Midgley, P.M., Eds.; Cambridge University Press: Cambridge, UK; New York, NY, USA, 2013.

46. Taylor, K. Summarizing multiple aspects of model performance in a single diagram. *J. Geophys. Res.-Atmos.* **2001**, *106*, 7183–7192. [CrossRef]

47. Abbaspour, K.C.; Yang, J.; Maximov, I.; Siber, R.; Bogner, K.; Mieleitner, J.; Zobrist, J.; Srinivasan, R. Modelling hydrology and water quality in the pre-alpine/alpine Thur watershed using SWAT. *J. Hydrol.* **2007**, *333*, 413–430. [CrossRef]

48. Williams, J.R. The EPIC model. In *Computer Models of Watershed Hydrology*; Singh, V.P., Ed.; Water Resources Publications: Highlands Ranch, CO, USA, 1995; pp. 909–1000.

49. Abbaspour, K.C.; Johnson, A.; van Genuchten, M.T. Estimating uncertain flow and transport parameters using a sequential uncertainty fitting procedure. *Vadose Zone J.* **2004**, *3*, 1340–1352. [CrossRef]

50. Abbaspour, K.C.; Vejdani, M.; Haghighat, S. SWAT-CUP calibration and uncertainty programs for SWAT. In Proceedings of the Modsim International Congress on Modelling & Simulation Land Water & Environmental Management Integrated Systems for Sustainability, Melbourne, Australia, 3–5 December 2007.

51. Abbaspour, K.C. SWAT-CUP: SWAT calibration and uncertainty programs—A user manual. *Eawag* **2015**, *6–69*, 72–80.

52. Moriasi, D.N.; Arnold, J.G.; Liew, M.W.V.; Bingner, R.L.; Harmel, R.D.; Veith, T.L. Model evaluation guidelines for systematic quantification of accuracy in watershed simulations. *Trans. ASABE* **2007**, *50*, 885–899. [CrossRef]

53. Santhi, C.; Arnold, J.G.; Williams, J.R.; Dugas, W.A.; Srinivasan, R.; Hauck, L.M. Validation of the SWAT model on a large river basin with point and nonpoint sources, Jawra. *J. Am. Water Resour. Assoc.* **2010**, *37*, 1169–1188. [CrossRef]

54. Moriasi, D.N.; Gitau, M.W.; Pai, N.; Daggupati, P. Hydrologic and water quality model: Performance measures and evaluation criteria. *Trans. ASABE* **2015**, *58*, 1763–1785.

55. Zhang, C.; Huang, Y.X.; Javed, A.; Arhonditsis, G.B. An ensemble modeling framework to study the effects of climate change on the trophic state of shallow reservoirs. *Sci. Total Environ.* **2019**, *697*, 134078. [CrossRef]

56. Guo, Y.X.; Fang, G.H.; Xu, Y.P.; Tian, X.; Xie, J.K. Identifying how future climate and land use/land cover changes impact streamflow in Xinanjiang Basin, East China. *Sci. Total Environ.* **2020**, *710*, 136275. [CrossRef]

57. Mann, H.B. Nonparametric tests against trend. *Econometrica* **1945**, *13*, 245–259. [CrossRef]

58. Kendall, M.G. *Rank Correlation Methods*, 4th ed.; Charles Griffin: London, UK, 1975.

59. Zhang, Y.L.; Shi, K.; Zhou, Y.Q.; Liu, X.H.; Qin, B.Q. Monitoring the river plume induced by heavy rainfall events in large, shallow, Lake Taihu using MODIS 250 m imagery. *Remote Sens. Environ.* **2016**, *173*, 109–121. [CrossRef]

60. Yang, Z.; Zhang, M.; Shi, X.L.; Kong, F.X.; Ma, R.H.; Yu, Y. Nutrient reduction magnifies the impact of extreme weather on cyanobacterial bloom formation in large shallow Lake Taihu (China). *Water Res.* **2016**, *103*, 302–310. [CrossRef]

water

Communication

The Feasibility of Monitoring Great Plains Playa Inundation with the Sentinel 2A/B Satellites for Ecological and Hydrological Applications

Hannah L. Tripp [1], Erik T. Crosman [1,*], James B. Johnson [1], William J. Rogers [1] and Nathan L. Howell [2]

[1] Department of Life, Earth and Environmental Sciences, West Texas A&M University, Canyon, TX 79015, USA; hannah.louise.tripp@gmail.com (H.L.T.); jbjohnson@wtamu.edu (J.B.J.); jrogers@wtamu.edu (W.J.R.)

[2] College of Engineering, West Texas A&M University, Eng & Computer Science Building (ECS), 2501 4th Ave, Canyon, TX 79015, USA; nhowell@wtamu.edu

* Correspondence: etcrosman@wtamu.edu; Tel.: +1-806-651-2294

Abstract: Playas are ecologically and hydrologically important ephemeral wetlands found in arid and semi-arid regions of the world. Urbanization, changes in agricultural land use and irrigation practices, and climate change all threaten playas. While variations in playa inundation on the Great Plains of North America have been previously analyzed by satellite using annual and decadal time scales, no study to our knowledge has monitored the Great Plains playa inundation area using sub-monthly time scales. Thousands of playas smaller than ~50 m in diameter, which were not previously identified by the Landsat satellite platform, can now be captured by higher resolution satellite data. In this preliminary study, we demonstrate monitoring spatial and temporal changes in the playa water inundation area on sub-monthly times scales between September 2018 and February 2019 over a region in West Texas, USA, using 10 m spatial resolution imagery from the Sentinel-2A/B satellites. We also demonstrate the feasibility and potential benefits of using the Sentinel-2A/B satellite retrievals, in combination with precipitation and evaporation data, to monitor playas for environmental, ecological, groundwater recharge, and hydrological applications.

Keywords: playa lakes; wetland; terminal basins; Sentinel-2A/B satellites; remote sensing; evaporation

Citation: Tripp, H.L.; Crosman, E.T.; Johnson, J.B.; Rogers, W.J.; Howell, N.L. The Feasibility of Monitoring Great Plains Playa Inundation with the Sentinel 2A/B Satellites for Ecological and Hydrological Applications. *Water* **2022**, *14*, 2314. https://doi.org/10.3390/w14152314

Academic Editors: Fei Zhang, Ngai Weng Chan, Xinguo Li and Xiaoping Wang

Received: 16 June 2022
Accepted: 22 July 2022
Published: 26 July 2022

Publisher's Note: MDPI stays neutral with regard to jurisdictional claims in published maps and institutional affiliations.

1. Introduction

Playas, which are generally defined as shallow, ephemeral wetlands located within closed basins, are found in many arid or semi-arid climates. Playas are generally more sensitive to changes in climate, land use, and irrigation practices than permanent bodies of water [1,2]. Thousands of small playas (both Great Plains playas and prairie potholes) are observed across the Great Plains of North America. Playa surfaces and their underlying soils are typically characterized by variable amounts of soluble salts, sand, clay, and silt that are deposited within these closed basins [3]. The frequency and duration of water inundation in playas also varies greatly.

Water is a critical resource on the Great Plains of North America, and the shallow, circular playa basins scattered across these environs are often hotspots of ecological diversity. These playa wetlands provide crucial habitats for many species of mammals, migratory birds, amphibians, and invertebrates [4]. Playas are also important in some regions for aquifer recharge. The Southern High Plains depend heavily on groundwater from the Ogallala Aquifer, which stretches from southern South Dakota to the Texas Panhandle. However, the magnitude of spatial variability of groundwater recharge from playas into the Ogallala Aquifer recharge remains somewhat uncertain and is an active area of research [5,6].

Many hundreds of thousands of playas dot the landscape across the Great Plains of North America [7]. Despite their ecological and hydrological importance to the region, to our knowledge, the use of frequent, high-resolution satellite data that has become available in the last several years has thus far been underutilized in monitoring these threatened and important ecosystems. While a number of satellite-derived analyses of playas have been conducted over the past 20 years, these have mainly used the Landsat satellite platform (e.g., Spain [8], California, USA [9], Texas, USA [10]) which provides a resolution that is too coarse to capture smaller playas, and the platform is infrequently available. On the Great Plains of North America, Starr and McIntyre [2] found a 77% decrease in the percentage of playa inundation (on playas resolved by Landsat) between 1980 and 2008 regarding a study region in West Texas, USA, as irrigation practices changed and warmer and drier conditions occurred.

In the last 10 years, the advent of new higher-resolution, frequent satellite imagery has revolutionized remote sensing for wetlands, lakes, and rivers. A number of studies have utilized the increased spatial and temporal resolution of the European Space Agency Sentinel-2 twin satellites: Sentinel-2A (launched in 2015) and Sentinel-2B (launched in 2017)) (e.g., [11–14]). Many previous playa studies were limited by lower spatial resolution and less frequent satellite imagery (Landsat spatial resolution is ~30 m, with a return period for satellite observations of ~16 days. For the Sentinel-2 satellites, the spatial resolution is ~10 m, and the return interval for observations is ~10 days (~5 days by utilizing both Sentinel-2A and 2B satellites). Thus, the Sentinel-2 satellite data effectively triples the spatial resolution and doubles the temporal resolution capabilities (compared to Landsat) to monitor changes in small, ephemeral lakes such as playas.

More recently, optical and radar satellite imagery have been combined to obtain a sophisticated analysis of vegetation, soils, and playa water inundation on seasonal scales from 1984–2019 over the Lordsburg Playa in New Mexico [15], while another recent study in Spain demonstrated using genetic programming to improve the reliability of algorithms to discern water versus non-water surfaces in complex shallow lakes and wetlands [16]. Improving satellite retrievals over the variable underlying soil surfaces and spectral characteristics of shallow, turbid waters with variable amounts of aquatic vegetation continues to be an active area of research.

Many different remote sensing techniques exist to identify water bodies by satellite, including image classification and derived water indices. It is beyond the scope of this short communication to review them all here. Because water has a distinct reflectance signature in the visible light wavelengths of the electromagnetic spectrum, the spectral contrast between land and water surfaces are generally pronounced. The normalized difference water index (NDWI) [17], and subsequent variations on this index are the most widely used water detection techniques, using visible and near-infrared (NIR) satellite spectral bands [18]. The goal of this short communication is not to test or recommend the most accurate techniques for retrieving water properties from satellite images taken over playas (which is certainly needed, and future work in this area is encouraged), but to demonstrate the general feasibility of monitoring the Great Plains playa inundation for hydrological and ecological applications using the freely available Sentinel 2A/B satellite data. Using an open-source remote sensing land surface classification tool, we demonstrate how playa inundation can be monitored on a sub-monthly basis, and then discuss the potential of utilizing the satellite-derived playa inundation area data, along with concurrent precipitation and water level imagery, to support ecological and hydrological applications, such as estimating groundwater recharge.

2. Materials and Methods

2.1. Study Area and 25 September 2018–17 February 2019 Case Study

The 900 km^2 (30 km × 30 km) study area analyzed in this paper was selected to represent a classic clay-lined playa environment in the Ogallala Aquifer region of the Southern High Plains (Figure 1). The time period between 25 September 2018 and 17 February 2019 was selected as a case study example of monitoring changes in playa inundation and water surface area for ecological and hydrological applications with Sentinel-2A/B satellite data. This time period was chosen to demonstrate how a sequence of Sentinel-2 satellite images is able to capture a rapid filling of the playas with water in late September through October 2018, followed by a slow decrease in the playa inundation between November 2018 and February 2019.

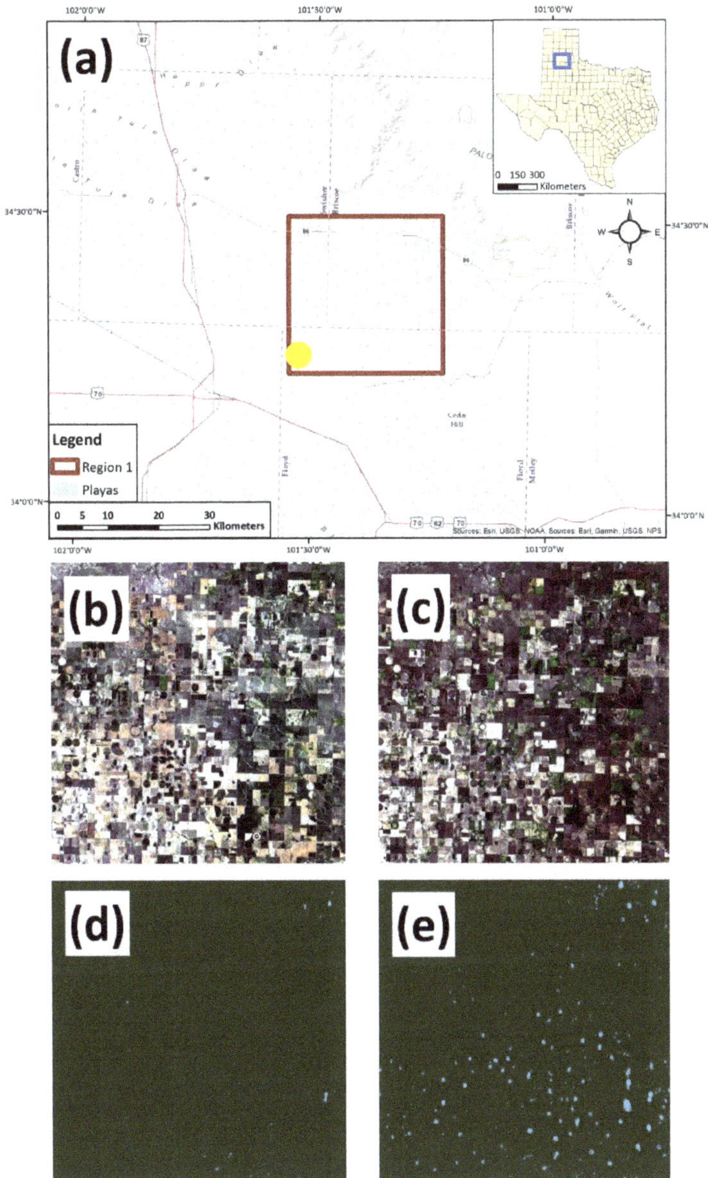

Figure 1. (**a**). Map of 30 km by 30 km study area (shown in brown square) in West Texas, USA (upper right inset). True color imagery of the study area for (**b**) 25 September 2018 and (**c**) 28 October 2018. The corresponding classified water surface area (water represented by blue colored areas, while all non-water surfaces are represented by green colored areas) for (**d**) 25 September 2018 and (**e**) 28 October 2018. The yellow dot in (**a**) represents the area shown in Figure 2.

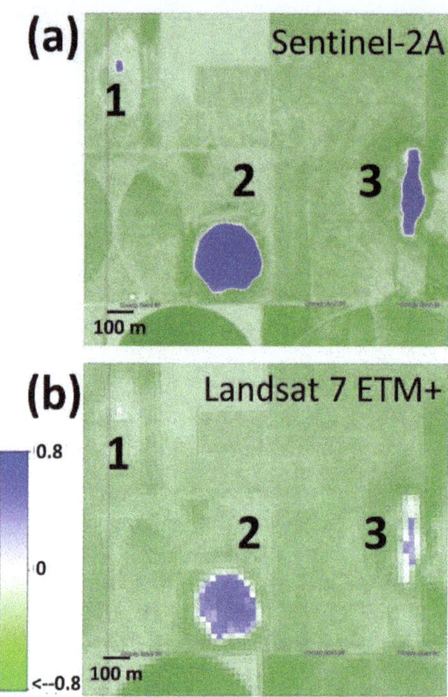

Figure 2. (**a**). Sentinel-2A imagery of the Normalized Difference Water Index (NDWI) for the location denoted with a yellow dot in Figure 1 on 22 September 2021, and (**b**) Landsat 7 ETM+ NDWI imagery from the same location on 19 September 2021. Blue areas represent water, while green areas indicate non-water areas. The 3 playas referenced in the text are indicated by the numbers 1–3. NDWI is processed by Sentinel Hub. The imagery contains modified European Space Agency Copernicus Sentinel data processed by Sentinel Hub, and Landsat 7 images are courtesy of the U.S. Geological Survey, processed by Sentinel Hub.

2.2. Sentinel-2A/B Satellite Imagery and Rainfall and Evaporation Data

The Sentinel-2 satellite mission is comprised of two sun-synchronous polar-orbiting satellites, phased at 180° to each other. The goal of the mission is to monitor land and ocean surface variability [19]. Sentinel-2A (Sentinel-2B) was launched by the European Space Agency (ESA) on 23 June 2015 (7 March 2017), each with a return frequency of ~10 days, such that data over any given area of the earth is obtained approximately every 5 days. The multispectral imager (MSI) on the Sentinel-2 satellite contains 13 spectral bands. These bands vary in wavelength from 442.7 nm to 2202.4 nm. The visible reflectance band images from these satellites (channels 2, 3, 4, 8) have a resolution of approximately 10 m, which is 3 times the resolution of previously widely utilized Landsat satellites [20].

For this study, the Sentinel-2A/B satellite data were downloaded through the Copernicus Open Access Hub maintained by the European Space Agency. A free account allowed easy downloading and retrieval of the satellite images. For this paper, 13 Sentinel-2A and Sentinel-2B satellite images from different days were analyzed between 25 September 2018 and 17 February 2019. A 2TB external hard drive provided adequate storage for storing the Sentinel-2 satellite images, which were 14.3 GB total in size (~1.1 GB per file). Each downloaded Sentinel-2A/B satellite data ZIP file (which contains 13 files for each of the 13 bands described below) was 10,980 pixels by 10,980 pixels, at 10 m resolution. A smaller 3000 by 3000 pixel subsection was selected for analysis in the selected 900 km² study area.

Several precipitation and evaporation estimate datasets were also used in this study, including daily precipitation data from the National Climate Data Center (NCDC) for Plainview, Texas (for Region 1), as well as radar precipitation estimates from the Advanced Hydrologic Prediction Service (AHPS) website: https://water.weather.gov/precip/ (accessed on 1 June 2022). Monthly evaporation estimates from the Texas Water Development Board (TWDB) were obtained from the online download portal at: https://waterdatafortexas.org/lake-evaporation-rainfall (accessed on 15 July 2022). The TWDB provides evaporation estimates on gridded one-degree latitude by one-degree longitude quadrangles for the entire state of Texas. The 'gross lake evaporation' rate, which is defined as the 'water loss caused by evaporation' was derived from Class A pan evaporation data.

2.3. Processing Methodoloy for Water Classification and Surface Area Calculations

Only images without clouds were used in this study. The images were loaded into QGIS software and subsected into 30 km by 30 km tiles over the region of interest, as previously discussed. Then, pixels in the image indicating water were identified through image classification by the free, open source QGIS Semi-Automatic Classification Plugin (SACP) [21]. Obtaining accurate retrieval of images showing shallow, turbid, and muddy waters, such as playas, remains an active area of research [22–26], and no efforts were made in this pilot study to evaluate the accuracy of the SACP for playa water retrieval. For this study, visual analysis of the water bodies after widespread heavy rains confirmed that the SACP did capture playa water inundation for hundreds of playas, but more rigorous evaluation is needed for future work.

The SACP, which is a Python tool in the QGIS environment, has been shown to be an effective application tool for land and water cover classification [27]. The SACP allows the user to download the images and perform both unsupervised and supervised classification, either manually or automatically. The SACP computes the spectral signatures of selected training features in the images and then compares these against the spectral signatures of other pixels in the image. In our study, we used supervised classification, in which we manually selected multiple lake surfaces that were known to have filled with water after heavy rains to define or 'train' the SACP regarding what the spectral characteristics of the water surfaces should be, and these characteristics were then used to define inundated playa surfaces throughout the remainder of the image.

The spectral signatures of data from the various bands, 2, 3, 4, 5, 6, 7, 8, 8A, 11, and 12, from the multispectral imager (MSI) on the Sentinel-2A and 2B satellites were evaluated by the Semi-Automatic Classification Plugin [21]. Briefly, the exact processing steps were as follows: (1) the satellite raw band data were loaded into QGIS, (2) the raw band images were converted to reflectance values appropriate for Sentinel-2, (3) simple atmospheric correction using the DOS1 method (dark object subtraction) was applied [21], (3) the spectral signatures of each of the satellite bands were manually selected by clicking on multiple known water surfaces (to train for classification according to the range of the spectral signatures of shallow water surfaces), as well as known non-water surfaces, in multiple regions of interest (ROIs). Then, (4) using the default minimum distance classification algorithm in the QGIS classification plug-in, the spectral signatures of the manually selected water and non-water surface ROIs were used to classify the entire image as either water or non-water. A total of 41 images obtained during this period were not classified, either because they were cloudy, or because they were within a week or two of another available non-cloudy image. The classification output accuracy and ability to identify water surfaces was manually checked through visual analysis of the satellite imagery and classification of several know water surface areas. Raster math calculations in QGIS were then used to compute the water surface area in each of the Sentinel-2 satellite images.

3. Results

The Sentinel-2A/B satellite data was collected and processed over the 900 km^2 study area to evaluate variations in water inundation of the several hundred observed playas during the 25 September 2018 to 17 February 2019 time period. Many of these playas would not be identified by the lower resolution Landsat Imagery used in many previous playa studies. Figure 2 provides a visual comparison of Sentinel-2A and Landsat 7 imagery, using the Normalized Water Difference Index (NDWI) [17], over a small region defined in Figure 1 with 3 different-sized playas. As can be clearly seen, a very small playa (playa 1 in Figure 2), with a diameter of less than 50 m, is clearly resolved by the 10 m resolution Sentinel-2A imagery (Figure 2a), but is not identified by the Landsat 7 imagery (Figure 2b). Similarly, the outline of a larger, elongated playa is well-characterized by Sentinel-2A, but poorly resolved by Landsat 7 (playa 3 in Figure 2). Finally, a large playa (diameter > 250 m, playa 2 in Figure 2) is resolved by both Sentinel-2A and Landsat 7; however, the full shape and edge regions are much better defined by Sentinel-2A. For determining the total surface area of playas, edge land contamination on the lower-resolution Landsat would also negatively bias the water inundation area estimates compared to the more defined playa edges resolved by the Sentinal-2 imagery.

The variations in playa inundation of the several hundred observed playas within the study region (Figure 1) during the 25 September 2018 to 17 February 2019 time period followed a realistic pattern based on the rainfall observed at a nearby weather station (Figure 3). Between 25 September and 20 October 2018, several weather systems each delivered 2.5–6.0 cm of rainfall (Figure 3). Prior to this rainfall, dry conditions resulted in only a few of the larger playas retaining water (Figure 1d). A rapid increase in the water area associated with playa inundation was observed across the region between late September and late October 2018 (Figures 1d,e and 3). Between 25 September and 3 October, the total surface water area in the 900 km^2 study area in West Texas, USA, increased from 1.1 to 8.2 square kilometers, with additional increases up to 14.5 km^2 by 10 October. These increases in water surface area are attributed to two main rainfall events—one in late September and another ~9 October (Figure 3).

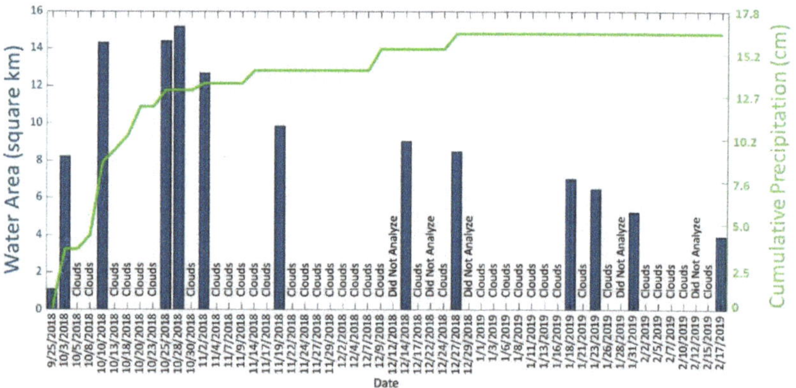

Figure 3. Observed surface water area (blue bar graph, in square kilometers) calculated for the 30 × 30 km square region shown in the brown square in Figure 1a for the period between 25 September 2018 and 17 February 2019. Missing dates are due to either cloud cover or data not being analyzed, as indicated on the plots. The corresponding cumulative precipitation at Plainview, Texas (in inches), is indicated by a green solid line (data courtesy of the National Weather Service through Mesowest https://mesowest.utah.edu (accessed on 15 March 2022).

These two wet periods were then followed by a generally dry weather pattern through February 2019. The absence of significant precipitation from November 2018 through February 2019 implies that the steady observed rates of change (decreases) in playa lake surface area would be primarily driven by evaporation or groundwater infiltration (Figure 3). The available retrievals indicate temporal variations in the rate of decrease in water surface area, e.g., between 28 October and 19 November, the water surface areas average rate of decrease was ~0.33 km^2 per day, whereas between 19 November and 17 February, the rate of decrease was ~0.08 square km^2 per day (Figure 3). The rates of change in playa inundation were also observed to vary between playas of different sizes (not shown).

The time-varying estimates of playa inundation can be combined with precipitation, evaporation, and playa depth estimates to provide potential added value for ecological and hydrological applications, such as the total playa water volume for ecological habitat or aquifer recharge estimates (Table 1).

Table 1. Approximation of total surface evaporation and playa volume in the study region between November 2018–January 2019. See text for description of assumptions used.

Month	Net Surface Evaporation (Rainfall Minus Evaporation (cm))	Estimated Playa Volume in Study Region [1]	Estimated Changes from Previous Month in Playa Volume Due to Evaporative Loss [1]	Estimated Changes from Previous Month in Playa Volume Due to Ground Infiltration [1]
November 2018	−7.9	2.25×10^6 m^3	–	–
December 2018	−5.5	1.35×10^6 m^3	7.9×10^5 m^3	1.1×10^5 m^3
January 2019	−6.4	0.8×10^6 m^3	4.3×10^5 m^3	1.2×10^5 m^3

[1] Note that this table is shown for illustrative purposes only, and all calculations in this table are based merely on approximations and assumptions.

For this study, estimates of lake surface net evaporation are obtained by the Texas Water Development Board and may or may not be representative for playa surfaces with varying salinity. Moreover, we do not have enough detailed measurements of (1) playa lake depth, and (2) playa basin size and runoff efficiency (the playa basin area controls how quickly a playa fills up for a given amount of rain [28]) across the many hundreds of playas to create accurate water budgets for the playas without the use of additional remote sensing and in situ datasets. However, if we analyze the period from November 2018–January 2019, the rainfall was negligible, so the playa basin rainfall catchment can be ignored, as water would not be entering the playas and the only escape mechanisms would be evaporation and ground infiltration. The average depth of the playas in the study region is not well-known, but based on general observations of depth [29], we estimate them here to be around 20 cm in our study region in early November 2018, and we assume that the depth of the playa water then decreases at a constant rate proportional to the playa water surface area (this may or may not be a good approximation). Based on these assumptions, we calculate the playa volumes in Table 1 using a simple relationship:

$$\text{Surface volume} = \text{surface area estimated from satellite data} \times \text{estimated mean playa depth.} \tag{1}$$

The changes in volume in the study region, due to either evaporation or ground infiltration, are then calculated using the following simple equation:

$$\text{Change in volume due to evaporation} = \text{monthly change in water surface area estimated from satellite data} \times \text{net surface evaporation.} \tag{2}$$

Finally, the difference between the total playa volume each month and the change due to evaporation can be inferred to be ground infiltration. The results of these simple illustrative calculations demonstrate that the changes in playa volume are mainly driven by evaporation (about 70–85%), followed by percolation into the ground (15–30%). Better known observations or estimates of the aforementioned variables (playa depth, basin size, evaporation rates, etc.) are needed to confidently close the hydrological budget of the playas and make more useful estimates of the groundwater infiltration rates using the satellite inundation surface area measurements, but the previous discussion lays out a realistic scenario, provided the assumptions we chose are reasonable.

Even within the relatively small 900 km^2 study area, large variations in playa inundation, driven by variations in rainfall, are observed. As an example, a period of heavy rainfall occurred in the study region (shown in Figure 1a) in West Texas, USA, between

15 June and 15 July 2021. The total rainfall in the region was highly variable, ranging between 5 and 18 cm over just 10 km (Figure 4). All playas in this region were observed to be dry in spring 2021 (not shown) with no surface water previous to the heavy rainfall in summer 2021. However, heavy thunderstorms in late June and early July 2021 resulted in significant increases in total playa inundation, and an increase in observed water surface area (15.23 km^2 of playa lake surface area) was noted in the study (Figure 4)). However, those areas in the southeastern third of the region of interest, where less rainfall was observed, saw only minimal increases in water surface area, compared to large increases in the regions where the heaviest rainfall occurred (Figure 4).

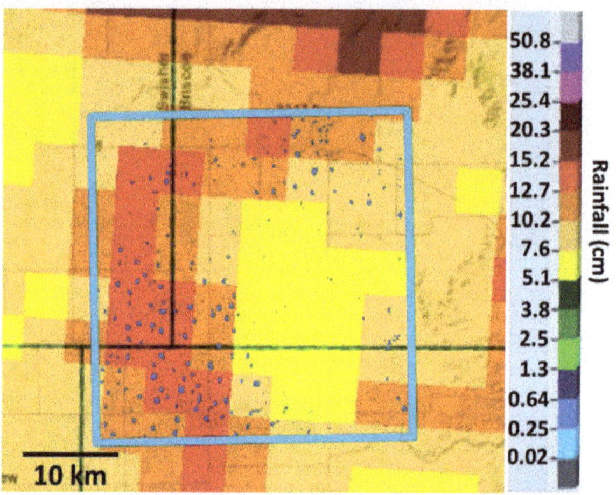

Figure 4. Playa lake surface area and radar-estimated precipitation for the 30 × 30 km square region shown in Figure 1a (and delineated here by the blue square). Playa water surface area for 9 July 2021 within the blue square is denoted in blue and calculated using the QGIS Semi-Automatic Classification Plugin [16,21]. The underlying map of radar-estimated rainfall between 15 June and 14 July 2021 is courtesy of the National Weather Service Advanced Hydrological Prediction Service https://water.weather.gov/precip/ (accessed on 15 March 2022).

4. Discussion

In this paper, we provide a preliminary framework for others to build upon. This short study leaves many questions unanswered for future research to consider. Future work will need to more carefully evaluate any limitations of using Sentinel-2 or other satellite imagery to monitor playas. In our study, many small playas had diameters around 50 m, with larger playa diameters of 300 m or more. Assuming a circular playa shape and edge detection errors of 10 m (one pixel) due to subpixel (half of pixel ground, half water, which will sometimes be detected as water, and sometimes as land) or other effects, this would result in possible area calculation errors in excess of 40% for small playas, with smaller estimated uncertainty (less than 10%) for larger diameter playas. To carefully evaluate the limitations of potential remote sensing errors, we recommend using in situ datasets to rigorously evaluate which water detection algorithms work best over the range of water depths, water colors, and water turbidity observed in shallow playas, to avoid biases or missed detections from the satellites of these complex surface features. A number of recent studies in the literature illustrate a wide range of potential errors and corrections that should be addressed, such as classification uncertainty [30], threshold detection errors [31], using frequency analyses of pixels rather than a single detection threshold (e.g., [32]), and combining the Sentinel-2 data with altimetric satellite missions to derive shoreline locations [33]. There are many possible avenues for future work, as there currently is no single universally approved detection algorithm for shallow lakes or wetlands, although the NDWI has been found to be problematic when used in shallow waters (e.g., [34]). The recent study by Jiang et al. [35] demonstrates a more sophisticated water detection methodology for use with water bodies of different clarity that could potentially be applied to playa lakes. Future work could also investigate blending the Sentinel-2 water surface area calculations with Sentinel-3 water level measurements, as demonstrated by recent studies [36], or using newly developed automated and hierarchical surface water fraction mapping developed by Wang et al. [37] for the thousands of playa lakes across the world. This limited demonstration study used a very simple manual classification detection algorithm, and made no attempt at evaluating uncertainties in either under- or over-detection, or at assessing the impacts of not resolving sub-pixel shoreline features, which would increase the potential errors in playa lake surface area calculations for small lakes nearing the detection limits (± 10 m). Some analysis of the impacts of aquatic plants and vegetation beneath the shallow water should also be considered [38–40]. Moreover, additional satellite platforms could also be considered in future work, including the proprietary Planet Scope and Worldview satellites (which have even higher temporal and spatial resolution than Sentinel-2). Recent studies have demonstrated using these platforms for the temporal classification of lake surfaces [41–43].

Evaluating spatial analyses of precipitation, in combination with the spatial analyses, may also help reveal which regions have adequate water in their playa lakes and which do not. The semi-arid climates that contain most of the world's playas are also regions with high variability in precipitation intensity and frequency. One single large localized rainstorm might mean the difference between a playa filling up with much-needed water for several months, or occasionally, even a year or more, or remaining dry for months or even years. The high-resolution spatial and temporal data from the Sentinel-2 satellites can also be used with precipitation data to determine playa inundation on sub-monthly time scales for hydrological and ecological applications.

In addition to the monitoring of playa lake surface area from the Sentinel-2 satellites, the satellite reflectance signatures could potentially be utilized to evaluate changes in playa lake water quality, as changes in water turbidity are sometimes correlated with water quality. Recent work has already developed regression algorithms for using Sentinel-2 satellites as proxies for the water quality monitoring of agricultural reservoirs in Oklahoma [43], as well as for utilizing remote sensing reflectance signatures for developing a water quality index [44]. Coupling the satellite reflectance data with in situ water quality sensors for a 'training set' of playa lakes could potentially be used to obtain water quality information for many thousands of playas where installing in situ measurements for playa lake water quality is not feasible. Because the satellite reflectance signatures differ between waters of differing turbidity and quality, properties of the water quality or wetland vegetation types could potentially be estimated; this is also recommended for future applications of Sentinel-2 or similar resolution satellite imagery for investigating playa lakes.

5. Conclusions

Over the last 5 years, the improved spatial (and temporal) resolution of Sentinel-2 satellite imagery has allowed many small, variable inland water bodies, such as rivers, glacial lakes, and rice paddies, to be studied globally for the first time. This study is the first to the authors' knowledge to demonstrate using Sentinel-2 satellite mission imagery to improve the monitoring of American Great Plains playa water inundation on sub-monthly time scales for ecological and hydrological applications. While many studies have documented the importance of the playa wetland ecosystems for wildlife in the American Great Plains, along with the variable infiltration and aquifer recharge rates occurring for different playas on the Great Plains, in situ monitoring of potential water inflow, evaporation, and infiltration rates for the many tens of thousands of North American Great Plains playas is simply not feasible due to the large number of small playas observed. The high-resolution Sentinel-2 mission satellite data, in combination with survey data on playa depth and nearby meteorological data (to determine evaporation rates), could be a major step forward in extensively and inexpensively quantifying the importance of playa lake groundwater recharge.

Evaluating the changes in specific playas across broad regions using satellite imaging on sub-monthly time scales may be useful for identifying playas with higher groundwater recharge potential (those that lose water more rapidly than others), and for analyzing periods when evaporation potential is low (such as during winter). Monitoring playas, as demonstrated in this short paper, would also be useful for ecological applications for determining the lakes that contain sufficient water to support wildlife, without requiring an airplane flyover or time-intensive field visits.

Author Contributions: Conceptualization, H.L.T., E.T.C. and W.J.R.; methodology, H.L.T., E.T.C., J.B.J. and N.L.H.; software, H.L.T. and E.T.C.; validation, H.L.T. and E.T.C.; formal analysis, H.L.T., E.T.C., J.B.J. and N.L.H.; investigation, H.L.T., E.T.C., J.B.J. and N.L.H.; resources, H.L.T. and E.T.C.; data curation, H.L.T. and E.T.C.; writing—original draft preparation, H.L.T., E.T.C., J.B.J. and N.L.H.; writing—review and editing, H.L.T., E.T.C., J.B.J. and N.L.H.; supervision, H.L.T., E.T.C., J.B.J. and N.L.H.; project administration, E.T.C. and H.L.T.; funding acquisition, E.T.C. and H.L.T. All authors have read and agreed to the published version of the manuscript.

Funding: This research was partially funded by a West Texas A&M University Kilgore graduate student research grant awarded to H.L.T. as part of her M.S. thesis research.

Institutional Review Board Statement: Not applicable.

Informed Consent Statement: Not applicable.

Data Availability Statement: Several publicly available datasets were used in this study. Satellite data from the Sentinel-2A and 2B satellites were freely downloaded from the European Space Agency (ESA) Copernicus Open Access Hub (https:scihub.copernicus.eu) (accessed on 1 June 2022). The QGIS software and Semi-Automatic Classification Plugin used in this study are also both freely available (https:qgis.org and https://plugins.qgis.org/plugins/SemiAutomaticClassificationPlugin/) (accessed on 1 June 2022). Radar derived precipitation estimate data was obtained from the National Weather Service Advanced Hydrological Prediction Service https://water.weather.gov/precip/ (accessed on 15 June 2022). Actual rainfall measurements from the National Weather Service Plainview Texas weather station were freely downloaded from Mesowest (https:mesowest.utah.edu) (accessed on 15 March 2022).

Acknowledgments: The authors are very thankful to the three anonymous expert reviewers whose time and effort improved this manuscript. The support of graduate school personnel and other faculty and staff at West Texas A&M University regarding the overall research efforts, including Angela Spaulding, Teresa Clemons, and Glenda Norton is acknowledged. West Texas A&M Information Technology also helped support the computing capabilities for research conducted at West Texas A&M. We also acknowledge the SentinelHub EO Browser for its help with creating Figure 2.

Conflicts of Interest: The authors declare no conflict of interest. The funders had no role in the design of the study; in the collection, analyses, or interpretation of data; in the writing of the manuscript, or in the decision to publish the results.

References

1. Covich, A.P.; Fritz, S.C.; Lamb, P.J.; Marzolf, R.D.; Matthews, W.J.; Poiani, K.A.; Prepas, E.E.; Richman, M.B.; Winter, T.C. Potential effects of climate change on aquatic ecosystems of the Great Plains of North America. *Hydrol. Process* **1997**, *11*, 993–1021. [CrossRef]
2. Starr, S.M.; McIntyre, N.E. Land-cover changes and influences on playa wetland inundation on the Southern High Plains. *J. Arid Environ.* **2020**, *175*, 104096. [CrossRef]
3. Last, W. Paleochemistry and paleohydrology of Ceylon Lake, a salt-dominated playa basin in the northern Great Plains, Canada. *J. Paleolimnol.* **1990**, *4*, 219–238. [CrossRef]
4. Smith, L.M. *Playas of the Great Plains: Peter T. Flawn Endowment in Natural Resource Management and Conservation;* University of Texas Press: Austin, TX, USA, 2003; Volume 3.
5. Gurdak, J.J.; Roe, C.D. Recharge rates and chemistry beneath playas of the High Plains aquifer, USA. *Hydrogeol. J.* **2010**, *18*, 1747–1772. [CrossRef]
6. McKenna, O.P.; Sala, O. Groundwater recharge in desert playas: Current rates and future effects of climate change. *Environ. Res. Lett.* **2018**, *13*, 014025. [CrossRef]

7. McIntyre, N.E.; Wright, C.K.; Swain, S.; Hayhoe, K.; Liu, G.; Schwartz, F.W.; Henebry, G.M. Climate forcing of wetland landscape connectivity in the Great Plains. *Front. Ecol. Environ.* **2014**, *12*, 59–64. [CrossRef]

8. Castañeda, C.; Herrero, J.; Casterad, M.A. Landsat monitoring of playa-lakes in the Spanish Monegros desert. *J. Arid Environ.* **2005**, *63*, 497–516. [CrossRef]

9. French, R.H.; Miller, J.J.; Dettling, C.; Carr, J.R. Use of remotely sensed data to estimate the flow of water to a playa lake. *J. Hydrol.* **2006**, *325*, 67–81. [CrossRef]

10. Collins, S.D.; Heintzman, L.J.; Starr, S.M.; Wright, C.K.; Henebry, G.M.; McIntyre, N.E. Hydrological dynamics of temporary wetlands in the southern Great Plains as a function of surrounding land use. *J. Arid Environ.* **2014**, *109*, 6–14. [CrossRef]

11. Walker, D.; Smigaj, M.; Jovanovic, N. Ephemeral sand river flow detection using satellite optical remote sensing. *J. Arid Environ.* **2019**, *168*, 17–25. [CrossRef]

12. Yang, X.; Chen, Y.; Wang, J. Combined use of Sentinel-2 and Landsat 8 to monitor water surface area dynamics using Google Earth Engine. *Remote Sens. Lett.* **2020**, *11*, 687–696. [CrossRef]

13. Li, M.; Hong, L.; Guo, J.; Zhu, A. Automated Extraction of Lake Water Bodies in Complex Geographical Environments by Fusing Sentinel-1/2 Data. *Water* **2022**, *14*, 30. [CrossRef]

14. Chen, Q.; Liu, W.; Huang, C. Long-Term 10 m Resolution Water Dynamics of Qinghai Lake and the Driving Factors. *Water* **2022**, *14*, 671. [CrossRef]

15. Eibedingil, I.G.; Gill, T.E.; Van Pelt, R.S.; Tong, D.Q. Combining Optical and Radar Satellite Imagery to Investigate the Surface Properties and Evolution of the Lordsburg Playa, New Mexico, USA. *Remote Sens.* **2021**, *13*, 3402. [CrossRef]

16. Doña, C.; Morant, D.; Picazo, A.; Rochera, C.; Sánchez, J.M.; Camacho, A. Estimation of Water Coverage in Permanent and Temporary Shallow Lakes and Wetlands by Combining Remote Sensing Techniques and Genetic Programming: Application to the Mediterranean Basin of the Iberian Peninsula. *Remote Sens.* **2021**, *13*, 652. [CrossRef]

17. Mcfeeters, S.K. The use of normalized difference water index (NDWI) in the delineation of open water features. *Int. J. Remote Sens.* **1996**, *17*, 1425. [CrossRef]

18. Huang, C.; Chen, Y.; Zhang, S.; Wu, J. Detecting, extracting, and monitoring surface water from space using optical sensors: A review. *Rev. Geophys.* **2018**, *56*, 333–360. [CrossRef]

19. Drusch, M.; Del Bello, U.; Carlier, S.; Colin, O.; Fernandez, V.; Gascon, F.; Hoersch, B.; Isola, C.; Laberinti, P.; Martimort, P.; et al. Sentinel-2: ESA's optical high-resolution mission for GMES operational services. *Remote Sens. Environ.* **2012**, *120*, 25–36. [CrossRef]

20. Mandanici, E.; Bitelli, G. Preliminary Comparison of Sentinel-2 and Landsat 8 Imagery for a Combined Use. *Remote Sens.* **2016**, *8*, 1014. [CrossRef]

21. Congedo, L. Semi-Automatic Classification Plugin: A Python tool for the download and processing of remote sensing images in QGIS. *J. Open Source Softw.* **2021**, *6*, 3172. [CrossRef]

22. Ghobadi, Y.; Pradhan, B.; Shafri, H.Z.M.; Bin Ahmad, N.; Kabiri, K. Spatio-temporal remotely sensed data for analysis of the shrinkage and shifting in the Al Hawizeh wetland. *Environ. Monit. Assess.* **2014**, *187*, 4156. [CrossRef] [PubMed]

23. Reif, M.; Frohn, R.C.; Lane, C.R.; Autrey, B. Mapping Isolated Wetlands in a Karst Landscape: GIS and Remote Sensing Methods. *GIScience Remote Sens.* **2009**, *46*, 187–211. [CrossRef]

24. Rokni, K.; Ahmad, A.; Selamat, A.; Hazini, S. Water Feature Extraction and Change Detection Using Multitemporal Landsat Imagery. *Remote Sens.* **2014**, *6*, 4173–4189. [CrossRef]

25. Sánchez-Espinosa, A.; Schröder, C. Land use and land cover mapping in wetlands one step closer to the ground: Sentinel-2 versus landsat 8. *J. Environ. Manag.* **2019**, *247*, 484–498. [CrossRef] [PubMed]

26. Wu, G.; Xiao, X.; Liu, Y. Satellite-Based Surface Water Storage Estimation: Its History, Current Status, and Future Prospects. *IEEE Geosci. Remote Sens. Mag.* **2022**, 2–23. [CrossRef]

27. Tempa, K.; Aryal, K.R. Semi-automatic classification for rapid delineation of the geohazard-prone areas using Sentinel-2 satellite imagery. *SN Appl. Sci.* **2022**, *4*, 141. [CrossRef]

28. Howell, N.L.; Butler, E.B.; Guerrero, B. Water quality variation with storm runoff and evaporation in playa wetlands. *Sci. Total Environ.* **2018**, *652*, 583–592. [CrossRef]

29. Anderson, J.T.; Smith, L.M. The effect of flooding regimes on decomposition of Polygonum pensylvanicum in playa wetlands (Southern Great Plains, USA). *Aquat. Bot.* **2002**, *74*, 97–108. [CrossRef]

30. Dronova, I.; Gong, P.; Wang, L. Object-based analysis and change detection of major wetland cover types and their classification uncertainty during the low water period at Poyang Lake, China. *Remote Sens. Environ.* **2011**, *115*, 3220–3236. [CrossRef]

31. Reis, L.G.d.M.; Souza, W.d.O.; Ribeiro Neto, A.; Fragoso, C.R., Jr.; Ruiz-Armenteros, A.M.; Cabral, J.J.d.S.P.; Montenegro, S.M.G.L. Uncertainties involved in the use of thresholds for the detection of water bodies in multitemporal analysis from Landsat-8 and Sentinel-2 Images. *Sensors* **2021**, *21*, 7494. [CrossRef]

32. Borro, M.; Morandeira, N.; Salvia, M.; Minotti, P.; Perna, P.; Kandus, P. Mapping shallow lakes in a large South American floodplain: A frequency approach on multitemporal Lansat TM/ETM data. *J. Hydrol.* **2014**, *512*, 39–52. [CrossRef]

33. Li, X.; Long, D.; Huang, Q.; Han, P.; Zhao, F.; Wada, Y. High-temporal-resolution water level and storage change data sets for lakes on the Tibetan Plateau during 2000–2017 using multiple altimetric missions and Landsat-derived lake shoreline positions. *Earth Syst. Sci. Data* **2019**, *11*, 1603–1627. [CrossRef]

34. Singh, K.; Ghosh, M.; Sharma, S.R. WSB-DA: Water surface boundary detection algorithm using Landsat 8 OLI data. *IEEE J. Sel. Top Appl. Earth Obs. Remote Sens.* **2015**, *9*, 363–368. [CrossRef]

35. Jiang, W.; Ni, Y.; Pang, Z.; Li, X.; Ju, H.; He, G.; Lv, J.; Yang, K.; Fu, J.; Qin, X. An Effective Water Body Extraction Method with New Water Index for Sentinel-2 Imagery. *Water* **2021**, *13*, 1647. [CrossRef]
36. Gourgouletis, N.; Bariamis, G.; Anagnostou, M.N.; Baltas, E. Estimating Reservoir Storage Variations by Combining Sentinel-2 and 3 Measurements in the Yliki Reservoir, Greece. *Remote Sens.* **2022**, *14*, 1860. [CrossRef]
37. Wang, Y.; Li, X.; Zhou, P.; Jiang, L.; Du, Y. AHSWFM: Automated and Hierarchical Surface Water Fraction Mapping for Small Water Bodies Using Sentinel-2 Images. *Remote Sens.* **2022**, *14*, 1615. [CrossRef]
38. Zhang, F.; Wang, J.; Wang, X. Recognizing the Relationship between Spatial Patterns in Water Quality and Land-Use/Cover Types: A Case Study of the Jinghe Oasis in Xinjiang, China. *Water* **2018**, *10*, 646. [CrossRef]
39. Jiang, X.; Wang, J.; Liu, X.; Dai, J. Landsat Observations of Two Decades of Wetland Changes in the Estuary of Poyang Lake during 2000–2019. *Water* **2022**, *14*, 8. [CrossRef]
40. Soria, J.; Ruiz, M.; Morales, S. Monitoring Subaquatic Vegetation Using Sentinel-2 Imagery in Gallocanta Lake (Aragón, Spain). *Earth* **2022**, *3*, 363–382. [CrossRef]
41. Tuzcu, A.; Taskin, G.; Musaoğlu, N. Comparison of Object Based Machine Learning Classifications of Planetscope and WORLDVIEW-3 Satellite Images for Land Use/Cover. *Int. Arch. Photogramm. Remote Sens. Spat. Inf. Sci.* **2019**, *42*, 1887–1892. [CrossRef]
42. Qayyum, N.; Ghuffar, S.; Ahmad, H.M.; Yousaf, A.; Shahid, I. Glacial Lakes Mapping Using Multi Satellite Planet Scope Imagery and Deep Learning. *ISPRS Int. J. Geo-Inf.* **2020**, *9*, 560. [CrossRef]
43. Mansaray, A.S.; Dzialowski, A.R.; Martin, M.E.; Wagner, K.L.; Gholizadeh, H.; Stoodley, S.H. Comparing PlanetScope to Landsat-8 and Sentinel-2 for Sensing Water Quality in Reservoirs in Agricultural Watersheds. *Remote Sens.* **2021**, *13*, 1847. [CrossRef]
44. Zhang, F.; Chan, N.W.; Liu, C.; Wang, X.; Shi, J.; Kung, H.-T.; Li, X.; Guo, T.; Wang, W.; Cao, N. Water Quality Index (WQI) as a Potential Proxy for Remote Sensing Evaluation of Water Quality in Arid Areas. *Water* **2021**, *13*, 3250. [CrossRef]

Article

Surface Water Quality Assessment and Contamination Source Identification Using Multivariate Statistical Techniques: A Case Study of the Nanxi River in the Taihu Watershed, China

Zhi-Min Zhang [1],*, Fei Zhang [2], Jing-Long Du [1] and De-Chao Chen [1]

[1] College of Geography Science and Geomatics Engineering, Suzhou University of Science and Technology, Suzhou 215009, China; jldu@mail.usts.edu.cn (J.-L.D.); dcchen@mail.usts.edu.cn (D.-C.C.)

[2] College of Resources and Environment Science, Xinjiang University, Urumqi 830046, China; zhangfei3s@xju.edu.cn

* Correspondence: zmzhang@mail.usts.edu.cn; Tel.: +86-137-7179-7831

Abstract: Understanding the spatiotemporal patterns of water quality is crucial because it provides essential information for water pollution control. The spatiotemporal variations in water quality for the Nanxi River in the Taihu watershed of China were evaluated by a water quality index (WQI) and multivariate statistical techniques; additionally, the potential sources of contamination were identified. The data set included 22 water quality parameters collected during the monitoring period from 2015 to 2020 for 14 monitoring stations. WQI assessment revealed that approximately 85% of monitoring stations were classified as "medium-low" water quality, and most showed continuous improvement in water quality. Cluster analysis divided the 14 monitoring stations into three clusters (low contamination, medium contamination and high contamination). Discriminant analysis identified pH, petroleum, volatile phenol, chemical oxygen demand, total phosphorus, F, S, fecal coliform, SO_4, Cl, NO_3-N, total hardness, NO_2-N and NH_3 as important parameters affecting spatial variations. Factor analysis identified four potential contamination source types: nutrient, organics, feces and oil. This study demonstrated the usefulness of multivariate statistical techniques in assessing large data sets, identifying contamination source types, and better understanding spatiotemporal variations in water quality to restore and protect water resources.

Keywords: Nanxi River; multivariate statistical techniques; water quality index; water quality assessment

Citation: Zhang, Z.-M.; Zhang, F.; Du, J.-L.; Chen, D.-C. Surface Water Quality Assessment and Contamination Source Identification Using Multivariate Statistical Techniques: A Case Study of the Nanxi River in the Taihu Watershed, China. *Water* **2022**, *14*, 778. https://doi.org/10.3390/w14050778

Academic Editor: Roko Andricevic

Received: 19 January 2022
Accepted: 27 February 2022
Published: 1 March 2022

Publisher's Note: MDPI stays neutral with regard to jurisdictional claims in published maps and institutional affiliations.

1. Introduction

Deterioration of the water environment is a prominent problem in worldwide watershed management and seriously threatens the security of the water ecological environment [1]. Natural factors (such as climate, topography, geology and soil) and human activities (such as urbanization, industrial production and agricultural practice) affect the surface water quality of an area [2–5]. The seasonal changes in precipitation, hydrological conditions and stream runoff have marked effects on stream flow and the consequent pollutant concentrations in surface water [1,6–8]. Dynamic spatiotemporal assessment of water quality can be used to analyze water contamination problems, identify potential contamination source types, and provide information support and reference to effectively manage water resources [3].

To effectively prevent and control surface water contamination, reliable water quality data for in-depth research is necessary. Considering the spatiotemporal variation in the physicochemical and biological characteristics of surface water, a long-term monitoring plan to accurately assess water quality should be developed [9]. Environmental protection departments in China have established sound water quality monitoring networks and continuous water quality monitoring procedures that monitor the physical properties (e.g., temperature, pH and electrical conductivity, etc.), total organic components, nutrients and inorganic components, as well as the biological and microbial conditions. In water quality assessments, multiple water quality parameters are typically collected at multiple monitoring stations in different monitoring periods, and this process generates a complex data matrix [10]. Due to the potential multivariable correlations among monitoring stations, monitoring periods and water quality parameters, this complex data set is often challenging to analyze and explain [11–13]. In a comprehensive assessment of water quality, the challenge is to determine whether the changes in water quality should be attributed to the contamination of rivers by human activities or biogeochemical changes in natural processes [14]. Furthermore, the water quality parameters that can best describe the spatiotemporal changes and identify contamination source types should be determined.

The water quality index (WQI) is a useful method for evaluating the change and trend of water environment quality by synthesizing multiple original parameters to a single index [10]. As a water quality assessment model, WQI determines the relative weight of each parameter based on its importance in water environment protection and integrates multiple variables into a dimensionless variable to represent the comprehensive water quality status and grade [15–18]. WQI has played an increasingly crucial role in the water quality assessment of rivers, lakes and groundwater [19–24].

With the increased number and dimension of measurement parameters in samples, the problem of allocating unknown samples and mining valuable information becomes increasingly complex. Therefore, using multivariate statistical techniques and data reduction simultaneously to obtain satisfactory results is necessary [10]. Multivariate statistical techniques, such as cluster analysis (CA), discriminant analysis (DA), principal component analysis (PCA) and factor analysis (FA) have been widely applied to evaluate water quality, can simplify data dimensions from complex water quality data matrices and remove redundant information without losing valuable information [8,25–29]. Multivariate statistical techniques can identify spatiotemporal patterns of water quality and analyze the possible factors causing spatiotemporal variations in water quality and affecting the health of water ecosystems [1,30].

The Nanxi River in the Taihu watershed, as a rapidly urbanized area in China, is experiencing high disturbance from human activities and serious water contamination problems [31]. The spatiotemporal variations in surface water quality and the identification of contamination source types are critical for sustainable watershed water quality management. However, studies focusing on the identification of contamination source types in the Nanxi River are limited. The primary aims of this study are as follows: (1) to evaluate the contamination levels of different monitoring stations and periods to examine the spatiotemporal distributions of water quality using WQI; (2) to extract the clustering information of monitoring stations, and determine the most important classification variables for the spatial variations in water quality; and (3) to analyze the potential impact factors of water quality in three regions with different contamination levels and explore possible contamination source types (natural processes or human activities).

2. Materials and Methods

2.1. Study Area

The Nanxi River (119°08′–119°36′ E, 31°1′–31°41′ N) is the main river in the western part of the Taihu watershed in China (Figure 1). The total extension of the study area is 1535.87 km^2 and includes 39% farmland, 23% water area, 22% forestland and 16% built-up land. It belongs to the subtropical monsoon climate zone, with an average annual temperature of 16 °C and an average annual precipitation of 1147 mm, 70% of which occurs in the rainy season from May to October. The area comprises low mountains, hills, plain polders and other landform types, with elevations of 1–702 m. The main types of soil are paddy soils, yellow-brown soils, and yellow cinnamon soils [32]. The regional zonal vegetation is an evergreen and deciduous broad-leaved mixed forest. The main crops are rice, rape, tea and sericulture, etc. [31].

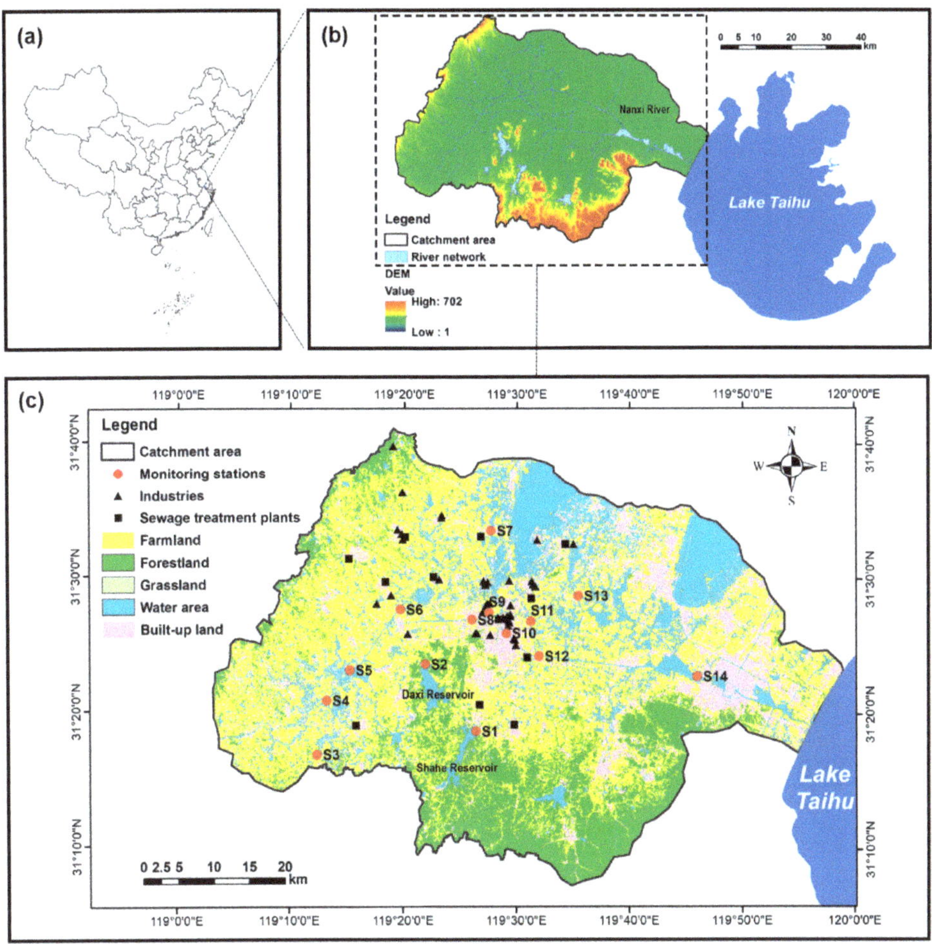

Figure 1. (**a**) Location of the study area in China; (**b**) Topography and river network of the study area; (**c**) Study area with 14 monitoring stations in the Nanxi River.

This area is relatively developed in the Taihu watershed, with a population of approximately 763,000. The area has many chemicals, synthetic materials, and mechanical, electronic and cement factories. The basin has fertile paddy soil, which is very suitable for various agricultural activities. The main contamination source types in this area include municipal and industrial wastewater, livestock and poultry breeding, planting and aquaculture. In 2019, the total amount of industrial and domestic sewage discharge was 2248.8×10^4 and 3244.9×10^4 tons, respectively, including 3043.5 tons of chemical oxygen demand (COD), 538.5 tons of ammonia nitrogen (NH_3-N), and 14.8% of the industrial sewage treatment rate; the COD, total nitrogen (TN) and total phosphorus (TP) of agricultural contamination sources were 3567.3, 338.2 and 202 tons, respectively. Thirteen sewage treatment plants (STPs) exist to treat domestic and industrial wastewater in the area. The maximum daily treatment capacity of these STPs is 15×10^4 m^3 [33].

2.2. Monitoring Stations and Water Quality Data

Water samples were collected from 14 monitoring stations (Figure 1) in the Nanxi River every month from January 2015 to November 2020. All the water samples were collected, stored, transported, and analyzed according to the Technical Specifications Requirements for Monitoring of Surface Water and Waste Water (HJ/T 91-2002) [34] and the Environmental Quality Standards for Surface Water (GB3838-2002) [35] to ensure the quality of the data. The final data were from the Liyang Environmental Protection Bureau.

These water quality data belonged to monthly routine sampling, which reflected the daily water quality status of each monitoring station, but cannot capture the dynamics of pollutants generated by episodic events (e.g., storms and pollution leakage accidents, etc.). Therefore, if possible, additional sampling of water quality before and after episodic events will be required in the future.

We selected 22 water quality parameters for this study. The abbreviations, units and descriptive statistics of them are summarized in Table 1.

Table 1. Water quality parameters, abbreviations, units and descriptive statistics.

Parameter	Abbreviation	Unit	Minimum	Maximum	Mean	S.D.	C.V.
Water temperature	Temp	°C	3.4	34.0	18.1	8.4	0.463
pH	pH		5.74	8.91	7.50	0.38	0.050
Electrical conductivity	EC	ms/m	9.4	181.0	38.0	16.6	0.437
Dissolved oxygen	DO	mg/L	1.8	13.0	6.5	1.9	0.291
Permanganate index	COD_{Mn}	mg/L	1.9	15.7	5.5	1.4	0.261
Biochemical oxygen demand	BOD_5	mg/L	1.8	9.0	3.6	1.0	0.287
Ammonia nitrogen	NH_3-N	mg/L	0.046	31.600	0.916	1.355	1.479
Petroleum	Petrol	mg/L	0.00	0.27	0.06	0.04	0.721
Volatile phenol	VP	mg/L	0.000	0.010	0.002	0.001	0.581
Chemical oxygen demand	COD	mg/L	5.0	87.9	19.1	6.8	0.354
Total nitrogen	TN	mg/L	0.24	37.40	2.87	2.01	0.699
Total phosphorus	TP	mg/L	0.010	0.444	0.131	0.071	0.546
Fluoride	F	mg/L	0.07	0.83	0.34	0.12	0.345
Sulfide	S	mg/L	0.000	0.173	0.046	0.033	0.715
Fecal coliform	F. Coli	CFU/L	360	9130	4226	2429	0.575
Sulphate	SO_4	mg/L	15.9	53.1	32.1	7.4	0.231
Chloride	Cl	mg/L	0	195	54	30	0.557
Nitrate nitrogen	NO_3-N	mg/L	0.020	1.200	0.294	0.127	0.434
Total suspended solids	TSS	mg/L	7	239	33	19	0.574
Total hardness	T-Hard	mg/L	41	255	147	37	0.251
Nitrite nitrogen	NO_2-N	mg/L	0.005	0.294	0.063	0.041	0.660
Nonionic ammonia	NH_3	mg/L	0.001	0.172	0.014	0.019	1.364

Mean represents the mean value; S.D. represents the standard deviation; C.V. represents the coefficient of variation.

2.3. Water Quality Index

WQI is an effective method for water quality assessments [36]. According to the impact of each water quality parameter on human water health and its relative importance in aquatic organisms [15], different weights were assigned in Table 2 [23]. Temp, pH, EC, DO, COD_{Mn}, COD, BOD_5, NH_3-N, TP, TN, Petrol, F. Coli, SO_4, CI, NO_2-N, NO_3-N, TSS and T-Hard were used, and the measured values were normalized.

The formula used to calculate WQI is as follows:

$$WQI = \frac{\sum_{i=1}^{n} C_i P_i}{\sum_{i=1}^{n} P_i} \tag{1}$$

where n is the total number of parameters involved in the calculation, C_i is the normalization factor of parameter i, and P_i is the relative weight of parameter i. The minimum value of P_i is 1, and the maximum weight specified is 4. These values were determined based on previous studies [23,37–39].

The calculated WQI is a dimensionless value from 0 to 100. Based on the WQI scores, surface water quality was divided into five categories [23]: excellent (90–100), good (70–89), moderate (60–69), low (40–59), and bad (0–39).

Table 2. Weights and normalization factors of the parameters used to calculate the water quality index.

Parameters	Units	Relative Weight (P_i)	Normalization Factor (C_i)										
			100	90	80	70	60	50	40	30	20	10	0
Temp	°C	1	21/16	22/15	24/14	26/12	28/10	30/5	32/0	36/−2	40/−4	45/−6	>45/<−6
pH		1	7	7–8	7–8.5	7–9	6.5–7	6–9.5	5–10	4–11	3–12	2–13	1–14
EC	ms/m	1	<75	<100	<125	<150	<200	<250	<300	<500	<800	≤1200	>1200
DO	mg/L	4	≥7.5	>7	>6.5	>6	>5	>4	>3.5	>3	>2	≥1	<1
COD_{Mn}	mg/L	3	<1	<2	<3	<4	<6	<8	<10	<12	<14	≤15	>15
COD	mg/L	3	<10	<15	<16	<18	<20	<25	<30	<34	<37	≤40	>40
BOD_5	mg/L	3	<2	<3	<3.4	<3.7	<4	<5	<6	<7	<9	≤10	>10
NH_3-N	mg/L	3	<0.1	<0.15	<0.3	<0.5	<1	<1.3	<1.5	<1.7	<1.9	≤2	>2
TP	mg/L	1	<0.01	<0.02	<0.05	<0.1	<0.15	<0.2	<0.25	<0.3	<0.35	≤0.4	>0.4
TN	mg/L	2	<0.1	<0.2	<0.35	<0.5	<0.75	<1	<1.25	<1.5	<1.75	≤2	>2
Petrol	mg/L	2	<0.01	<0.02	<0.03	<0.04	<0.05	<0.3	<0.5	<0.7	<0.9	≤1	>1
F. Coli	CFU/L	3	<100	<200	<1000	<2000	<10,000	<15,000	<20,000	<30,000	<35,000	≤40,000	>40,000
SO_4	mg/L	2	<25	<50	<75	<100	<150	<250	<400	<600	<1000	≤1500	>1500
CI	mg/L	1	<25	<50	<100	<150	<200	<300	<500	<700	<1000	≤1500	>1500
NO_2-N	mg/L	2	<0.005	<0.01	<0.03	<0.05	<0.1	<0.15	<0.2	<0.25	<0.5	≤1	>1
NO_3-N	mg/L	2	<0.5	<2	<4	<6	<8	<10	<15	<20	<50	≤100	>100
TSS	mg/L	4	<20	<40	<60	<80	<100	<120	<160	<240	<320	≤400	>400
T-Hard	mg/L	1	<25	<100	<200	<300	<400	<500	<600	<800	<1000	≤1500	>1500

Normalization factors are according to GB3838-2002 and weight, as proposed by Wu et al. (2018) [23].

2.4. Statistical Analysis

The water quality data sets over six years (2015–2020) were checked to eliminate possible missing and abnormal values [40]. The parameter F. Coli had missing values, which were replaced by sequence mean values. Data conforming to normal distribution are needed for most multivariate statistical techniques. Therefore, kurtosis and skewness statistics were analyzed to test whether each water quality parameter conformed to a normal distribution [5,13]. The original data showed that the kurtosis value was between −1.888 and 357.738, and the skewness value was between −1.215 and 16.671, indicating that the raw data were far from the normal distribution. Because most kurtosis and skewness values were greater than 0, the raw data were logarithmically converted ($x = \log10(x)$) [41]. After logarithmic transformation, kurtosis and skewness were in the ranges of −0.640 to 4.577 and −2.279 to 0.096 respectively. To minimize the influence of different units and variances on the parameters, Z-scale standardization (mean value is 0, variance is 1) was performed on the data.

Cluster analysis (CA) was performed on the standardized data to explore the spatial similarity and clustering information on water quality. Principal component analysis/factor analysis (PCA/FA) was performed on the standardized data to explore possible contamination source types [9,28]. Discriminant analysis (DA) was performed on the raw data to extract the important variables reflecting the variations between groups [5]. STATISTICA 10 was used for statistical analysis.

Please refer to Supplementary Materials for the detail about the above multivariate statistical methods.

3. Results and Discussion

The descriptive statistics of 22 water quality parameters are summarized in Table 1. The pH values ranged from 5.74 to 8.91, which were basically within the standard limit of 6–9 allowed by GB3838-2002. The mean values of F, S, F. Coli, and VP in most water samples were far lower than the class III standard (GB3838-2002), while that of Petrol (0.058 mg/L) was slightly higher than the class III standard (0.05 mg/L). Among nutrients, the mean value of TN was 2.87 and far higher than the class III standard (1.0 mg/L); the mean values of NO_3-N and NO_2-N were 0.294 and 0.063 respectively, which were far lower than the class III standard (10 mg/L); the one of NH_3-N was 0.916 and lower than the class III standard (1.0 mg/L). TN is the sum of NO_3-N, NO_2-N, NH_3-N and organic nitrogen, which is the main indicator of water eutrophication. Thus, the main nutrient in the study area was organic nitrogen. The concentration levels of COD_{Mn}, BOD_5 and COD deserve attention because these parameters represent the levels of biological, chemical and organic contamination in surface water, respectively. The maximum values of these parameters were 15.7, 9.0 and 87.9 mg/L, respectively, all exceeding the class III standard (6, 4 and 20 mg/L, respectively). Therefore, the study area had a relatively high contamination level. The coefficients of variation for NH_3-N, NH_3, Petrol, S, TN and NO_2-N were relatively high, indicating significant temporal and spatial differences in the distributions of these water quality parameters.

3.1. Water Quality Assessment Using WQI

The water quality of most monitoring stations was classified as "medium-low", accounting for approximately 84.52% (of which "medium" accounted for 64.29% and "low" accounted for 20.24%). Additionally, 13.10% of the water quality was "good", and only 2.38% was "excellent" (Figure 2). The water quality of S1 and S2 was always above "good", especially the water quality of S2, which was "excellent" in 2015 and 2018. Because these two monitoring stations are in the Daxi Reservoir and Shahe Reservoir within the urban centralized drinking water protection area, their water quality has been maintained in good condition due to the good natural ecological environment and strict contamination control measures. Other monitoring stations are in urbanized or agricultural areas.

From the interannual change trend of WQI (Figure 3), about half of the monitoring stations showed an increasing trend, most of which were generally stable. The water quality of monitoring stations S1 and S2 decreased slightly. Due to rapid urbanization and population growth, water environment security is facing increased pressure, and the protection of water sources should be further strengthened. The water quality of other monitoring stations showed continuous improvement, especially from 2016 to 2019. Since 2017, Changzhou city has adopted special actions: "two reductions" (reducing total coal consumption and backward chemical production), "six governance" (governing the Taihu Lake water environment, domestic garbage, black and smelly water bodies, livestock and poultry breeding contamination, volatile organic compound contamination, and hidden environmental dangers), and "three improvements" (improving the level of ecological protection, environmental-economic policy regulation, and environmental law enforcement and supervision) [42]. The environmental quality has been significantly improved, the total discharge of major pollutants has been markedly reduced, and the environmental risks have been effectively controlled.

3.2. Spatial Similarities and Clustering

Spatial CA generated a dendrogram, dividing the 14 monitoring stations into 3 clusters at (Dlink/Dmax) × 100 < 40 (Figure 4). According to the physical, chemical and microbiological characteristics of water quality, each cluster was classified into its own contamination category. Cluster A included stations S1 and S2 and corresponded to low contamination. Cluster B contained six monitoring stations (S5, S6, S8, S9, S11 and S13) and was classified as medium contamination. Cluster C comprised six monitoring stations (S3, S4, S7, S10, S12 and S14) and was classified as high contamination.

In cluster A, S1 and S2 are in the Daxi Reservoir and Shahe Reservoir. The contamination of the six monitoring stations in cluster B mainly derives from nonpoint source contamination, such as agricultural runoff, livestock and poultry breeding, and fishpond drainage. The monitoring stations of cluster C are mainly located in urban areas and downstream reaches, and the possibility of water contamination is higher because of the comprehensive impacts of domestic sewage, industrial wastewater and upstream inflow water [5,7,10].

The above spatial CA results coincided with the average WQI of the monitoring stations. The WQI values of S1 and S2 were the highest; those of S3, S4, S7, S10, S12 and S14 were relatively low; those of S5, S6, S8, S9, S11 and S13 were at a medium level (Figure 5). Thus, CA can be used to provide reliable water quality classification throughout monitoring stations; however, designing optimal spatial sampling strategies is warranted in the future [10,28,41,43].

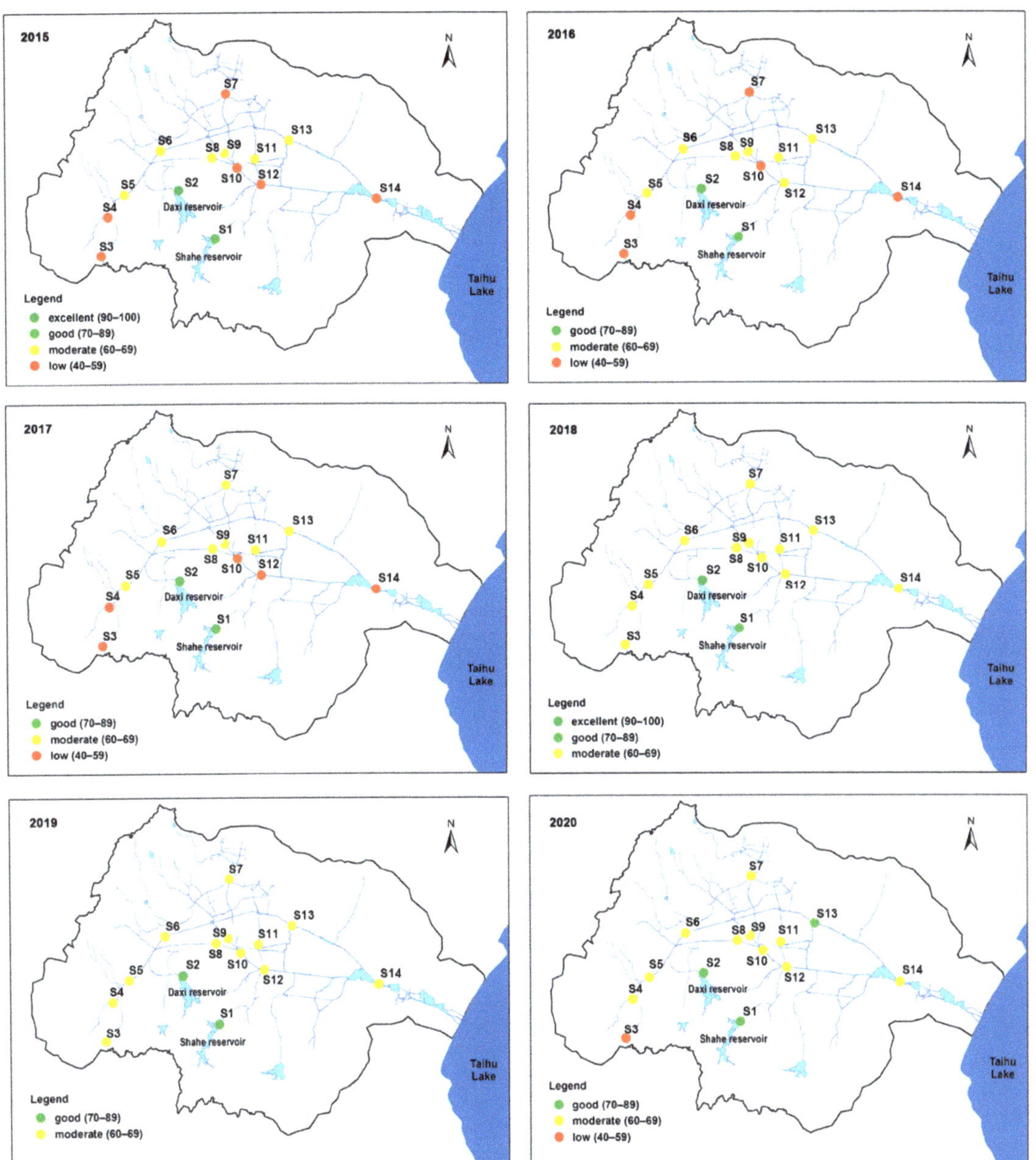

Figure 2. Spatial distribution pattern of the average annual WQI categories in the Nanxi River from 2015 to 2020.

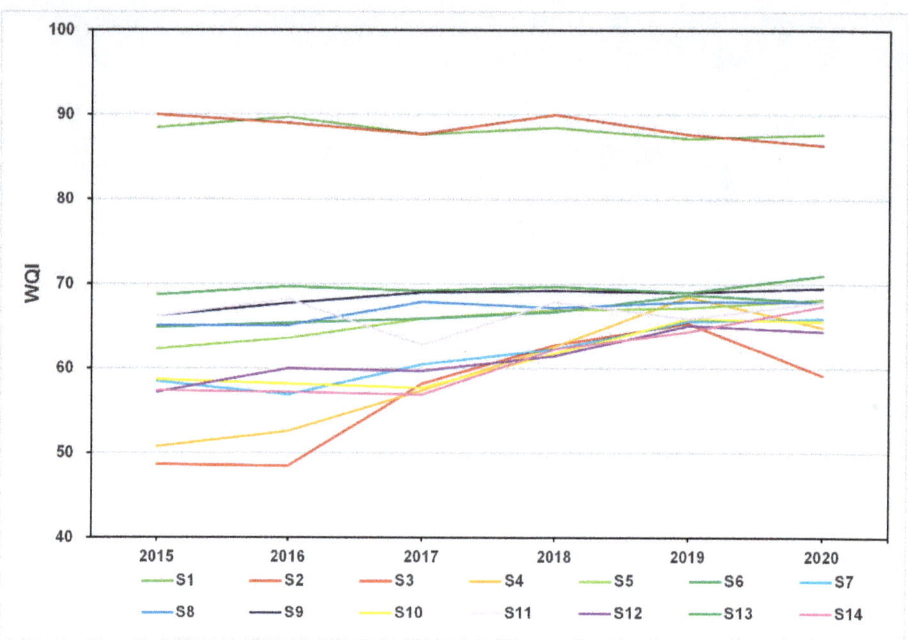

Figure 3. The average annual WQI scores of 14 monitoring stations in the Nanxi River from 2015 to 2020.

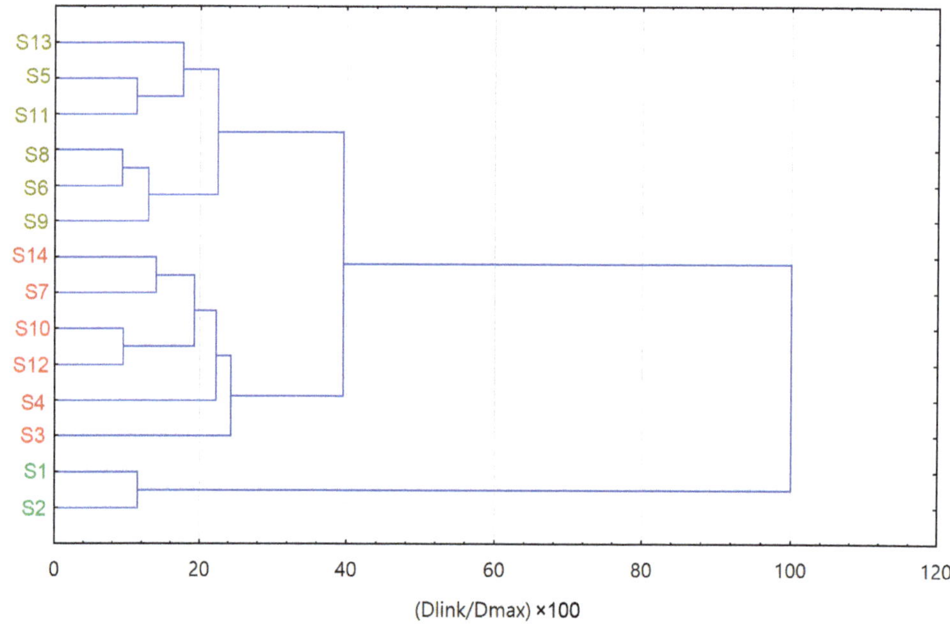

Figure 4. Dendrogram of spatial similarities and clustering of monitoring stations (S1–S14) in the Nanxi River.

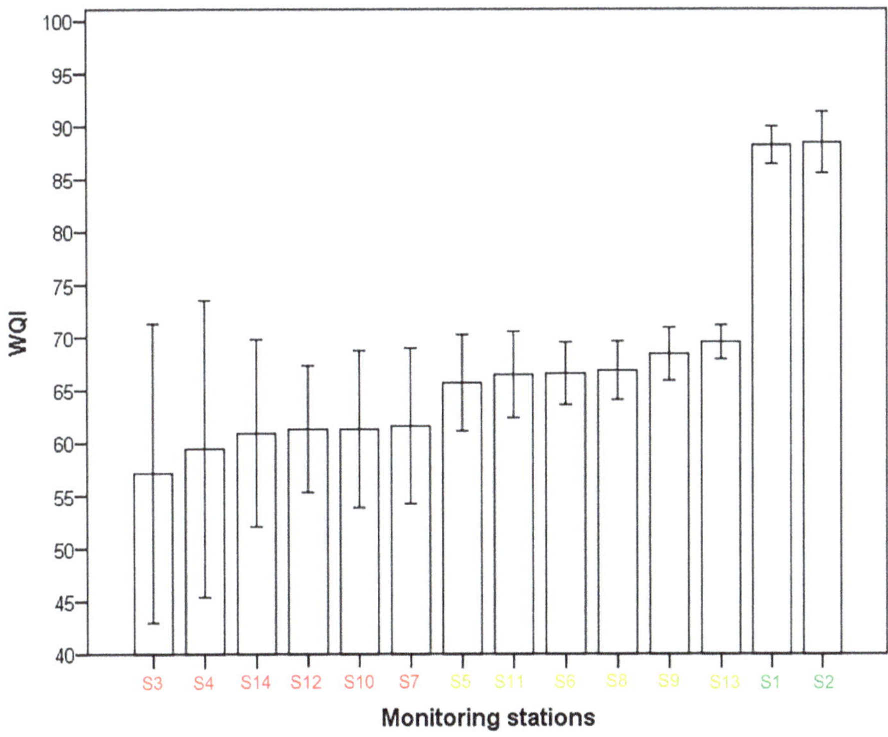

Figure 5. Average WQI of 14 monitoring stations in the Nanxi River.

3.3. Spatial Variations in Water Quality

Based on the CA data, discriminant analysis was used to detect the significance of the discriminant function and to identify the important variables reflecting the variation between clusters. The Wilks' lambda and chi-square values in all discriminant functions were in the range of 0.036–0.509 and 504.269–2479.317, respectively, and the *p* values were all less than 0.01 (Table 3), indicating that the spatial DA was valid [13].

Table 3. Spatial discriminant analysis results for spatial variations in water quality.

Modes	Discriminant Function	R	Wilks' Lambda	Chi-Square	*p* Value
Standard	1	0.963697	0.035502	2478.586	0.00
	2	0.708513	0.498009	517.624	0.00
Forward	1	0.963686	0.035547	2479.317	0.00
	2	0.708173	0.498491	517.255	0.00
Backward	1	0.960275	0.039628	2409.856	0.00
	2	0.700788	0.508897	504.269	0.00

Tables 4 and 5 show the discriminant function and classification matrix generated from the standard, forward stepwise and backward stepwise modes of DA. The standard and forward stepwise models of the discriminant function used 22 and 21 discriminant variables, respectively, and obtained the corresponding classification matrix, which correctly assigned approximately 88% of cases. However, in the backward stepwise mode, DA generated nearly 87% of the correct allocation to the classification matrix using only 14 discrimination parameters. Spatial DA showed that pH, Petrol, VP, COD, TP, F, S, F. Coli, SO_4, Cl, NO_3-N, T-Hard, NO_2-N, and NH_3 were the critical variables to distinguish the water quality of the three spatial clusters and explained most of the spatial variations in expected water quality.

Table 4. Discriminant function coefficients of discriminant analysis for spatial variations in water quality.

Parameters	Standard Mode			Forward Stepwise Mode			Backward Stepwise Mode		
	Cluster A	Cluster B	Cluster C	Cluster A	Cluster B	Cluster C	Cluster A	Cluster B	Cluster C
Temp	2.19	2.15	2.16						
pH	115.00	107.97	107.52	102.81	95.983	95.470	95.99	89.346	88.763
EC	−0.09	−0.10	−0.08	−0.02	−0.035	−0.016			
DO	5.19	4.17	4.10	1.01	0.066	−0.021			
COD_{Mn}	−1.36	1.34	1.28	−2.35	0.373	0.314			
BOD_5	12.67	11.16	11.64	12.40	10.899	11.372			
NH_3-N	9.04	4.81	5.01	8.72	4.494	4.688			
Petrol	−87.09	−63.79	3.49	−107.72	−84.054	−16.893	−29.04	−15.614	53.431
VP	−5928.45	−1499.07	41.86	−3233.54	1148.149	2704.845	−3985.81	552.192	2072.728
COD	1.09	0.60	0.75	0.56	0.081	0.232	1.09	0.769	0.921
TN	−2.78	−0.32	−0.53	−3.21	−0.752	−0.964			
TP	38.74	93.48	99.55	21.98	77.019	82.983	73.23	130.491	137.176
F	7.53	40.57	37.79	8.13	41.154	38.375	8.79	42.410	39.641
S	−50.45	90.85	97.30	−101.99	40.229	46.377	−89.33	50.326	53.812
F. Coli	0.00	0.00	0.00	0.00	0.002	0.002	0.00	0.001	0.001
SO_4	−0.27	0.47	0.46	0.35	1.070	1.065	0.38	1.196	1.188
Cl	0.10	0.17	0.16	−0.07	0.006	−0.005	−0.08	0.005	−0.002
NO_3-N	−3.63	19.19	22.51	−17.28	5.789	9.028	−9.40	13.946	17.819
TSS	0.22	0.26	0.25	0.23	0.268	0.259			
T-Hard	0.44	0.76	0.76	0.30	0.624	0.627	0.33	0.627	0.630
NO_2-N	−46.64	50.99	33.76	−11.68	85.329	68.309	22.86	110.897	95.396
NH_3	−1329.83	−1190.28	−1191.98	−936.82	−804.223	−803.626	−838.04	−706.073	−697.089
Constant	−522.73	−554.90	−562.82	−433.50	−468.801	−475.693	−397.88	−434.420	−440.556

Table 5. Classification matrix of discriminant analysis for spatial variations in water quality.

Monitoring Stations	Percent Correct	Stations Assigned by Discriminant Analysis		
		Cluster A	Cluster B	Cluster C
Standard mode				
Cluster A	100.0000	108	0	0
Cluster B	95.6790	0	310	14
Cluster C	77.4691	0	73	251
Total	88.4921	108	383	265
Forward stepwise mode				
Cluster A	100.0000	108	0	0
Cluster B	95.3704	0	309	15
Cluster C	76.8518	0	75	249
Total	88.0952	108	384	264
Backward stepwise mode				
Cluster A	100.0000	108	0	0
Cluster B	94.7531	0	307	17
Cluster C	75.3086	0	80	244
Total	87.1693	108	387	261

Based on the discriminant parameters analyzed by DA, box and whisker plots of three clusters (cluster A, cluster B, cluster C) were constructed to evaluate the spatial variations in water quality (Figure 6). Most of the parameters showed significant differences between clusters. Overall, the average concentration of cluster A was much lower than that of clusters B and C, and the average concentration of cluster C was slightly higher than that of cluster B. Higher Petrol, COD and TP values were found in cluster C, indicating that organic contamination and eutrophication were the most serious water environment problems in cluster C. Additionally, lower pH values were found at the monitoring stations of cluster C, likely because of the hydrolysis of acidic substances (ammonia and organic acids) [5]. In conclusion, the water contamination of cluster C was more serious than that of the other two clusters. Thus, the prevention and control of contamination sources and treatment capacity of point source contamination must be strengthened, such as strengthening the construction and treatment capacity of STPs.

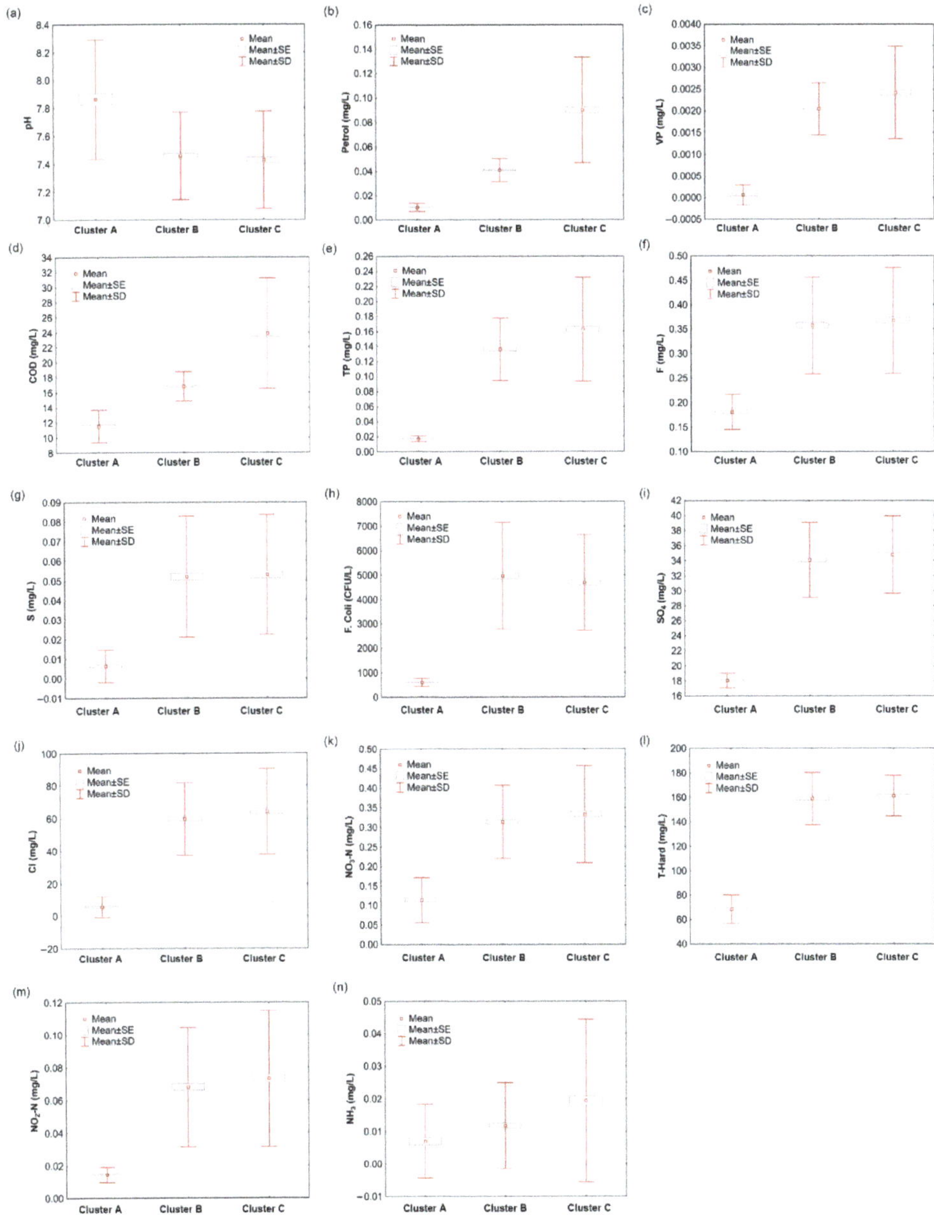

Figure 6. Spatial variations in water quality in three spatial clusters: (**a**) pH, (**b**) Petrol, (**c**) VP, (**d**) COD, (**e**) TP, (**f**) F, (**g**) S, (**h**) F. Coli, (**i**) SO_4, (**j**) Cl, (**k**) NO_3-N, (**l**) T-Hard, (**m**) NO_2-N, and (**n**) NH_3 in the Nanxi River.

3.4. Principal Component Determination and Contamination Source Identification

Because the contamination levels of the three spatial clusters (clusters A, B, and C) were significantly different, PCA/FA was used to identify the water contamination source types for the normalized data sets of the three spatial clusters.

PCA/FA of the three data matrices obtained six, eight and seven variance factors (VFs) with eigenvalues ≥ 1, explaining 71.5%, 66.8% and 67.9% of the total variance in the corresponding data sets, respectively (Tables 6–8). Additionally, the loadings of parameters on VFs were categorized as "high", "medium" and "low" based on absolute loading values of > 0.75, 0.75–0.50 and 0.50–0.30 [44].

Table 6. Loadings of 22 water quality parameters on VFs for cluster A in the Nanxi River.

Parameters	VF1	VF2	VF3	VF4	VF5	VF6
Temp	0.145	−0.110	0.168	0.351	0.665	0.214
pH	0.043	0.182	0.026	0.075	**0.764**	0.386
EC	0.073	0.190	0.168	0.113	**0.744**	−0.335
DO	−0.368	0.047	−0.319	−0.286	−0.223	0.287
COD_{Mn}	−0.198	−0.084	−0.024	**0.808**	0.243	−0.017
BOD_5	−0.068	0.103	0.130	**0.852**	0.101	−0.015
NH_3-N	0.091	−0.313	−0.489	0.093	−0.049	**−0.568**
Petrol	**0.829**	−0.066	0.060	−0.128	0.134	0.011
VP	**0.891**	0.022	0.119	−0.160	0.130	0.060
COD	−0.180	−0.376	0.546	−0.032	−0.189	−0.091
TN	**0.724**	0.014	0.199	−0.144	0.038	−0.273
TP	0.074	−0.107	−0.618	0.347	0.141	−0.024
F	**0.767**	−0.141	0.006	0.092	−0.081	0.014
S	0.003	0.091	**0.735**	0.248	−0.138	−0.153
F. Coli	0.566	0.529	0.321	0.048	−0.004	−0.306
SO_4	−0.011	**−0.750**	−0.115	0.239	0.144	−0.084
Cl	0.238	0.046	**0.889**	0.061	0.041	0.084
NO_3-N	0.133	0.092	**0.910**	0.007	0.074	0.022
TSS	0.589	−0.061	−0.162	0.403	−0.188	0.395
T-Hard	0.205	**−0.826**	0.096	−0.128	−0.243	0.022
NO_2-N	0.038	**0.748**	0.255	0.045	0.273	0.126
NH_3	0.019	0.097	−0.220	0.062	**0.884**	−0.113
Eigenvalue	4.554	3.373	3.067	2.266	1.380	1.089
% Total variance	20.699	15.332	13.939	10.302	6.271	4.952
Cumulative % variance	20.699	36.031	49.969	60.271	66.542	71.494

VFs represent the variance factors after varimax raw rotation for principal components; bold values represent medium-high loadings.

Among the six VFs of cluster A, VF1 explained 20.7% of the total variance and had high positive loadings on Petrol, VP, TN and F. This factor indicated toxic organic contamination from farmland drainage, oily sewage discharge from ship operation, domestic sewage, industrial wastewater, atmospheric deposition and precipitation leaching. VF2 (15.3% of the total variance) had high negative loadings on SO_4 and TSS, and high positive loadings on NO_2-N. The presence of nitrite in water indicated that the decomposition process of organic matter continued, and the risk of organic matter contamination persisted. VF3 (13.9%) had high positive loadings on NO_3-N, Cl and S, indicating nutrients from agricultural runoff and atmospheric deposition and the natural source of soil erosion and salt ions (CI, S) in the watershed [45]. VF4 (10.3%) had high positive loadings on BOD_5 and COD_{Mn}, representing organic contamination in sewage [6]. VF5 (6.3%) had high positive loadings on NH_3, pH and EC. Generally, EC indicates natural contamination, which may be due to soil erosion or an increase in the number of salt ions in water [44]. Additionally, VF6 (only 5.0%) had a medium negative loading on NH_3-N.

Table 7. Loadings of 22 water quality parameters on VFs for cluster B in the Nanxi River.

Parameters	VF1	VF2	VF3	VF4	VF5	VF6	VF7	VF8
Temp	−0.063	0.109	**−0.865**	0.041	−0.052	−0.100	0.099	0.153
pH	0.042	0.021	0.220	−0.234	−0.117	0.125	0.077	**0.787**
EC	−0.214	0.036	−0.180	−0.032	−0.062	**0.823**	−0.024	0.046
DO	0.024	0.073	**0.750**	0.086	−0.126	−0.378	0.066	0.055
COD_{Mn}	0.191	**−0.847**	0.041	−0.076	0.028	−0.006	−0.074	0.015
BOD_5	−0.032	**−0.863**	0.013	0.174	0.108	−0.020	0.041	−0.075
NH_3-N	0.211	−0.222	−0.088	−0.038	**0.703**	0.085	−0.019	−0.038
Petrol	−0.216	−0.357	0.073	0.067	0.629	−0.159	0.131	0.083
VP	0.074	0.141	−0.134	**−0.647**	−0.073	−0.134	0.358	0.181
COD	−0.161	0.132	0.157	0.225	0.520	0.004	0.082	0.104
TN	−0.021	0.009	0.193	0.014	0.142	0.090	**0.719**	0.061
TP	0.049	−0.135	−0.160	**0.621**	0.164	0.046	−0.064	−0.073
F	**−0.810**	0.169	−0.006	−0.117	−0.011	0.086	0.061	−0.010
S	**0.706**	0.171	0.121	−0.335	0.093	−0.062	−0.090	0.199
F. Coli	**0.707**	−0.178	0.010	0.400	−0.226	−0.052	0.095	0.061
SO_4	−0.006	0.028	−0.166	−0.111	−0.103	0.007	**0.786**	0.066
Cl	0.392	0.026	0.354	0.086	0.161	0.568	0.291	0.061
NO_3-N	−0.185	0.013	0.064	**−0.664**	0.038	0.314	−0.173	−0.140
TSS	−0.141	0.218	−0.169	−0.047	0.544	0.052	−0.206	−0.103
T-Hard	−0.093	−0.130	0.050	−0.158	0.035	0.202	0.131	−0.663
NO_2-N	0.165	0.085	0.189	0.572	−0.046	0.089	−0.490	0.143
NH_3	0.081	−0.024	−0.355	0.119	0.135	0.045	0.092	**0.812**
Eigenvalue	2.837	2.440	2.038	1.980	1.548	1.419	1.297	1.146
% Total variance	12.897	11.089	9.262	8.998	7.038	6.452	5.897	5.209
Cumulative % variance	12.897	23.986	33.248	42.246	49.284	55.735	61.632	66.841

VFs represent the variance factors after varimax raw rotation for principal components; bold values represent medium-high loadings.

Regarding the data set of cluster B, among the eight VFs, VF1, which accounted for 12.9% of the total variance, represented a high negative loading on F but medium positive loadings on F. Coli and S, indicating microbial contamination from municipal sewage, livestock and poultry breeding. VF2 (11.1% of the total variance) represented high negative loadings on BOD_5 and COD_{Mn}, indicating organic contamination in urban sewage and industrial wastewater. VF3 (9.3%) represented a high positive loading on DO but a high negative loading on Temp. VF4 (9.0%) represented only a moderate positive loading on TP, revealing nutrient contamination (e.g., P), especially from sewage containing detergents, industrial wastewater and fertilizer. Point source contamination (such as wastewater from the phosphorus chemical industry) and nonpoint source contamination (such as animal breeding and agricultural fertilizer) from P, constitute common eutrophication-causing contamination in this area [46]. VF5 (7.0%) applied only a moderate positive loading on NH_3-N, representing the contamination of animal feces and agricultural fertilizers. VF6 (only 6.5%) presented a high positive loading only on EC, likely because of the mineral composition in river water [6]. VF7 (only 5.9%) presented a high positive loading on SO_4 and a medium positive loading on TN, representing industrial wastewater using sulfate or sulfuric acid. Finally, VF8 (only 5.2%) had a high positive loading on NH_3 and pH, likely because of industrial wastewater containing alkaline substances, such as NH_3.

Regarding the seven VFs of cluster C, VF1 (20.5% of the total variance) showed high positive loadings on NH_3-N and TN, representing nutrient contamination from agricultural runoff, municipal sewage and fertilizer plant wastewater (e.g., N). VF2 (12.2%) showed a high positive loading on F, representing industrial wastewater containing fluoride. VF3 (10.2%) showed a high positive loading on pH. VF4 (7.5%) showed a high positive loading on Temp and a moderate negative loading on DO, contrasting the results for VF3 of cluster B. VF5 (6.7%) showed a high positive loading on Petrol, representing contamination from oily sewage discharge from ship operations and wastewater from the petrochemical industry. VF6 (5.7) showed moderate positive loadings on TSS and EC. Agricultural runoff, wastewater discharge, solid waste disposal and irrigation return increased the suspended solids loading in streams [45]. VF7 (5.0%) showed a high positive loading on SO_4, similar to VF7 of cluster B.

Table 8. Loadings of 22 water quality parameters on VFs for cluster C in the Nanxi River.

Parameters	VF1	VF2	VF3	VF4	VF5	VF6	VF7
Temp	−0.189	0.073	0.090	**0.862**	−0.116	−0.034	0.125
pH	−0.158	0.007	**0.781**	−0.105	−0.132	−0.052	0.135
EC	0.250	0.211	−0.077	0.206	−0.068	**0.704**	−0.077
DO	−0.216	0.030	−0.051	**−0.743**	−0.118	−0.155	0.090
COD_{Mn}	0.593	−0.065	0.051	0.032	0.414	−0.285	0.186
BOD_5	0.534	−0.030	0.056	0.142	0.655	−0.255	0.098
NH_3-N	**0.829**	−0.001	−0.085	0.025	0.175	0.135	−0.189
Petrol	0.114	−0.105	0.086	−0.108	**0.819**	0.055	0.064
VP	0.001	−0.100	0.155	0.004	−0.632	−0.283	0.139
COD	0.546	−0.013	0.294	−0.245	0.472	0.075	0.071
TN	**0.827**	0.127	−0.065	−0.055	0.075	0.099	−0.001
TP	0.302	0.110	−0.201	0.061	0.648	0.167	−0.116
F	0.115	**0.759**	0.076	−0.063	0.044	0.214	0.200
S	−0.082	−0.673	0.355	−0.245	−0.133	0.154	−0.048
F. Coli	0.045	**−0.721**	−0.077	0.028	0.255	−0.249	0.071
SO_4	0.053	0.197	0.074	0.080	−0.178	0.022	**0.818**
Cl	0.182	−0.435	0.133	−0.133	0.157	0.489	0.315
NO_3-N	0.262	0.268	−0.213	−0.041	−0.291	−0.013	−0.632
TSS	−0.106	0.174	−0.067	−0.033	0.153	**0.706**	0.091
T-Hard	0.148	0.105	−0.659	−0.187	−0.140	0.086	0.031
NO_2-N	0.191	−0.147	0.046	0.013	0.631	−0.043	−0.438
NH_3	0.312	−0.001	0.689	0.404	0.104	0.007	0.135
Eigenvalue	4.513	2.685	2.250	1.661	1.469	1.252	1.098
% Total variance	20.515	12.202	10.227	7.549	6.678	5.693	4.989
Cumulative % variance	20.515	32.717	42.944	50.493	57.171	62.863	67.852

VFs represent the variance factors after varimax raw rotation for principal components; bold values represent medium-high loadings.

We have identified four contamination source types—nutrient, organics, feces and oil. Specifically, nutrient represented point source contamination, such as urban domestic wastewater and industrial wastewater from chemical fertilizer plants, and nonpoint source contamination, such as that related to agricultural activities and aquaculture. Second, organics were mainly derived from oxygen consumption and toxic organic matter from municipal sewage and industrial sewage. Third, feces were mainly derived from animal fecal drainage in the fishery and livestock breeding industries. Finally, oil represented the contamination characters from the petroleum chemical industry and oily sewage discharge from ship operation.

4. Conclusions

In the Nanxi River of the Taihu watershed in China, WQI and multivariate statistical techniques were used to assess the spatiotemporal variations in water quality and to identify contamination source types.

(1) The WQI findings indicated that the water quality of most monitoring stations was classified as "medium-low" and presented a continuous improvement trend. The water quality of S1 and S2 was always above "good", especially the water quality of S2, which was "excellent" in 2015 and 2018.

(2) Cluster analysis divided the 14 monitoring stations into 3 clusters of low contamination, medium contamination and high contamination.

(3) Discriminant analysis used 14 parameters (pH, Petrol, VP, COD, TP, F, S, F. coli, SO_4, Cl, NO_3-N, T-Hard, NO_2-N, and NH_3) for important data reduction and provided an 87% correct allocation in the spatial variation analysis for the 3 clusters.

(4) PCA/FA was used to analyze the data sets of three spatial clusters and obtained six, eight and seven potential factors. The study showed that the sources of water contamination were mainly related to nutrients (livestock and poultry breeding, agricultural activities), salt ions (natural) and toxic organic contamination (urban sewage, industrial wastewater and ship operation) in cluster A; fecal coliform (livestock and poultry breeding), organic contamination (industrial and domestic sewage), temperature (natural), nutrients (point source: industrial wastewater and domestic sewage; nonpoint sources: livestock and poultry breeding, agricultural fertilizer) in cluster B; and fluoride (industrial wastewater), pH and temperature (natural), and petroleum (ship operation and industrial wastewater) in cluster C.

Supplementary Materials: The following supporting information can be downloaded at: https://www.mdpi.com/article/10.3390/w14050778/s1, Supplementary Material.docx (the explanation about the statistical methods) and Supplementary Material.xlsx (the raw water quality data).

Author Contributions: Conceptualization, Z.-M.Z. and F.Z.; methodology, Z.-M.Z.; software, Z.-M.Z.; validation, F.Z.; formal analysis, Z.-M.Z.; investigation, D.-C.C.; resources, D.-C.C.; data curation, D.-C.C.; writing—Original draft preparation, Z.-M.Z.; writing—Review and editing, F.Z.; visualization, Z.-M.Z.; supervision, J.-L.D.; project administration, J.-L.D.; funding acquisition, F.Z. All authors have read and agreed to the published version of the manuscript.

Funding: This research was funded by the National Natural Science Foundation of China, grant number U1603241, U2003205.

Institutional Review Board Statement: Not applicable.

Informed Consent Statement: Not applicable.

Data Availability Statement: The data presented in this study are available in Supplementary Material.xlsx.

Acknowledgments: The authors would like to thank the reviewers for providing helpful suggestions to improve this manuscript.

Conflicts of Interest: The authors declare no conflict of interest.

References

1. Singh, K.; Malik, A.; Sinha, S. Water quality assessment and apportionment of pollution sources of Gomti river (India) using multivariate statistical techniques—A case study. *Anal. Chim. Acta* **2005**, *538*, 355–374. [CrossRef]
2. Carpenter, S.R.; Caraco, N.F.; Correll, D.L.; Howarth, R.W.; Sharpley, A.N.; Smith, V.H. Nonpoint pollution of surface waters with phosphorus and nitrogen. *Ecol. Appl.* **1998**, *8*, 559–568. [CrossRef]
3. Zhai, X.; Xia, J.; Zhang, Y. Water quality variation in the highly disturbed Huai River Basin, China from 1994 to 2005 by multi-statistical analyses. *Sci. Total Environ.* **2014**, *496C*, 594–606. [CrossRef] [PubMed]
4. Sundaray, K.S.; Panda, U.C.; Nayak, B.B.; Bhatta, D. Multivariate statistical techniques for the evaluation of spatial and temporal variations in water quality of the Mahanadi river–estuarine system (India)—A case study. *Environ. Geochem. Health* **2006**, *28*, 317–330. [CrossRef]
5. Xu, Y.; Xie, R.; Wang, Y.; Jian, S. Spatio-temporal variations of water quality in Yuqiao Reservoir Basin, North China. *Front. Env. Sci. Eng.* **2015**, *9*, 649–664. [CrossRef]
6. Singh, K.P.; Malik, A.; Mohan, D.; Sinha, S. Multivariate statistical techniques for the evaluation of spatial and temporal variations in water quality of Gomti River (India)—A case study. *Water Res.* **2004**, *38*, 3980–3992. [CrossRef]
7. Shrestha, S.; Kazama, F. Assessment of surface water quality using multivariate statistical techniques: A case study of the Fuji river basin, Japan. *Environ. Modell. Softw.* **2007**, *22*, 464–475. [CrossRef]
8. Vega, M.; Pardo, R.; Barrado, E.; Debn, L. Assessment of seasonal and polluting effects on the quality of river water by exploratory data analysis. *Water Res.* **1998**, *32*, 3581–3592. [CrossRef]
9. Muangthong, S.; Shrestha, S. Assessment of surface water quality using multivariate statistical techniques: Case study of the Nampong river and Songkhram river, Thailand. *Environ. Monit. Assess.* **2015**, *187*, 548. [CrossRef]
10. Alberto, W.D.; Pilar, D.D.; Valeria, A.M.; Fabiana, P.S.; Cecilia, H.A.; De, B.M. Pattern recognition techniques for the evaluation of spatial and temporal variations in water quality. A case study: Suquía River Basin (Córdoba-Argentina). *Water Res.* **2001**, *35*, 2881–2894. [CrossRef]
11. Dixon, W.; Chiswell, B. Review of aquatic monitoring program design. *Water Res.* **1996**, *30*, 1935–1948. [CrossRef]

12. Zhou, F.; Guo, H.; Liu, Y.; Jiang, Y. Chemometrics data analysis of marine water quality and source identification in Southern Hong Kong. *Mar. Pollut. Bull.* **2007**, *54*, 745–756. [CrossRef] [PubMed]

13. Zhang, Y.; Guo, F.; Meng, W.; Wang, X. Water quality assessment and source identification of Daliao river basin using multivariate statistical methods. *Environ. Monit. Assess.* **2009**, *152*, 105–121. [CrossRef]

14. He, W.; Li, F.; Yu, J.; Chen, M.; Deng, Y.; Li, J.; Tang, X.; Chen, Z.; Yan, Z. Risk assessment and source apportionment of trace elements in multiple compartments in the lower reach of the Jinsha River, China. *Sci. Rep.* **2021**, *11*, 20041. [CrossRef] [PubMed]

15. Pesce, S.F.; Wunderlin, D.A. Use of water quality indices to verify the impact of Córdoba City (Argentina) on Suqua River. *Water Res.* **2000**, *34*, 2915–2926. [CrossRef]

16. Adimalla, N.; Li, P.; Venkatayogi, S. Hydrogeochemical evaluation of groundwater quality for drinking and irrigation purposes and integrated interpretation with water quality index studies. *Environ. Process.* **2018**, *5*, 363–383. [CrossRef]

17. Alver, A. Evaluation of conventional drinking water treatment plant efficiency according to water quality index and health risk assessment. *Environ. Sci. Pollut. R.* **2019**, *26*, 27225–27238. [CrossRef]

18. Hurley, T.; Sadiq, R.; Mazumder, A. Adaptation and evaluation of the Canadian Council of Ministers of the Environment Water Quality Index (CCME WQI) for use as an effective tool to characterize drinking source water quality. *Water Res.* **2012**, *46*, 3544–3552. [CrossRef]

19. Şener, Ş.; Şener, E.; Davraz, A. Evaluation of water quality using water quality index (WQI) method and GIS in Aksu River (SW-Turkey). *Sci. Total Environ.* **2017**, *584–585*, 131–144. [CrossRef]

20. Hoseinzadeh, E.; Khorsandi, H.; Wei, C.; Alipour, M. Evaluation of Aydughmush River water quality using the National Sanitation Foundation Water Quality Index (NSFWQI), River Pollution Index (RPI), and Forestry Water Quality Index (FWQI). *Desalin. Water Treat.* **2015**, *54*, 2994–3002. [CrossRef]

21. Suratman, S.; Sailan, M.I.M.; Hee, Y.Y.; Bedurus, E.A.; Latif, M.T. A preliminary study of water quality index in Terengganu River Basin, Malaysia. *Sains Malays.* **2015**, *44*, 67–73. [CrossRef]

22. Sun, W.; Xia, C.; Xu, M.; Guo, J.; Sun, G. Application of modified water quality indices as indicators to assess the spatial and temporal trends of water quality in the Dongjiang River. *Ecol. Indic.* **2016**, *66*, 306–312. [CrossRef]

23. Wu, Z.; Wang, X.; Chen, Y.; Cai, Y.; Deng, J. Assessing river water quality using water quality index in Lake Taihu Basin, China. *Sci. Total Environ.* **2018**, *612*, 914–922. [CrossRef] [PubMed]

24. Wu, H.; Yang, W.; Yao, R.; Zhao, Y.; Zhao, Y.; Zhang, Y.; Yuan, Q.; Lin, A. Evaluating surface water quality using water quality index in Beiyun River, China. *Environ. Sci. Pollut. R.* **2020**, *27*, 35449–35458. [CrossRef]

25. Helena, B.; Pardo, R.; Vega, M.; Barrado, E.; Fernandez, J.M.; Fernandez, L. Temporal evolution of groundwater composition in an alluvial aquifer (Pisuerga River, Spain) by principal component analysis. *Water Res.* **2000**, *34*, 807–816. [CrossRef]

26. Liu, C.; Lin, K.; Kuo, Y. Application of factor analysis in the assessment of groundwater quality in a blackfoot disease area in Taiwan. *Sci. Total Environ.* **2003**, *313*, 77–89. [CrossRef]

27. Reghunath, R.; Murthy, T.R.S.; Raghavan, B.R. The utility of multivariate statistical techniques in hydrogeochemical studies: An example from Karnataka, India. *Water Res.* **2002**, *36*, 2437–2442. [CrossRef]

28. Simeonov, V.; Stratis, J.A.; Samara, C.; Zachariadis, G.; Voutsa, D.; Anthemidis, A.; Sofoniou, M.; Kouimtzis, T. Assessment of the surface water quality in Northen Greece. *Water Res.* **2003**, *37*, 4119–4124. [CrossRef]

29. Bu, H.; Tan, X.; Li, S.; Zhang, Q. Temporal and spatial variations of water quality in the Jinshui River of the South Qinling Mts., China. *Ecotox. Environ. Safe.* **2010**, *783*, 907–913. [CrossRef]

30. Alexakis, D. Assessment of water quality in the Messolonghi–Etoliko and Neochorio region (West Greece) using hydrochemical and statistical analysis methods. *Environ. Monit. Assess.* **2011**, *182*, 397–413. [CrossRef]

31. Zhang, Z.; Zhang, F.; Du, J.; Chen, D.; Zhang, W. Impacts of land use at multiple buffer scales on seasonal water quality in a reticular river network area. *PLoS ONE* **2021**, *16*, e0244606. [CrossRef] [PubMed]

32. Standardization Administration of China. *Classification and Codes for Chinese Soil (GB/T 17296-2009)*, 3rd ed.; Standards Press of China: Beijing, China, 2009; pp. 2–113.

33. Liyang Statistical Yearbook. Available online: http://www.liyang.gov.cn/class/KPAQQDDJ#ly4 (accessed on 11 February 2022).

34. Technical Specifications Requirements for Monitoring of Surface Water and Waste Water. Available online: https://www.mee.gov.cn/ywgz/fgbz/bz/bzwb/jcffbz/200301/t20030101_66890.htm (accessed on 11 February 2022).

35. Environmental Quality Standards for Surface Water. Available online: https://www.mee.gov.cn/ywgz/fgbz/bz/bzwb/shjbh/shjzlbz/200206/t20020601_66497.shtml (accessed on 11 February 2022).

36. Zhao, Y.; Song, Y.; Cui, J.; Gan, S.; Yang, X.; Wu, R.; Guo, P. Assessment of water quality evolution in the Pearl River estuary (South Guangzhou) from 2008 to 2017. *Water* **2020**, *12*, 59. [CrossRef]

37. Debels, P.; Figueroa, R.; Urrutia, R.; Barra, R.; Niell, X. Evaluation of water quality in the Chillán river (Central Chile) using physicochemical parameters and a modified water quality index. *Environ. Monit. Assess.* **2005**, *110*, 301–322. [CrossRef] [PubMed]

38. Kocer, M.A.T.; Sevgil, H. Parameters selection for water quality index in the assessment of the environmental impacts of land-based trout farms. *Ecol. Indic.* **2014**, *36*, 672–681. [CrossRef]

39. Zhao, P.; Tang, X.; Tang, J.; Wang, C. Assessing water quality of three gorges reservoir, China, over a five-year period from 2006 to 2011. *Water Resour. Manag.* **2013**, *27*, 4545–4558. [CrossRef]

40. Li, B.; Wan, R.; Yang, G.; Wang, S.; Wagner, P. Exploring the spatiotemporal water quality variations and their influencing factors in a large floodplain lake in China. *Ecol. Indic.* **2020**, *115*, 106454. [CrossRef]

41. Zhang, X.; Wang, Q.; Liu, Y.; Wu, J.; Yu, M. Application of multivariate statistical techniques in the assessment of water quality in the Southwest New Territories and Kowloon, Hong Kong. *Environ. Monit. Assess.* **2011**, *173*, 17–27. [CrossRef]
42. Changzhou "263" Special Action Topic. Available online: http://www.changzhou.gov.cn/ns_class/cz263zt (accessed on 12 January 2022).
43. Bouza-Deao, R.; Ternero-Rodríguez, M.; Fernández-Espinosa, A.J. Trend study and assessment of surface water quality in the ebro river (spain). *J. Hydrol.* **2008**, *361*, 227–239. [CrossRef]
44. Ogwueleka, T.C. Use of multivariate statistical techniques for the evaluation of temporal and spatial variations in water quality of the Kaduna River, Nigeria. *Environ. Monit. Assess.* **2015**, *187*, 137. [CrossRef]
45. Varol, M.; Gökot, B.; Bekleyen, A.; Şen, B. Water quality assessment and apportionment of pollution sources of Tigris River (Turkey) using multivariate statistical techniques—A case study. *River Res. Appl.* **2012**, *28*, 1428–1438. [CrossRef]
46. Wang, Y.; Wang, P.; Bai, Y.; Tian, Z.; Li, J.; Shao, X.; Mustavich, L.; Li, B. Assessment of surface water quality via multivariate statistical techniques: A case study of the Songhua River Harbin region, China. *J. Hydro-Environ. Res.* **2013**, *7*, 30–40. [CrossRef]

Article

Long-Term 10 m Resolution Water Dynamics of Qinghai Lake and the Driving Factors

Qianqian Chen [1,2], Wanqing Liu [1,2] and Chang Huang [1,2,3,*]

1   Shaanxi Key Laboratory of Earth Surface System and Environmental Carrying Capacity, Northwest University, Xi'an 710127, China; 202021042@stumail.nwu.edu.cn (Q.C.); liuwqing@nwu.edu.cn (W.L.)

2   College of Urban and Environmental Sciences, Northwest University, Xi'an 710127, China

3   Institute of Earth Surface System and Hazards, Northwest University, Xi'an 710127, China

*   Correspondence: changh@nwu.edu.cn; Tel.: +86-29-8830-8412

Abstract: As the largest inland saltwater lake in China, Qinghai Lake plays an important role in regional sustainable development and ecological environment protection. In this study, we adopted a spatial downscaling model for mapping lake water at 10 m resolution through integrating Sentinel-2 and Landsat data, which was applied to map the water extent of Qinghai Lake from 1991 to 2020. This was further combined with the Hydroweb water level dataset to establish an area-level relationship to acquire the 30-year water level and water volume. Then, the driving factors of its water dynamics were analyzed based on the grey system theory. It was found that the lake area, water level, and water volume decreased from 1991 to 2004, but then showed an increasing trend afterwards. The lake area ranges from 4199.23 to 4494.99 km^2. The water level decreased with a speed of ~0.05 m/a before 2004 and then increased with a speed of 0.22 m/a thereafter. Correspondingly, the water volume declined by 5.29 km^3 in the first 13 years, and rapidly increased by 15.57 km^3 thereafter. The correlation between climatic factors and the water volume of Qinghai Lake is significant. Precipitation has the greatest positive impact on the water volume variation with the relational grade of 0.912, while evaporation has a negative impact.

Keywords: water level; water volume; spatial downscaling; water dynamics; climate change

Citation: Chen, Q.; Liu, W.; Huang, C. Long-Term 10 m Resolution Water Dynamics of Qinghai Lake and the Driving Factors. *Water* **2022**, *14*, 671. https://doi.org/10.3390/w14040671

Academic Editors: Fei Zhang, Ngai Weng Chan, Xinguo LI and Xiaoping Wang

Received: 25 December 2021
Accepted: 18 February 2022
Published: 21 February 2022

1. Introduction

Lakes are an important part of global hydrological and ecological processes [1–3], providing humans with indispensable resources and services, including drinking water supply, agricultural production, transportation, recreation, fishery, etc. [4,5]. Ongoing global warming and climatic change [6] is enhancing the global hydrological cycle and affecting water availability. As a result, efficient management of water resources is needed [7,8]. Warming-induced hydrological cycle intensification and its impacts on local and global ecosystems have brought increasing attention to the links between climatic change/variability, hydrological processes, and water resources across various temporal and spatial scales during the last few decades [9,10]. Therefore, understanding the hydrological changes of lakes and their potential driving factors can provide insights into lake conservation and water resource management [11,12]. As the largest inland saltwater lake in China, Qinghai Lake is located at the northeastern part of the Tibetan Plateau, which is extremely sensitive to climate change and plays a crucial role in maintaining the regional hydrological cycle [13]. Therefore, monitoring the long-term dynamics of Qinghai Lake and analyzing its driving factors are of great significance for local sustainable development and ecological environment protection.

Remote sensing provides an effective way of monitoring surface water, mainly in the forms of microwave remote sensing and optical remote sensing. Microwave remote sensing is powerful due to its less atmospheric effect and all-weather observation [14], while optical remote sensing is widely used because of the data availability and appropriate spatial and temporal resolutions [15]. For example, high temporal resolution multispectral data, including MODIS and AVHRR, have been widely used to detect the seasonal and inter-annual changes of lakes in the Tibetan Plateau [16], bearing in mind that the coarse resolution may cause a lack of water extraction details and low accuracy at a regional scale [17,18], while higher spatial resolution remote sensing data (e.g., Landsat imagery) make it possible to accurately detect and delineate the water body information [19–22]. For example, Cui et al. [23] analyzed the coastline change of Qinghai Lake and its surrounding lakes from 1973 to 2015 by utilizing multitemporal Landsat imagery. Zhang et al. [24] estimated the water balances of the ten largest lakes in China using ICESat and Landsat data between 2003 and 2009. They proved that satellite remote sensing could serve as a fast and effective tool for estimating lake water balance. Although Landsat imagery has higher spatial resolution in comparison with MODIS or AVHRR, the accuracy of water body extraction was still limited by its 30 m resolution. Sentinel-2 satellites are able to obtain multispectral remote sensing data with a higher spatial resolution of up to 10 m, which is assumed to be better for mapping surface water [25]. Existing research, such as Du et al. [26] and Yang et al. [27], has demonstrated that Sentinel-2 data can provide more explicit and accurate surface water information with the advantages of intensively and continuously monitoring the surface of the Earth and higher spatial resolution. However, as this is a recent satellite mission, its data have a relatively short time series, which fails to meet the requirements of long-term analysis of lake water dynamics.

The mixed pixel issue usually hinders the accurate drawing and monitoring of lake water. There are two popular methods to alleviate mixed pixel issues, pixel unmixing and reconstruction, and spatial and temporal fusion [25]. The purpose of pixel unmixing and reconstruction is to achieve higher resolution land cover mapping from coarse-resolution data under the assumption that each mixed pixel can be expressed in the form of certain combinations of a number of pure spectral signatures [25]. Spatial and temporal fusion (spatio-temporal fusion) aims to blend high spatial resolution data with high temporal resolution data to achieve both high spatial and high temporal resolutions [28–31], so that the mixed pixel issue of the coarse spatial resolution data can be alleviated. Wu et al. [32] proposed a downscaling algorithm that established a statistical regression model between MODIS and Landsat data for generating a higher resolution inundation map from MODIS. Through this downscaling process, they managed to generate 30 m water maps from coarse resolution MODIS data while keeping their high temporal resolution. It was proved that the downscaled water maps provide more spatial details and have higher accuracy.

The rapid development of remote sensing technology also brings new ideas for monitoring lake water volume changes. This can be achieved by combing the lake area derived from optical remote sensing and water level estimated by satellite altimetry data. Satellite radar/laser altimeters such as TOPEX/POSEIDON, ENVISAT, JASON-1, and ICESat/GLAS have been successfully applied for monitoring lake level variations [33–36]. For example, Zhang et al. [37] utilized Landsat and ICESat datasets to examine annual changes in lake area, level, and volume of the Tibetan Plateau and explored the reasons for the lake water volume changes from the 1970s to 2015. The Hydroweb, maintained by LEGOS/GOHS in France, provides water level/area information derived from a combination of multiple altimetry satellite observations of more than 150 inland lakes and reservoirs [38], which serves as a useful data source for lake monitoring. For example, Liu et al. [39] combined the Hydroweb and Landsat data recorded from 1975 to 2015 to evaluate water volume variations and the water balance of Taihu Lake.

In this study, we aim to achieve a long-term and high-resolution analysis of the water variation of Qinghai Lake in the past 30 years. To fulfil this objective, we adopt Wu et al.'s [32] downscaling method to generate 10 m resolution water maps from a long-term Landsat image series, with Sentinel-2 data as the auxiliary. To facilitate the computation, we implement this method on Google Earth Engine (GEE) [40], an advanced remote sensing cloud computing platform for large-scale and long-term remote sensing analysis and processing. We also want to combine the long-term water area variation with water level information to estimate the water volume dynamics of Qinghai Lake, and ultimately analyze the driving factors.

2. Study Area and Materials

2.1. Study Area

Qinghai Lake is the largest plateau inland saltwater lake in China, located in the northeast corner of the Tibet Plateau (36°32′–37°15′ N, 99°36′–100°47′ E) (Figure 1) at an altitude of 3196 m. It belongs to the semi-arid climate on a continental plateau, with large evaporation, great temperature difference between day and night, and a short frost-free and long freezing period [41]. The annual precipitation in the lake area is about 357 mm, and the annual average temperature is approximately 1.2 °C [42]. More than 40 rivers (or streams) flow into Qinghai Lake, with the two largest rivers, the Buha River and Shaliu River, accounting for 63% of the total recharge volume [43]. As a closed inland lake, the variations of Qinghai Lake water are closely related to, and highly affected by the climate, while human activities contribute little [41,42,44], probably because it is a salt lake.

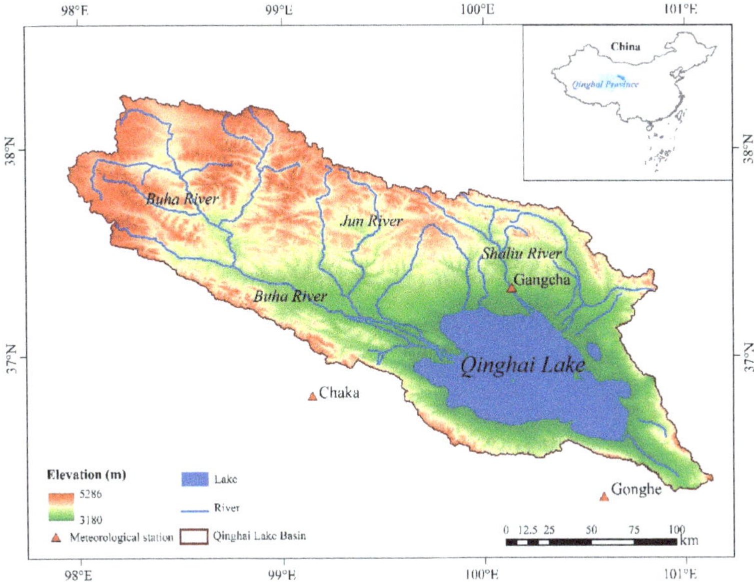

Figure 1. Qinghai Lake Basin.

2.2. Materials

Data used in this study include Landsat imagery, Sentinel-2 imagery, water level data from the Hydroweb, and meteorological data (Table 1). Landsat 5 TM, Landsat 7 ETM+, and Landsat 8 OLI data were employed together to implement a long-term earth observation from 1991–2020. Sentinel-2 MSI imagery, with a spatial resolution up to 10 m, was employed to establish the downscaling model to generate 10 m water extent from Landsat data. Both Landsat and Sentinel-2 data were obtained and pre-processed on GEE. Considering the interference of clouds, Landsat images from May to November were mosaiced to generate a cloud-free image for each year. In order to reduce distortion caused by projection, Sentinel-2 data were reprojected to the same coordinate system as Landsat data (WGS 84/UTM zone 47N). The Hydroweb dataset (http://Hydroweb.theia-land.fr, accessed on 20 October 2020) provides long-term water level, area, and water storage estimations of major lakes globally. Its water level dataset is a fusion of multiple altimetry satellites with different service years, including Topex-Poseidon (1992–2005), Jason-1 (2001–2013), ICESat (2003–2009), Jason-2 (2008-), Jason-3 (2016-), Sentinel-3A (2016-), ICESat-2 (2018-), and so on [35]. Meteorological data were obtained from the China Surface Climate Data Daily Value Dataset (V3.0) published by the China Meteorological Data Service Center (https://data.cma.cn/en/?r=data/index, accessed on 20 October 2020). We acquired temperature, evaporation, and precipitation of Gangcha, Chaka, and Gonghe stations near Qinghai Lake from this dataset, and used them to analyze the driving factors of Qinghai Lake's water dynamics.

Table 1. Materials used in this study.

	Year	Selected Bands	Spatial Resolution (m)	Purpose
Landsat 5 TM	1991–2011	B2, B4	30	Water extraction
Landsat 7 ETM+	2012	B2, B4	30	Water extraction
Landsat 8 OLI	2013–2020	B3, B5	30	Downscaling Model & water extraction
Sentinel-2 MSI	2015–2019	B3, B8	10	Downscaling Model
Hydroweb dataset	1995–2020	-	-	Water volume estimation
Meteorological dataset	1991–2017	-	-	Driving factor analysis

3. Methodology

We utilized Landsat and Sentinel-2 images in the overlapping period (2015–2019) on GEE to establish the statistical regression downscaling model as developed by Wu et al. [32]. This model was then applied to generate long-term (1991–2020) and high-resolution (10 m) water maps from Landsat imagery. Through integrating with the water level from the Hydroweb dataset, the water volume variation in the past 30 years was analyzed based on the area-level relationship. Finally, the meteorological dataset was used to analyze the driving factors of lake volume changes. The flowchart of the methodology of this study is shown in Figure 2.

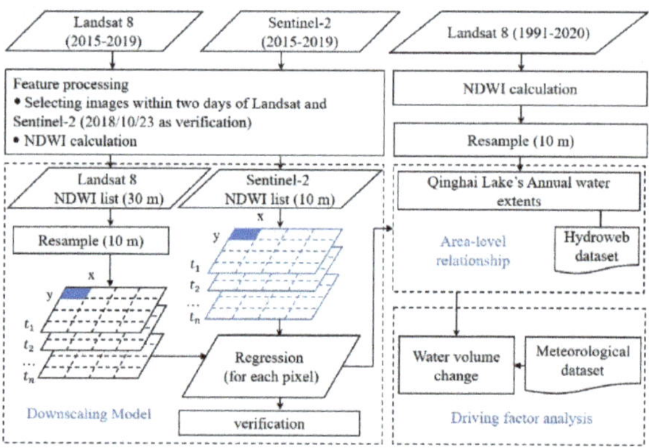

Figure 2. Workflow of this study.

3.1. Downscaled Mapping of Surface Water

We adopted the statistical regression model proposed by Wu et al. [32] to downscale Landsat imagery from 30 m to 10 m resolution, with the assistance of 10 m resolution Sentinel-2 data. This model is based on regressing water index images derived from Landsat and Sentinel-2 (Equation (1)). Specifically, Landsat 8 and Sentinel-2 with close dates (less than 3 days) from 2015 to 2019 were selected to construct the regression model (Table 2). Among the selected 11 pairs of Landsat-8 and Sentinel-2 images, the one on

23 October 2018 was selected to validate the downscaled results only, while the remaining were selected for regression.

$$\text{NDWI}^{\text{Fine}}_{i,j,t} = a_{i,j} \cdot \text{NDWI}^{\text{Coarse}}_{i,j,t} + b_{i,j}, \tag{1}$$

Table 2. Selected Landsat 8 and Sentinel-2 imagery for establishing the regression model. The bolded pair was for validation only.

Sequence	The Date of Landsat 8	The Date of Sentinel-2
1	2016/07/29	2016/07/30
2	2016/10/17	2016/10/18
3	2017/07/16	2017/07/15
4	2017/11/05	2017/11/07
5	2017/12/07	2017/12/07
6	2018/02/09	2018/02/10
7	2018/02/25	2018/02/25
8	2018/03/13	2018/03/12
9	**2018/10/23**	**2018/10/23**
10	2019/01/11	2019/01/11
11	2019/04/17	2019/04/16

In Equation (1), $a_{i,j}$ and $b_{i,j}$ are the fitted regression coefficients, $\text{NDWI}^{\text{Fine}}_{i,j,t}$ and $\text{NDWI}^{\text{Coarse}}_{i,j,t}$ are the normalized difference water index (NDWI) [45] of fine and coarse resolution images at time t and pixel location (x, y), respectively. NDWI was calculated as the normalized difference of GREEN and near-infrared (NIR) bands (Equation (2)).

$$\text{NDWI} = (\text{GREEN} - \text{NIR})/(\text{GREEN} + \text{NIR}), \tag{2}$$

We first resampled the coarse resolution NDWI image to the same resolution as the fine resolution NDWI imagery using the NEAREST interpolation method, i.e., resampled the 30 m Landsat NDWI to 10 m resolution, and then established the regression model based on the resampled Landsat NDWI and Sentinel-2 NDWI on a pixel-by-pixel basis. Using this model, higher resolution (10 m) NDWI images can be generated from any input of Landsat NDWI image. OTSU thresholding [46] was then applied to the resultant NDWI images to extract the surface water extent.

3.2. Water Volume Estimation

To calculate the relative water volume variation, the lake was assumed to be circular with a regular shape. In this study, we adopted the method used in [47] to estimate the lake volume change (ΔV), as shown in Equation (3).

$$\Delta V = \frac{1}{3}(H_1 - H_2) \cdot (A_1 + A_2 + \sqrt{A_1 \cdot A_2}), \tag{3}$$

where H_1 and A_1 represent the corresponding lake water level and area at time 1, and H_2 and A_2 are the water level and area at time 2, respectively.

3.3. Driving Factor Analysis

As the human activities had limited impacts on the water volume variation of Qinghai Lake [48], we assume there is no impact caused by anthropogenic factors and only analyze the climatic driving factors for lake water dynamics. Due to the complexity of climatic change and the diversity of influencing factors of lakes, nonlinear constraints and uncertainties are involved in the consideration of the impact of climate elements on the lake dynamics, which causes extensive greyness [49]. Therefore, the Grey Relation Analysis (GRA) [50] was applied to analyze the response of the water volume to climate factors.

The GRA uses the correlation of two sequences to characterize the degree of association between them, called the relational grade, which is calculated as:

$$R_{ij} = \frac{1}{N} \sum_{t=1}^{N} R_{ij}(t), \tag{4}$$

where R_{ij} represents the relational grade between the sequences of i and j, N is the length of the sequence, and $R_{ij}(t)$ is the correlation coefficient between the sequences of i and j at time t, calculated as Equation (5):

$$R_{ij}(t) = \frac{\Delta_{min} + \rho\Delta_{max}}{\Delta_{ij}(t) + \rho\Delta_{max}}, \tag{5}$$

where Δ_{min} and Δ_{max} denote the minimum and maximum of the absolute difference of two sequences at each time, respectively, $\Delta_{ij}(t)$ represents the absolute error between sequences at time t, and ρ is the resolution coefficient ($\rho \in (0,1)$), usually set to 0.5 [49].

In addition, we adopted three different methods to calculate the correlation coefficient, namely Pearson [51], Spearman [51], and Kendall [51], to compare with the GRA analysis results. The Pearson correlation coefficient was also used to investigate the climate influence on water volume, which was calculated as Equation (6):

$$r = \frac{n\sum_{x_i y_i} - \sum_{x_i} \sum_{y_i}}{\sqrt{n\sum_{x^2} - \left(\sum_{x_i}\right)^2} \sqrt{n\sum_{y^2} - \left(\sum_{y_i}\right)^2}}, \tag{6}$$

where r is the correlation coefficient ranging from -1 to 1, x and y are the values of the two variables, and n is the number of samples. While the absolution value of r is closer to 1, the correlation between variables is stronger.

4. Results

4.1. Validation of Downscaled Water Maps

We utilized a pair of Landsat 8 and Sentinel-2 images on 23 October 2018, which were not employed for establishing the downscaling model but to validate the downscaling method. A 10 m resolution water map was generated from a downscaled NDWI image derived from the Landsat 8 image using OTSU thresholding. Another 10 m resolution water map derived directly from the Sentinel-2 image was employed as the reference to validate the downscaled result. Two maps were generated by overlaying the water map derived from the original Landsat image and the downscaled result with the Sentinel-2 derived referencing water map, respectively (Figure 3). From these maps, it is obvious that the Landsat 8 image can accurately extract the major water body of Qinghai Lake, either with or without the downscaling process. The extraction differences are mainly distributed along the boundary, especially in Haixi Island, the estuaries of the Buha River and Shaliu River, the sandy area of Shadao Lake, and Haiyan Bay. Compared with the referencing Sentinel-2 water map, the water map derived from the original Landsat 8 image has many misclassified pixels, shown as red color for omission errors and green color for commission errors. The water map derived from downscaled Landsat 8 data showed some improvement, with more detailed features and small water bodies successfully being extracted, for example in the sandy area.

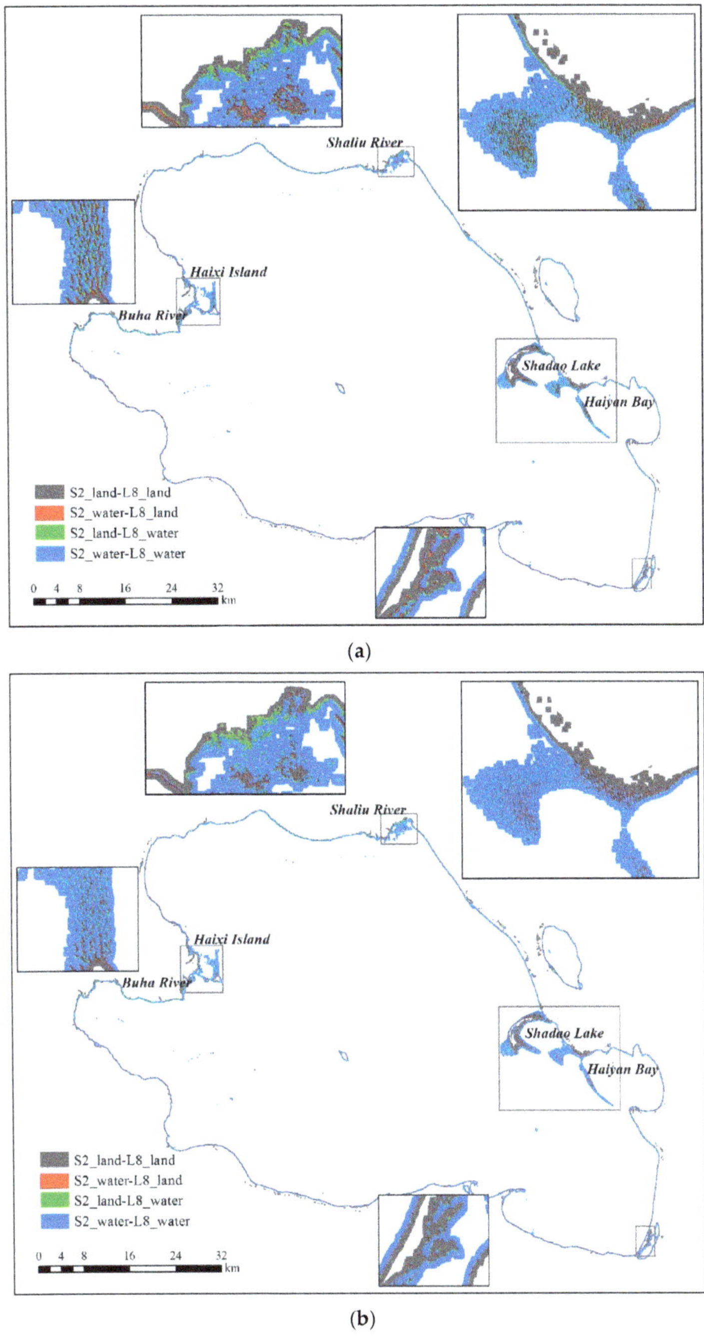

Figure 3. Comparison between Sentinel-2 and Landsat 8 water maps, (**a**) original Landsat 8 image, and (**b**) downscaled Landsat 8 image. Grey color (S2_land-L8_land) stands for pixels that were identified

as Land on Sentinel-2 image and Land on original/downscaled Landsat 8 image. Red color (S2_warer-L8_land) stands for pixels that were identified as Water on Sentinel-2 image and Land on original/downscaled Landsat 8 image (omission error). Green color (S2_land-L8_water) stands for pixels that were identified as Land on Sentinel-2 image and Water on original/downscaled Landsat 8 image (commission error). Blue color (S2_water-L8_water) stands for pixels that were identified as Water on Sentinel-2 image and Water on original/downscaled Landsat 8 image.

Based on these overlaying results (Figure 3), we calculated a confusion matrix by counting the number of four types of overlay map pixels. In this process, as both the reference and verification object are raster data, we took all the pixels as the samples to construct the confusion matrix, based on which accuracy indicators including commission error, omission error, overall accuracy, and Kappa coefficient were calculated (Table 3). It was found that the overall accuracy was clearly improved from 88.35% to 92.10%, and the commission error decreased by 2.46% and omission error by 1.94%. The Kappa coefficient was increased from 0.77 to 0.84. These accuracy indices suggest that the lake water was mapped more accurately by the downscaling method.

Table 3. The accuracy of water maps derived from the original Landsat 8 image and downscaled Landsat 8 image.

Accuracy Indicators	Landsat 8 Image	Downscaled Landsat 8 Image
commission error (%)	6.18	3.72
omission error (%)	5.47	3.53
overall accuracy (%)	88.35	92.10
Kappa coefficient	0.77	0.84

4.2. Lake Area and Shoreline Dynamics

We applied the downscaling model to generate 10 m resolution water maps from selected Landsat images for Qinghai Lake from 1991 to 2020. The lake water area exhibits a two-phase changing pattern as shown in Figure 4a. Taking 2004 as a turning point, the water area showed an overall downward trend at the first stage, dropping from 4316.20 km^2 in 1991 to 4199.23 km^2 in 2004. Since 2004, the water area of Qinghai Lake has been increasing gradually, reaching 4494.99 km^2 in 2020, with an annual growth rate of 18.49 km^2/a.

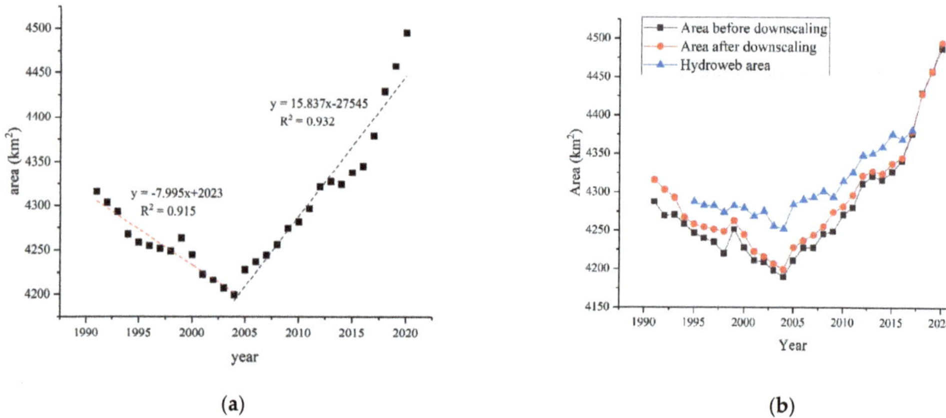

Figure 4. (**a**) Annual water area of Qinghai Lake; (**b**) annual lake area derived from the Landsat image before and after downscaling, in comparison with that extracted from Hydroweb dataset (available for 1995–2017).

We compared the lake area derived from Landsat before and after downscaling with that extracted from the Hydroweb dataset (Figure 4b). Hydroweb only provides area estimation of Qinghai Lake from 1995 to 2017 through a combination of multiple satellite data such as Landsat and CBERS-2 [35]. The annual area was taken from the average value from May to November. It is shown in Figure 4b that the annual lake water areas are consistent among the three data sources. The Hydroweb area is overall slightly higher than the area derived from Landsat images, which may be accounted for by the area integrated by different remote sensing satellites. It is also observed that through the downscaling process, the Qinghai Lake area extracted by the Landsat images is closer to the observations of Hydroweb.

We took 1991, 2004, and 2020 to elaborate the spatial dynamics of Qinghai Lake shoreline (Figure 5). It can be seen clearly that the shoreline at the west, east, and north banks shrank in 2004 in comparison with 1991, particularly in the east bank. In 2004, Shadao Lake was separated from the main body of Qinghai Lake due to water receding. Compared with 2004, the water extent of Qinghai Lake in 2020 was much larger. The Shadao Lake and Haiyan Bay on the east was integrated with the main body of Qinghai Lake. The Tiebuka Bay, Buha River, and Haixi Island also expended significantly, but the Gahai Lake has not changed significantly. In addition, the shoreline on the south bank also had an apparent expansion from 2004 to 2020.

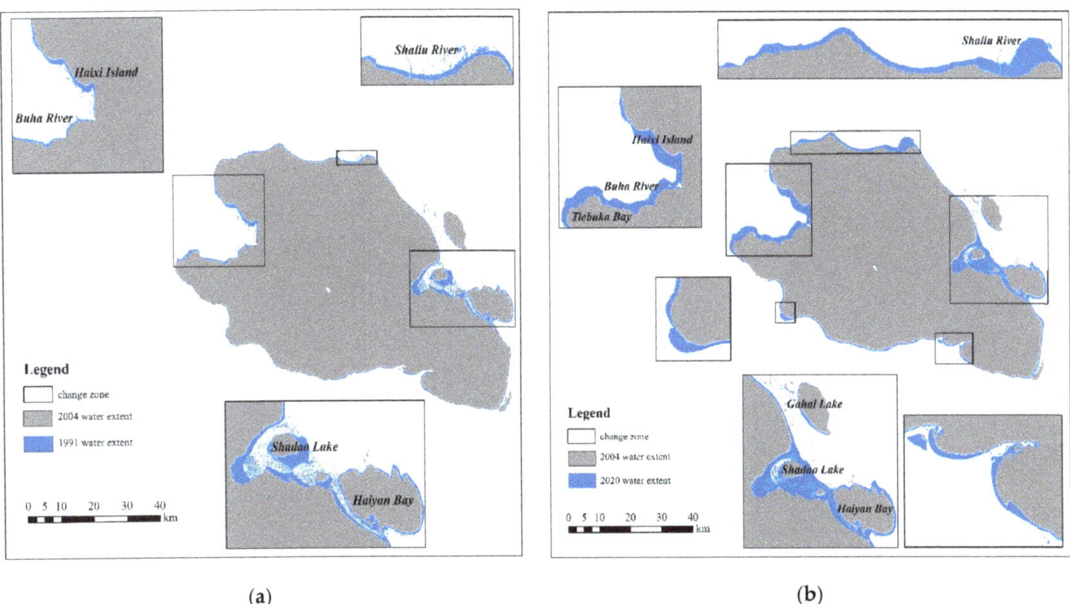

Figure 5. Shoreline change of Qinghai Lake for (**a**) 1991–2004, and (**b**) 2004–2020.

4.3. Lake Water Volume Variation

We extracted the annual average water level of Qinghai Lake from the Hydroweb dataset by taking the average water level from May to November each year. Due to the data availability, we only have the water level record from 1995–2020. The water level dropped from 3194.22 m in 1995 to 3193.62 m in 2004 with an average descending speed of 0.05 m/a, and then raised to 3197.20 m in 2020, with an average rate of 0.22 m/a. A significant correlation between the area and water level of Qinghai Lake was identified (R^2 = 0.976, RMSE = 11.67, Figure 6a). Based on the regression model of water level and area, we estimated the water level of Qinghai Lake from 1991 to 1994 (red dots in Figure 6b) and made a full time series of the water level for 1991–2020 (Figure 6b). Similar to the area variation, the water level variation of Qinghai Lake also exhibits a first-decline-then-increase pattern. We fit a linear regression for the water level of 1991–2004 and 2004–2020, respectively, and found that both periods have significant linear trends, with R^2 both greater than 0.8.

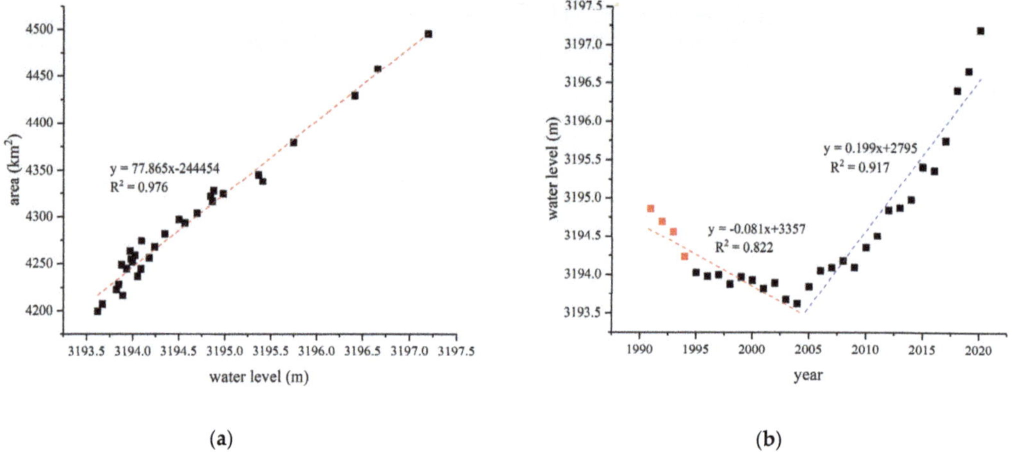

(a)

(b)

Figure 6. (a) The area-level correlation; (b) The water level of the Qinghai Lake in 1991–2020.

Taking the water volume of 1991 as the baseline, the water volume dynamics in the past 30 years were calculated from water area and water level using Equation (3). As shown in Figure 7, it is clear that the water volume also shows a first-decline-then-increase pattern. We also fit a linear regression for the water volume variation in 1991–2004 and 2004–2020, respectively. It was found that both regression models have a high R^2, suggesting significant linear trends. From these models, it is obvious that the water volume decreased from 1991 to 2004, with a fitted rate near to 0.38 km^3/a, while it increased from 2004 to 2020, with a fitted rate of 0.89 km^3/a.

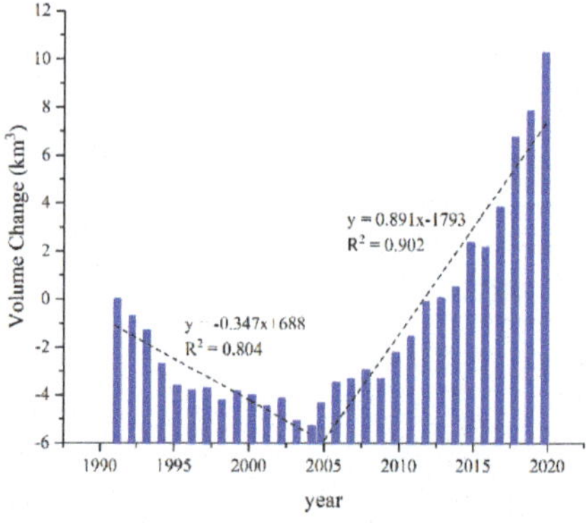

Figure 7. 30-year water volume dynamics of Qinghai Lake based on water volume of 1991.

4.4. Driving Factors of Qinghai Lake Water Variation

In this paper, we calculated the annual accumulated temperature by selecting the daily temperature greater than 10 °C, which is proven to be increasingly important for assessing the impact of climate change [52].We adopted the Mann–Kendall (M–K) [53] trend analysis to identify the tipping point and trend of accumulated temperature, precipitation, and evaporation from 1991 to 2017 in Qinghai Lake Basin (Figure 8). The M–K method is a nonparametric analysis method that has been extensively used for time-series hydrological analysis [54]. The results show that the annual accumulated temperature, precipitation, and evaporation in the Qinghai Lake Basin was overall increasing gradually. The tipping points of accumulated temperature and precipitation are 2005 and 2003, respectively, which is close to the turning point of the lake water volume. The average accumulated temperature in 1991–2005 is 1374.47 °C, which jumps to 1520.15 °C in 2005–2017. The average precipitation changes from 285.31 mm in 1991–2003 to 336.67 mm in 2003–2017. However, the change point of evaporation occurs in 1995. The average evaporation of 1991–1995 and 1995–2017 are 1669.86 mm and 1736.36 mm, respectively.

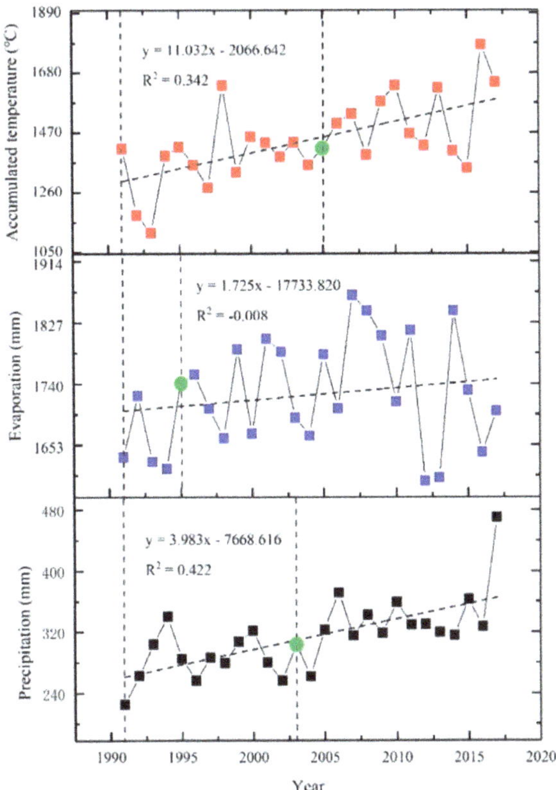

Figure 8. The change point and trend of accumulated temperature, precipitation, and evaporation from 1991 to 2017 in Qinghai Lake Basin.

We employed the GRA to investigate how the climatic elements have affected the relative water volume of Qinghai Lake in the past 30 years. The relational grade (Table 4) between the annual accumulated temperature, precipitation, and evaporation, and the water volume of Qinghai Lake was obtained through Equations (4) and (5). In addition, three different correlation analysis methods (i.e., Pearson, Kendall, and Spearman) were adopted for cross comparison.

Table 4. The correlation of annual mean values of accumulated temperature, precipitation, and evaporation with the annual water volume.

	Accumulated Temperature	Precipitation	Evaporation
Pearson	0.25	0.46 *	−0.26
Kendall	0.12	0.28	−0.14
Spearman	0.14	0.40 *	−0.23
Relational grade	0.56	0.95	0.51

Note: * $p < 0.05$.

According to the results of three different correlation analyses, the correlation of precipitation is the highest no matter which method is applied. The correlation between water volume and accumulated temperature is relatively low, while that with evaporation is negative. The relational grade of GRA also suggests that precipitation has the greatest impact on the water volume, with a relational grade of 0.95. The accumulated temperature has a value of 0.56, and evaporation exerts the weakest effect on water volume dynamics, with a relational grade of 0.51.

To further explore the relationship between climate factors and Qinghai Lake water volume, we performed the Pearson analysis in 1991–2004 and 2004–2017 separately (Table 5). During the period of 1991–2004, it seems that the accumulated temperature is the major factor affecting the decline of Qinghai Lake water volume. For the period of 2004–2017, the increase of water volume seems to be mainly positively affected by the precipitation, with the correlation coefficient close to 0.6 and $p < 0.05$.

Table 5. Pearson's r between climate factors and water volume for period 1991–2004 and 2004–2017.

Period	Accumulated Temperature	Precipitation	Evaporation
1991–2004	−0.70 **	0.12	−0.24
2004–2017	0.36	0.60 *	−0.32

Note: * $p < 0.05$; ** $p < 0.01$.

5. Discussion

As the largest inland saline lake on the plateau in China, Qinghai Lake not only regulates the local climate through the "lake effect", but also directly affects the wetlands and sandy land around the lake. This study made full use of the continuity of the medium-to high-resolution Landsat imagery and combined them with higher-resolution Sentinel-2 imagery for more accurate and long-term monitoring of Qinghai Lake water area dynamics. Meanwhile, the water level data acquired by satellite altimetry were employed to transform the Landsat-based water area dynamics to water volume dynamics. The results show that the water area, water level, and water volume of Qinghai Lake from 1991 to 2020 all exhibit a first-decline-then-increase pattern. The turning point occurred in 2004, when the water level and area reached the minimum. Since then, Qinghai Lake has entered into a period of stable expansion. Overall, our findings were found to be consistent with previous studies [23,24,48,55,56]. However, compared with the annual average water level obtained from gauge stations of Qinghai Lake by Li et al. [42], the water level of the Hydroweb dataset is relatively higher. Due to the lack of lake bathymetry dataset, the water volume estimated in this study only represents the water volume change relative to 1991, instead of the real water volume change. Moreover, different altimetry data have different uncertainties due to their different data quality. In the future, we will consider combining the lake bathymetry and fusing different altimetry satellite data to deepen the research on water level and water volume.

Existing studies have proven that local climate change in the Qinghai Lake Basin in recent years leads to gradual increases in temperature and precipitation and decreases in evapotranspiration [48,57]. Zhang et al. [37] found that increased net precipitation contributes the majority of the water supply (74%) for the lake volume increase, followed by glacier mass loss (13%) and ground ice melt due to permafrost degradation (12%) on the Tibetan Plateau from 2003–2009. Song et al. [58] also pointed out that the meltwater from mountainous glaciers and snow cover have become important water sources for Qinghai Lake, supported by the work of Zhang et al. [33]. Considering the increasing contribution of glaciers and precipitation to the water balance, it is anticipated that the water volume of the plateau inland lakes will continue to increase in the next few decades. We also found that the increasing precipitation had a major contribution to the increase of Qinghai Lake's water volume, indicating possible continuous water increasing in the near future [59]. Continuous rising of water level and expansion of water area may breed a better ecological environment and richer biodiversity, which would be beneficial for local ecological protection and desertification prevention [48].

6. Conclusions

We integrated Landsat and Sentinel-2 remote sensing imagery to construct a long-term 10 m resolution lake water area variation series, which was further associated with the Hydroweb water level dataset to estimate the water volume change. Through this process, we were able to provide the highest resolution long-term Qinghai Lake water monitoring results to date. The driving factors of lake water variation were further analyzed through the grey theory. Based on the results, we draw the following conclusions.

(1) The spatial downscaling method that was incorporated with the Sentinel-2 and Landsat imagery can effectively take advantage of Landsat's long time series and Sentinel-2's high spatial resolution and thus achieve long-term and high-resolution lake monitoring. The resultant water extent was proven to have an improved overall accuracy of 92.10% and Kappa coefficient of 0.84.

(2) The area, water level, and water volume of the Qinghai Lake exhibit the same first-decline-then-increase pattern, with 2004 as the turning point. The minimum lake area that occurred in 2004 is 4199.23 km^2, and the maximum is 4494.99 km^2 in 2020. The water level dropped from 3194.22 m in 1995 to 3193.62 m in 2004 with an average descending speed of 0.05 m/a, and then raised to 3197.20 m in 2020, with an average rate of 0.22 m/a. The water volume decreased between 1991 and 2004, with a fitted rate of 0.34 km^3/a, while it increased between 2004 and 2020, with a fitted rate of 0.89 km^3/a.

(3) The results of the GRA and three correlation analyses all indicate that precipitation has the greatest impact on the water volume variation of Qinghai Lake, followed by accumulated temperature and evaporation. From 1991–2004, the Pearson correlation analysis indicates that accumulated temperature is the primary factor affecting the decline of Qinghai Lake water volume, while the increase of water volume from 2004–2017 seems to be mainly positively affected by precipitation.

Author Contributions: Conceptualization, W.L. and C.H.; Data curation, Q.C.; Formal analysis, Q.C.; Funding acquisition, C.H.; Methodology, Q.C. and C.H.; Resources, C.H.; Software, Q.C.; Supervision, C.H.; Writing—original draft, Q.C.; Writing—review and editing, W.L. and C.H. All authors have read and agreed to the published version of the manuscript.

Funding: This work was funded by the Shaanxi Natural Science Foundation (2021JM-314), and the National Key R&D Program of China (2017YFC0404302).

Institutional Review Board Statement: Not applicable.

Informed Consent Statement: Not applicable.

Data Availability Statement: Not applicable.

Conflicts of Interest: The authors declare no conflict of interest.

References

1. Sobek, S.; Algesten, G.; BergstrÖM, A.-K.; Jansson, M.; Tranvik, L.J. The catchment and climate regulation of pCO2 in boreal lakes. *Glob. Change Biol.* **2003**, *9*, 630–641. [CrossRef]
2. Verpoorter, C.; Kutser, T.; Seekell, D.A.; Tranvik, L.J. A global inventory of lakes based on high-resolution satellite imagery. *Geophys. Res. Lett.* **2014**, *41*, 6396–6402. [CrossRef]
3. Wang, J.; Song, C.; Reager, J.T.; Yao, F.; Famiglietti, J.S.; Sheng, Y.; MacDonald, G.M.; Brun, F.; Schmied, H.M.; Marston, R.A.; et al. Recent global decline in endorheic basin water storages. *Nat. Geosci.* **2018**, *11*, 926–932. [CrossRef]
4. Alsdorf, D.E.; Rodríguez, E.; Lettenmaier, D.P. Measuring surface water from space. *Rev. Geophys.* **2007**, *45*. [CrossRef]
5. Lehner, B.; Döll, P. Development and validation of a global database of lakes, reservoirs and wetlands. *J. Hydrol.* **2004**, *296*, 1–22. [CrossRef]
6. Hansen, J.; Ruedy, R.; Sato, M.; Lo, K. Global surface temperature change. *Rev. Geophys.* **2010**, *48*, RG4004. [CrossRef]
7. Jung, M.; Kim, H.; Mallari, K.J.B.; Pak, G.; Yoon, J. Analysis of effects of climate change on runoff in an urban drainage system: A case study from Seoul, Korea. *Water Sci. Technol.* **2014**, *71*, 653–660. [CrossRef]
8. Li, B.; Yu, Z.; Liang, Z.; Song, K.; Li, H.; Wang, Y.; Zhang, W.; Acharya, K. Effects of Climate Variations and Human Activities on Runoff in the Zoige Alpine Wetland in the Eastern Edge of the Tibetan Plateau. *J. Hydrol. Eng.* **2014**, *19*, 1026–1035. [CrossRef]
9. Qin, B.; Huang, Q. Evaluation of the Climatic Change Impacts on the Inland Lake—A Case Study of Lake Qinghai, China. *Clim. Change* **1998**, *39*, 695–714. [CrossRef]
10. Rouhani, H.; Jafarzadeh, M.S. Assessing the climate change impact on hydrological response in the Gorganrood River Basin, Iran. *J. Water Clim. Change* **2017**, *9*, 421–433. [CrossRef]
11. Zhu, J.; Song, C.; Wang, J.; Ke, L. China's inland water dynamics: The significance of water body types. *Proc. Natl. Acad. Sci. USA* **2020**, *117*, 13876–13878. [CrossRef] [PubMed]
12. Luo, S.; Song, C.; Zhan, P.; Liu, K.; Chen, T.; Li, W.; Ke, L. Refined estimation of lake water level and storage changes on the Tibetan Plateau from ICESat/ICESat-2. *Catena* **2021**, *200*, 105177. [CrossRef]
13. Wang, H.; Long, H.; Li, X.; Yu, F. Evaluation of changes in ecological security in China's Qinghai Lake Basin from 2000 to 2013 and the relationship to land use and climate change. *Environ. Earth Sci.* **2014**, *72*, 341–354. [CrossRef]
14. Schumann, G.J.P.; Moller, D.K. Microwave remote sensing of flood inundation. *Phys. Chem. Earth Parts A/B/C* **2015**, *83–84*, 84–95. [CrossRef]
15. Huang, C.; Chen, Y.; Wu, J.; Li, L.; Liu, R. An evaluation of Suomi NPP-VIIRS data for surface water detection. *Remote. Sens. Lett.* **2015**, *6*, 155–164. [CrossRef]
16. Song, C.; Huang, B.; Ke, L. Inter-annual changes of alpine inland lake water storage on the Tibetan Plateau: Detection and analysis by integrating satellite altimetry and optical imagery. *Hydrol. Process.* **2014**, *28*, 2411–2418. [CrossRef]
17. Lyons, E.A.; Sheng, Y.; Smith, L.C.; Li, J.; Hinkel, K.M.; Lenters, J.D.; Wang, J. Quantifying sources of error in multitemporal multisensor lake mapping. *Int. J. Remote. Sens.* **2013**, *34*, 7887–7905. [CrossRef]
18. Li, S.; Sun, D.; Goldberg, M.; Stefanidis, A. Derivation of 30-m-resolution water maps from TERRA/MODIS and SRTM. *Remote. Sens. Environ.* **2013**, *134*, 417–430. [CrossRef]
19. Smith, L.C. Satellite remote sensing of river inundation area, stage, and discharge: A review. *Hydrol. Process.* **1997**, *11*, 1427–1439. [CrossRef]
20. Jain, S.K.; Singh, R.D.; Jain, M.K.; Lohani, A.K. Delineation of Flood-Prone Areas Using Remote Sensing Techniques. *Water Resour. Manag.* **2005**, *19*, 333–347. [CrossRef]
21. Frazier, P.S.; Page, K.J. Water body detection and delineation with Landsat TM data. *Photogramm. Eng. Remote. Sens.* **2000**, *66*, 1461–1468.
22. Du, Z.; Li, W.; Zhou, D.; Tian, L.; Ling, F.; Wang, H.; Gui, Y.; Sun, B. Analysis of Landsat-8 OLI imagery for land surface water mapping. *Remote. Sens. Lett.* **2014**, *5*, 672–681. [CrossRef]
23. Cui, B.-L.; Xiao, B.; Li, X.-Y.; Wang, Q.; Zhang, Z.-H.; Zhan, C.; Li, X.-D. Exploring the geomorphological processes of Qinghai Lake and surrounding lakes in the northeastern Tibetan Plateau, using Multitemporal Landsat Imagery (1973–2015). *Glob. Planet. Change* **2017**, *152*, 167–175. [CrossRef]
24. Zhang, G.; Xie, H.; Yao, T.; Kang, S. Water balance estimates of ten greatest lakes in China using ICESat and Landsat data. *Chin. Sci. Bull.* **2013**, *58*, 3815–3829. [CrossRef]
25. Huang, C.; Chen, Y.; Zhang, S.; Wu, J. Detecting, Extracting, and Monitoring Surface Water from Space Using Optical Sensors: A Review. *Rev. Geophys.* **2018**, *56*, 333–360. [CrossRef]
26. Du, Y.; Zhang, Y.; Ling, F.; Wang, Q.; Li, W.; Li, X. Water Bodies' Mapping from Sentinel-2 Imagery with Modified Normalized Difference Water Index at 10-m Spatial Resolution Produced by Sharpening the SWIR Band. *Remote Sens.* **2016**, *8*, 354. [CrossRef]
27. Yang, X.; Zhao, S.; Qin, X.; Zhao, N.; Liang, L. Mapping of Urban Surface Water Bodies from Sentinel-2 MSI Imagery at 10 m Resolution via NDWI-Based Image Sharpening. *Remote Sens.* **2017**, *9*, 596. [CrossRef]
28. Emelyanova, I.V.; McVicar, T.R.; Van Niel, T.G.; Li, L.T.; van Dijk, A.I.J.M. Assessing the accuracy of blending Landsat–MODIS surface reflectances in two landscapes with contrasting spatial and temporal dynamics: A framework for algorithm selection. *Remote. Sens. Environ.* **2013**, *133*, 193–209. [CrossRef]
29. Feng, G.; Masek, J.; Schwaller, M.; Hall, F. On the blending of the Landsat and MODIS surface reflectance: Predicting daily Landsat surface reflectance. *IEEE Trans. Geosci. Remote. Sens.* **2006**, *44*, 2207–2218. [CrossRef]

30. Zhu, X.; Chen, J.; Gao, F.; Chen, X.; Masek, J.G. An enhanced spatial and temporal adaptive reflectance fusion model for complex heterogeneous regions. *Remote. Sens. Environ.* **2010**, *114*, 2610–2623. [CrossRef]
31. Wu, P.; Shen, H.; Zhang, L.; Göttsche, F.-M. Integrated fusion of multi-scale polar-orbiting and geostationary satellite observations for the mapping of high spatial and temporal resolution land surface temperature. *Remote. Sens. Environ.* **2015**, *156*, 169–181. [CrossRef]
32. Wu, G.; Liu, Y. Downscaling Surface Water Inundation from Coarse Data to Fine-Scale Resolution: Methodology and Accuracy Assessment. *Remote Sens.* **2015**, *7*, 15989–16003. [CrossRef]
33. Zhang, G.; Xie, H.; Kang, S.; Yi, D.; Ackley, S.F. Monitoring lake level changes on the Tibetan Plateau using ICESat altimetry data (2003–2009). *Remote. Sens. Environ.* **2011**, *115*, 1733–1742. [CrossRef]
34. Medina, C.E.; Gomez-Enri, J.; Alonso, J.J.; Villares, P. Water level fluctuations derived from ENVISAT Radar Altimeter (RA-2) and in-situ measurements in a subtropical waterbody: Lake Izabal (Guatemala). *Remote. Sens. Environ.* **2008**, *112*, 3604–3617. [CrossRef]
35. Crétaux, J.F.; Arsen, A.; Calmant, S.; Kouraev, A.; Vuglinski, V.; Bergé-Nguyen, M.; Gennero, M.C.; Nino, F.; Abarca Del Rio, R.; Cazenave, A.; et al. SOLS: A lake database to monitor in the Near Real Time water level and storage variations from remote sensing data. *Adv. Space Res.* **2011**, *47*, 1497–1507. [CrossRef]
36. Birkett, C.M. The contribution of TOPEX/POSEIDON to the global monitoring of climatically sensitive lakes. *J. Geophys. Res. Ocean.* **1995**, *100*, 25179–25204. [CrossRef]
37. Zhang, G.; Yao, T.; Shum, C.K.; Yi, S.; Yang, K.; Xie, H.; Feng, W.; Bolch, T.; Wang, L.; Behrangi, A.; et al. Lake volume and groundwater storage variations in Tibetan Plateau's endorheic basin. *Geophys. Res. Lett.* **2017**, *44*, 5550–5560. [CrossRef]
38. Yue, H.; Liu, Y.; Wei, J. Dynamic change and spatial analysis of Great Lakes in China based on Hydroweb and Landsat data. *Arab. J. Geosci.* **2021**, *14*, 149. [CrossRef]
39. Liu, Y.; Li, Y.; Lu, Y.; Yue, H. Remote Sensing Analysis of Volume in Taihu Lake: Application for Icesat/Hydroweb and Landsat Data. *Int. Arch. Photogramm. Remote. Sens. Spat. Inf. Sci.* **2018**, *XLII-3*, 1161–1163. [CrossRef]
40. Gorelick, N.; Hancher, M.; Dixon, M.; Ilyushchenko, S.; Thau, D.; Moore, R. Google Earth Engine: Planetary-scale geospatial analysis for everyone. *Remote. Sens. Environ.* **2017**, *202*, 18–27. [CrossRef]
41. Chang, B.; He, K.-N.; Li, R.-J.; Sheng, Z.-P.; Wang, H. Linkage of Climatic Factors and Human Activities with Water Level Fluctuations in Qinghai Lake in the Northeastern Tibetan Plateau, China. *Water* **2017**, *9*, 552. [CrossRef]
42. Li, X.-Y.; Cui, B.-L. The impact of climate changes on water level of Qinghai Lake in China over the past 50 years. *Hydrol. Res.* **2016**, *47*, 532–542. [CrossRef]
43. Cui, B.-L.; Li, X.-Y. Stable isotopes reveal sources of precipitation in the Qinghai Lake Basin of the northeastern Tibetan Plateau. *Sci. Total. Environ.* **2015**, *527–528*, 26–37. [CrossRef] [PubMed]
44. Li, X.-Y.; Xu, H.-Y.; Sun, Y.-L.; Zhang, D.-S.; Yang, Z.-P. Lake-Level Change and Water Balance Analysis at Lake Qinghai, West China during Recent Decades. *Water Resour. Manag.* **2006**, *21*, 1505–1516. [CrossRef]
45. McFeeters, S.K. The use of the Normalized Difference Water Index (NDWI) in the delineation of open water features. *Int. J. Remote. Sens.* **1996**, *17*, 1425–1432. [CrossRef]
46. Guo, Q.; Pu, R.; Li, J.; Cheng, J. A weighted normalized difference water index for water extraction using Landsat imagery. *Int. J. Remote. Sens.* **2017**, *38*, 5430–5445. [CrossRef]
47. Zhang, G.; Chen, W.; Xie, H. Tibetan Plateau's Lake Level and Volume Changes from NASA's ICESat/ICESat-2 and Landsat Missions. *Geophys. Res. Lett.* **2019**, *46*, 13107–13118. [CrossRef]
48. Zhang, M.; Song, Y.; Dong, H. Hydrological trend of Qinghai Lake over the last 60 years: Driven by climate variations or human activities? *J. Water Clim. Change* **2019**, *10*, 524–534. [CrossRef]
49. Allen, R.G.P.; Luis, S.; Raes, D.; Smith, M. Crop evapotranspiration-Guidelines for computing crop water requirements-FAO Irrigation and drainage paper 56. *Food Agric. Organ. United* **1998**, *300*, D05109.
50. Chan, J.W.K.; Tong, T.K.L. Multi-criteria material selections and end-of-life product strategy: Grey relational analysis approach. *Mater. Des.* **2007**, *28*, 1539–1546. [CrossRef]
51. Bolboaca, S.-D.; Jäntschi, L. Pearson versus Spearman, Kendall's tau correlation analysis on structure-activity relationships of biologic active compounds. *Leonardo J. Sci.* **2006**, *5*, 179–200.
52. Hallett, S.H.; Jones, R.J.A. Compilation of an accumulated temperature database for use in an environmental information system. *Agric. For. Meteorol.* **1993**, *63*, 21–34. [CrossRef]
53. Wang, M.; Du, L.; Ke, Y.; Huang, M.; Zhang, J.; Zhao, Y.; Li, X.; Gong, H. Impact of Climate Variabilities and Human Activities on Surface Water Extents in Reservoirs of Yongding River Basin, China, from 1985 to 2016 Based on Landsat Observations and Time Series Analysis. *Remote Sens.* **2019**, *11*, 560. [CrossRef]
54. Choi, W.; Nauth, K.; Choi, J.; Becker, S. Urbanization and Rainfall–Runoff Relationships in the Milwaukee River Basin. *Prof. Geogr.* **2016**, *68*, 14–25. [CrossRef]
55. Fu, C.; Wu, H.; Zhu, Z.; Song, C.; Xue, B.; Wu, H.; Ji, Z.; Dong, L. Exploring the potential factors on the striking water level variation of the two largest semi-arid-region lakes in northeastern Asia. *Catena* **2021**, *198*, 105037. [CrossRef]
56. Wang, L.; Chen, C.; Thomas, M.; Kaban, M.K.; Güntner, A.; Du, J. Increased water storage of Lake Qinghai during 2004–2012 from GRACE data, hydrological models, radar altimetry and in situ measurements. *Geophys. J. Int.* **2018**, *212*, 679–693. [CrossRef]

57. Fang, J.; Li, G.; Rubinato, M.; Ma, G.; Zhou, J.; Jia, G.; Yu, X.; Wang, H. Analysis of Long-Term Water Level Variations in Qinghai Lake in China. *Water* **2019**, *11*, 2136. [CrossRef]
58. Song, C.; Huang, B.; Ke, L. Modeling and analysis of lake water storage changes on the Tibetan Plateau using multi-mission satellite data. *Remote. Sens. Environ.* **2013**, *135*, 25–35. [CrossRef]
59. Fan, C.; Song, C.; Li, W.; Liu, K.; Cheng, J.; Fu, C.; Chen, T.; Ke, L.; Wang, J. What drives the rapid water-level recovery of the largest lake (Qinghai Lake) of China over the past half century? *J. Hydrol.* **2021**, *593*, 125921. [CrossRef]

 *water*

Article

A Novel Early Warning System (EWS) for Water Quality, Integrating a High-Frequency Monitoring Database with Efficient Data Quality Control Technology at a Large and Deep Lake (Lake Qiandao), China

Liancong Luo [1,*], Jia Lan [2], Yucheng Wang [2], Huiyun Li [3,*], Zhixu Wu [2], Chrisopher McBridge [4], Hong Zhou [5], Fenglong Liu [6], Rufeng Zhang [1], Falu Gong [1], Jialong Li [1], Lan Chen [1] and Guizhu Wu [1]

[1] Institute for Ecological Research and Pollution Control of Plateau Lakes, School of Ecology and Environmental Science, Yunnan University, Kunming 650504, China; hangrufeng1997@126.com (R.Z.); gongfl2022@163.com (F.G.); jlli@mail.yun.edu.cn (J.L.); 15111718997@163.com (L.C.); wuguizhu0223@163.com (G.W.)

[2] Chunan Branch, Hangzhou Municipal Ecology and Environment Bureau, Chunan 311700, China; lanika0916@163.com (J.L.); iceman0011@163.com (Y.W.); caepb@126.com (Z.W.)

[3] Nanjing Institute of Geography and Limnology, Chinese Academy of Sciences, Nanjing 210008, China

[4] Environmental Research Institute, The University of Waikato, Hamilton 3240, New Zealand; christopher.mcbride@waikato.ac.nz

[5] Waikato Institute of Technology, Hamilton 3204, New Zealand; hong.zhou@wintec.ac.nz

[6] Financial Department, Hunan University of Humanities, Science and Technology, Loudi 417000, China; 13907384452@139.com

[*] Correspondence: billluo@ynu.edu.cn (L.L.); hyli@niglas.ac.cn (H.L.); Tel.: +86-(0)-181-1290-7530 (L.L.); +86-(0)-159-0515-7570 (H.L.)

Citation: Luo, L.; Lan, J.; Wang, Y.; Li, H.; Wu, Z.; McBridge, C.; Zhou, H.; Liu, F.; Zhang, R.; Gong, F.; et al. A Novel Early Warning System (EWS) for Water Quality, Integrating a High-Frequency Monitoring Database with Efficient Data Quality Control Technology at a Large and Deep Lake (Lake Qiandao), China. *Water* **2022**, *14*, 602. https://doi.org/10.3390/w14040602

Academic Editors: Fei Zhang, Ngai Weng Chan, Xinguo Li and Xiaoping Wang

Received: 31 December 2021
Accepted: 13 February 2022
Published: 16 February 2022

Publisher's Note: MDPI stays neutral with regard to jurisdictional claims in published maps and institutional affiliations.

Abstract: To assess water quality (WQ) online for assuring the safety of drinking water, a novel early warning system integrating a high-frequency monitoring system (HFMS) and data quality control (QC) was developed at Lake Qiandao. The HFMS was designed for monitoring water quality, nutrient inputs by main tributaries, water currents and meteorology at different sites at Lake Qiandao. The EWS focused on data availability, a QC method, a statistical analysis method and data applications instead of technological aspects for sondes, wireless data transfer and interface software development. QC was implemented before use to delete the abnormal values of outliers, to detect change points, to analyse the change trend, to interpolate discrete missing measurements, and find continuous missing or wrong observations caused by technical problems with the sonde. For demonstrating advantages and data availability, surface and profiling measurements at two sites were plotted. The plots show obvious seasonal and diel variations, demonstrating the success of integration of the system with advanced automated technology and good QC. This successfully developed system is now not only giving early warning signals, but also providing critical WQ information for the security of drinking water diverted to Hangzhou city through a tunnel of 110 km length. The automatic monitoring data with QC is also being used to produce initial conditions for WQ prediction based on a three dimensional hydrodynamic-ecosystem model.

Keywords: early warning system; high-frequency monitoring; data quality control; water quality; Lake Qiandao

1. Introduction

Lake eutrophication is a long-term global problem caused by excess nutrient inputs [1], and exacerbated by long water residence times that delay WQ responses to management actions. It is a common problem in the lakes located at the middle and lower catchment of the Yangtze River even in the "good WQ" lakes classified by the Ministry of Ecology and Environment of the People's Republic of China (e.g., Lake Qiandao). Impairments associated with eutrophication include poor water clarity, harmful algal blooms (HABs) [2–4], and the loss of biodiversity, which affect drinking water supplies and the recreational use of lakes. Some impairments are highly dynamic (e.g., HABs, loss of dissolved oxygen, etc.), which has resulted in the rapid proliferation of EWS for monitoring key variables that can cause rapid changes in the water quality of coastal water [5] and freshwater systems [6–10].

An EWS is an integrated system consisting of in situ autonomous sensors for continuous rapid monitoring. The measured data are analysed and interpreted for the purpose of forecasting changes in water quality by the system. It provides a fast and accurate way to distinguish abnormal/abrupt variations in WQ due to biochemical and physical interactions over short time scales. EWS requires the fast detection of abnormalities in WQ parameters, which calls for high-frequency real-time monitoring technologies, wireless communication and appropriate data storage and analysis. A new generation of online monitoring tools based on sensor sonde technology and satellite-based remote sensoring (RS) has emerged in recent years [11–16]. However, the effective implementation of these tools has not been fully realised due to their limitations relating to meeting practical utility needs, high costs, unsatisfied reliability, hardware maintenance demands, and cumbersome data management and analysis approaches, with respect to practical operations.

Conventional sample collection and laboratory-based methods are too slow to achieve operational response and temporal–spatial continuity. There is a clear and increasing need to rapidly detect WQ parameters to ensure an appropriate and timely response to instances of accidental or deliberate contamination [13]. For the past two decades, Wireless Sensor Networks (WSNs) technology has been applied increasingly to environmental monitoring for providing high-frequency scientific data. These high-tech smart devices have offered a vital approach to environmental monitoring and have monitored some lesser-studied fundamental processes, due to their inaccessibility [12]. Generally, the sensor nodes acquire data autonomously, process them locally, and transfer the information to a base station with an internet connection [17].

WSN technology integrated with floating buoys has been widely used to acquire high-frequency WQ data for lakes in the world. Due to severe environmental problems, the Chinese great lakes, which were or are currently supplying drinking water, have been the focus of many efforts to build dense buoy monitoring networks. For example, on Lake Taihu (Jiangsu Province, surface area 2338 km^2, mean water depth 1.9 m), 18 WSN buoy stations are operated by the Chinese Academy of Science, 21 stations by the Jiangsu Environmental Monitoring Station [18], and one each by the Suzhou Meteorological Bureau and Nanjing Normal University. At Lake Dianchi (Yunnan Province, surface area ~300 km^2), there are 30 monitoring buoy stations (MBSs) [19]. Lake Taihu and Lake Dianchi are both key lakes which have been invested in tremendously by the Chinese central government for ecological restoration over the past two decades. At Lake Qiandao, which is a drinking water supply reservoir with an area of 580 km^2 [20], there are four buoy profilers with a meteorological station, 10 MBSs for surface WQ detection with a meteorological station, 13 MBSs for inflow river WQ detection, and four for both surface WQ detection and current measurements by an Acoustic Doppler Current Profiler (ADCP).

Key issues that need to be addressed for a EWS in order to assure the accuracy and precision of measurements, are data quality control (QC) and quality assurance (QA). QC and QA are fundamental for decision making based on reliable data analysis. For a specific water quality parameter, QC generally involves a number of internal consistency tests, a threshold test, a step change point and trend detections for finding potential outliers at a particular station [21]. Measured data at a given site may also be compared with measurements from surrounding sites for an accuracy assessment. An effective QC and QA system is critical to the success of any environmental project, which has been successfully applied to the fields of climatology, oceanography and other geosciences. However, there has been limited application to developing an EWS with real-time high-frequency monitoring. Therefore, the aims of this study were to introduce a comprehensive

EWS. developed for Lake Qiandao, and the corresponding QC method for real-time high-frequency monitoring data.

2. Methodology

2.1. Study Area

Built in 1959, Lake Qiandao is located at Chun'an County, which is at the west of Zhejiang Province, China (29°22′–9°50′N, 118°34′–119°15′E, Figure 1) [22]. It is one of the largest reservoirs in China with a surface water area of 573 km^2 and a water capacity of 178.4×10^8 m^3, when the water level is 108 m [23]. The mean water depth is 34 m and the maximum depth is 100 m. It is used to supply drinking water for the 450,000 people in Chun'an County. Now, it is also providing drinking water for five million people in Hangzhou City, through a tunnel with a length of ~110 km. There are 34 inflow tributaries around the lake. The largest one (Xinan River, Figure 1) is from the northwest, carrying 51.4% of the total inflow to the lake from all sources, not including rainfall and ground water. It carries 34.3% of the total phosphorus (TP) loading and 63.7% of the total nitrogen (TN) loading [19]. The multi-year average inflow and outflow are 103.44×10^8 m^3 and 97.45×10^8 m^3, respectively [23], with a residence time of ~668 days.

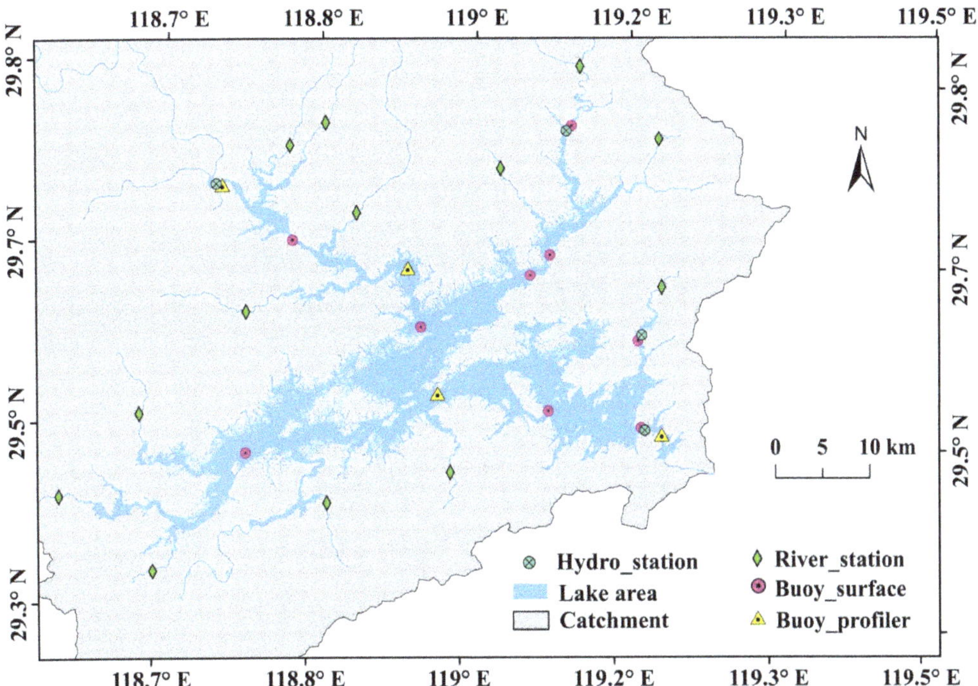

Figure 1. Monitoring sites at Lake Qiandao and its location within China.

2.2. Monitoring Stations

The HFMS at Lake Qiandao includes thirteen river stations; fourteen buoy stations—including ten buoys measuring the surface WQ (Buoy_surface, model EMM700, YSI Incorporated, Yellow Springs, USA); four 'profiler' buoys (Buoy_profiler, model EMM2500, Yellow Springs, USA); four hydrological stations (Hydro_station)—these were deployed at the three main tributaries and the only river outflow and measure water current speed; and four flux stations (Flux_station, model Tenghai HZF3, Tenghai Science & Technology Ltd., Hangzhou, China), located alongside the hydrological stations, measuring WQ parameters. There are also thirteen river stations (River_station, model EMM700, YSI

Incorporated, Yellow Springs, USA) measuring WQ, deployed at the main inflow tributaries. Meteorological sensors (Met_station) are deployed at the top of each buoy station, except for one site (Figure 1). The Buoy_profilers were deployed at sites 1–4. The profiling information at site 3 is representative of the lake because it is at the centroid, while site 4 is the deepest monitoring point, which is located at the biggest outflow channel. Sites 5 and 6 with Buoy_surface are near the middle, capturing surface WQ variation.

2.3. Sensor Information and Alert Range

Table 1 shows all the sensor metadata. The alert range for a specific sensor, in the seventh column of Table 1, was decided by analysing historical data manually, or by using the sonde measurements. The alert thresholds are equal to the corresponding reasonable "minimum-maximum-range" (MMR) of each sonde. When the measured value is out of the alert range, an alert report will be recorded and the EWS can find the report by searching alert report once an hour. Once alert information for a specific sonde is found, a message will be sent to the EWS manager's cellphone. The integrated sonders with a metal protective cage move up and down through the water column at a constant speed The average return times for sonders moving at site 1, site 2, site 3 and site 4, are 55 min through a water column of 65 m, 45 min (water column 40 m), 50 min (water column 46 m), and 30 min (water column 16 m), respectively. The measurement values are recorded every minute at all the four sites. All the buoy systems are solar-powered and the data are transferred to a computer server at the Chunan Branch of Hangzhou Ecology and Environment Bureau by 4G wireless telemetry. The whole monitoring system is being maintained by Hangzhou Tenghai Science and Technology Limited, with the sondes cleaned to wipe bio-fouling once a month and calibrated once every three months for data assurance. The power supply system with a solar panel and wireless data transfer are also regularly checked and maintained by this company.

Table 1. Sensor metadata for all monitoring buoys at Lake Qiandao.

Buoy Type	Number of Buoys	Measured Parameter	Unit	Sensor Model	Measurement Range	Alert Range	Monitoring Frequency
Buoy_surface	10	WT	°C		−5–+50	9–35	
		PH			0–14	6–9	
		ORP	mV		−999–+999	−30–+500	
		COND	mS cm^{-1}		0–200	80–170	
		DO_con	mg L^{-1}	YSI	0–50	4–13	
		DO_sat	%	EXO2	0–500	39–170	30 min
Buoy_profiler	4	TURB	NTU		0–4000	0–43	
		CHLA	µg L^{-1}		0–400	0–25	
		PC	µg L^{-1}		0–100	0–7	
		FDOM	QSE		0–300	0–7	
River_ station (EMM700)	13	WT	°C	YSI EXO2	−5–+50		2 h
		COND	µS cm^{-1}		0–200		
		TURB	NTU		0–4000		
		FDOM	QSE		0–300		
Met_ station	13	RH	%	VAISALA WXT520	0–100	35–90	30 min
		BP	hpa		600–1100	970–1030	
		Wind_spd	m s^{-1}		0–60	0–13	
		Wind_dir	°C		0–360	0–360	
		TEMP	°C		−52–+60	1–33	
		RAIN	mm		0–200 mm h^{-1}	0–162 mm day^{-1}	
Flux_ station	4	TN	mg L^{-1}	TriOS OPUS		0.4–1.5	1 h
		TP	mg L^{-1}			0–0.03	
		COD	mg L^{-1}		0–500	4–11	
		NO$_3$-N	mg L^{-1}		0–100	0–0.08	
		TOC	mg L^{-1}		0–500		

Table 1. *Cont.*

Buoy Type	Number of Buoys	Measured Parameter	Unit	Sensor Model	Measurement Range	Alert Range	Monitoring Frequency
Hydro_Station	4	Current Speed	$m\ s^{-1}$	ADCP TRDI WHR600k	0–5	0–1	30 min

Abbreviations: water temperature (WT), oxidation reduction potential (ORP), electrical conductivity (COND), dissolved oxygen concentration (DO_con), dissolved oxygen saturation (DO_sat), turbidity (TURB), chlorophyll *a* (CHLA), phycocyanin (a pigment specific to cyanobacteria, PC), fluorescent dissolved organic matter (FDOM), total nitrogen (TN), total phosphorus (TP), chemical oxygen demand (COD), nitrate (NO_3^-N), total organic carbon (TOC), relative humidity (RH), air pressure (BP), wind speed (Wind_spd), wind direction (Wind_dir) and air temperature (TEMP).

2.4. Data Quality Control

The monitoring stations produce large volumes of data, requiring specialised tools to facilitate quality control and to ensure that data are fit-for-purpose. We developed bespoke software in Fortran, employing two principle methods of quality control. Firstly, an MMR was adopted whereby the minimum and maximum values of the raw data measured by each sensor were specified, by assessing the range of previous observations and defining a 'reasonable range' (larger than or equal to the alert range at Table 1) for each variable based on a large volume of historical measurements from the lake area and inflows/outflows. The lowest value from both historical observations was adopted as the minimum value for MMR, with the maximum value defined with a similar method. Table 2 shows all the maximum (Max)/minimum (Min)/average (Avg) values for WQ measurements including WT, pH, DO, permanganate index (PI), chemical oxygen demand (COD), five-day biochemical oxygen demand (BOD_5), ammonia (NH_4-N), TP, TN, CHLA and Secchi depth (SD). Unfortunately, only WT, pH and DO were observed by the monitoring buoys. Subsequently, data outside of the specified range for each variable were quarantined with a unique flag number (e.g., '8888') and will be further investigated.

Table 2. Statistical value of measured water quality parameters at the four sites of Lake Qiandao from April 2001 to May 2021.

Site	Number of Samples	Statistical Value	WT (°C)	pH	DO (mg L^{-1})	PI (mg L^{-1})	COD (mg L^{-1})	BOD$_5$ (mg L^{-1})	NH$_4$-N (mg L^{-1})	TP (mg L^{-1})	TN (mg L^{-1})	CHLA (µg L^{-1})	SD (m)
Site 1	195	Max	32.8	9.1	14.3	3.53	17.0	3.50	0.51	0.173	2.47	72	6.0
		Min	9.0	6.4	4.9	1.23	5.0	0.34	0.01	0.002	0.59	0.6	0.1
		Avg	20.5	7.7	8.7	1.96	7.0	1.05	0.08	0.029	1.24	9.5	2.39
Site 2	184	Max	34.3	8.8	14.9	2.67	14.0	2.70	0.18	0.050	1.63	47.0	7.8
		Min	9.3	6.7	6.7	0.78	1.1	0.28	0.01	0.002	0.33	0	0.8
		Avg	20.9	7.8	9.3	1.54	5.1	1.04	0.03	0.013	1.04	7.0	3.81
Site 3	230	Max	33.5	8.8	11.8	2.00	12.0	1.80	0.09	0.027	1.45	20.2	11.0
		Min	9.6	6.7	6.4	0.71	0.7	0.22	0.01	0.002	0.42	0.3	1.7
		Avg	20.7	7.8	8.9	1.37	4.1	0.85	0.02	0.009	0.84	4.0	5.48
Site 4	230	Max	32.9	8.5	11.6	2.13	12.0	1.60	0.03	0.025	1.48	15.6	11.0
		Min	6.7	6.6	6.0	0.61	0.0	0.19	0.01	0.002	0.40	0.0	2.4
		Avg	20.7	7.7	8.6	1.27	4.0	0.80	0.01	0.007	0.82	3.3	6.02

The second approach is an "abnormal" value detection method, as follows:

(1) Suspected abnormal value judgement. For a target value, not including the first and last ones (e.g., 'x_i' in Equations (1)–(3)), if it is either larger or smaller than its adjacent values, then the target value will be regarded as a suspected abnormal value and flagged.

$$f_1 = x_i - x_{i-1} \tag{1}$$

$$f_2 = x_{i+1} - x_i \tag{2}$$

$$ff = f_1 \times f_2 \tag{3}$$

where x_i ($I = 2, 3, 4, \ldots n - 1$) represents the time series of buoy measurements, excluding the first and last values. So if $ff < 0$, the measurement at the ith time will be regarded as a suspected abnormal value.

(2) Abnormal value confirmation. We calculate the average value $\bar{x}$ of raw data after MMR control, the anomaly $|x_{max} - \bar{x}|$ between the maximum value x_{max} and $\bar{x}$, and the anomaly $|x_{min} - \bar{x}|$ between the minimum value x_{min} and $\bar{x}$. The larger value $|x - \bar{x}|$ between $|x_{max} - \bar{x}|$ and $|x_{min} - \bar{x}|$ will be chosen to compare with the absolute value $|ff|$ of ff. If $|ff|$ is larger than, or equal to, $|x - \bar{x}|^2$, then the measurement at the ith time will be confirmed as an abnormal value.

2.5. Change Point and Trend Detections

The Pettitt test was used to automatically detect change points in the data series once a week. Pettitt's test is a nonparametric test to detect a single change point in a time series with continuous data. Its calculation procedures can be found in detail in [24]. The identified change points were then compared to the minimum and maximum values for each sensor. If their values are all in MMR, the validity of change point values will be confirmed. Otherwise the values will be removed from the time series or marked for further check. An exploratory analysis was also carried out to detect the trend of hourly and daily data for all the parameters using the Mann-Kendall method once a week. If the serial data kept increasing or decreasing for more than one week, its validity would be manually and carefully investigated.

2.6. Data Availability, Daily and Hourly Data Calculation

Most of the buoy monitoring datasets at the lake area (Buoy_surface, Buoy_profiler, Met_station) commenced in September 2015. The River_station data collection began in August 2016 and the Flux_station and Hydro_station data collection began in April 2017. A software developed by the authors is used to analyse and summarise the high-frequency data, including the calculation of daily and hourly values, based on quality-controlled raw data. Small data gaps without measurements ($\leq$days) are interpolated by the software and the large data gaps are arbitrarily set up with a unique flag number (e.g., '8888'), which will be not included for calculating daily and hourly values.

3. Results

3.1. Buoy Photographs

Figure 2 shows photographs of the Buoy_surface system at site 5 (Figure 2A), located at the mouth of largest tributary (Xinan River, Figure 1), and the Buoy_profiler system at site 4 (Figure 2B), located at the deepest area in front of the dam for the power station (Figure 1), which is the only outflow. The web interface, which dynamically updates all station data from the database, allows the user to make requests for time periods of interest, review data from specific sites, visualise data as a function of time, and perform simple statistical analyses of the real-time data. All historical data from the monitoring system can be downloaded through the web interface by authorised users.

Figure 2. Pictures of (**A**) the Buoy_surface system at site 5, (**B**) the Buoy_profiler at site 4.

3.2. Surface Measurements

The measurements from site 5 are presented here as an example, and show the daily and hourly variations in surface WQ measured by the buoy probes. Figure 3A shows the time series of daily surface WT from 30 September 2015 to 1 August 2020, and daily surface DO, CHLA and PC from 27 January 2016 to 1 August 2020. The values of maximum (Max), minimum (min), average and standard deviation (Stdev) for WT, DO, CHLA and PC at site 5 are given in Table 3. All the maximum values for WT, CHLA and PC occurred in summer, but their lowest observations occurred in winter (PC) or spring (WT, DO and CHLA). CHLA and PC showed higher variability over time than those of WT and DO during the study period, based on their statistical Stdevs compared to their average values.

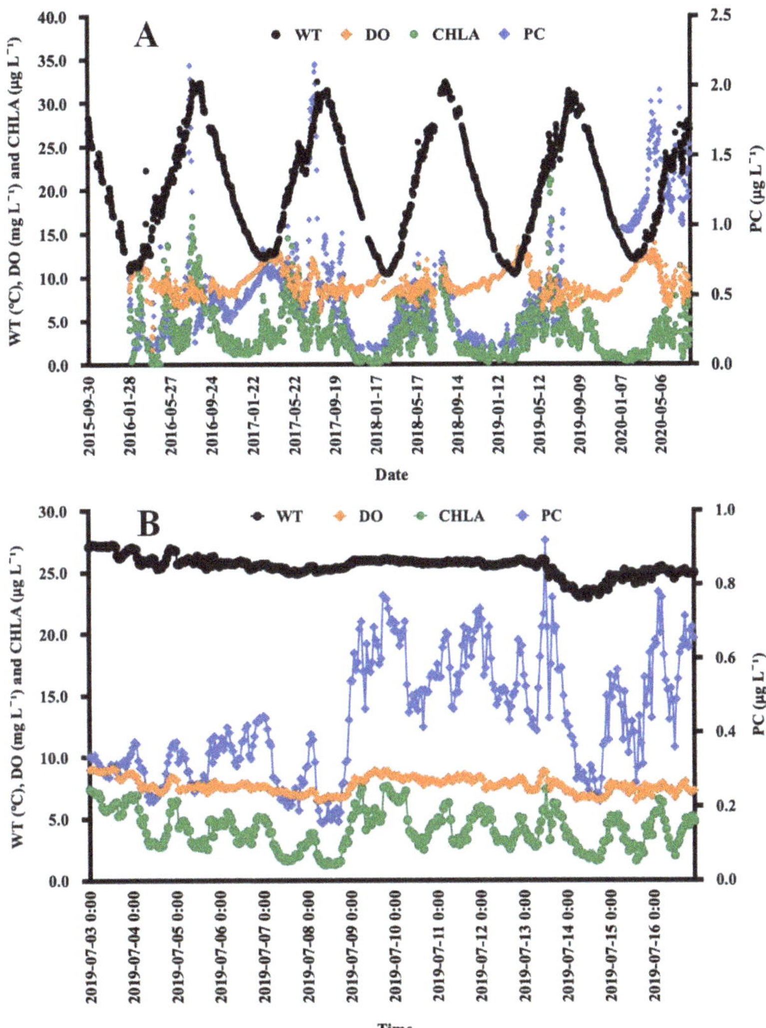

Figure 3. Time serials of (**A**) daily and (**B**) hourly WT (black dots, °C), DO (orange diamonds, mg L^{-1}), CHLA (green dots, µg L^{-1}) and PC (blue diamonds, µg L^{-1}) at site 5. Left Y-axes is for WT, DO and CHLA, and right Y-axes is for PC. X-axes is for date in format of year-month-day.

Table 3. Statistical summaries of surface WT, DO, CHLA and PC measured by monitoring buoys at site 5 and site 6.

Site	Parameter	Max		Min		Avg.	Stdev	*n*
		Value	Date	Value	Date			
Site 5	WT (°C)	32.4	30 July 2017	10.3	1 March 2019	20.2	6.3	1617
	DO (mg L^{-1})	13.9	18 April 2020	1.7	3 April 2016	9.3	1.6	1546
	CHLA (µg L^{-1})	24.3	17 June 2019	0.1	9 April 2016	3.6	3.0	1530
	PC (µg L^{-1})	2.15	23 July 2017	0.1	10 February 2016	0.5	0.4	1318

To show diel variation, hourly data of WT, DO, CHLA and PC at site 5 for the period of 00:00 a.m. 3 July 2019–11:00 p.m. 16 July 2019, without data gaps, are presented as an example in Figure 3B. CHLA and PC show obvious diel variation with higher values in daytime relative to night time, while WT and DO keep more constant than CHLA and PC, showing no diel variation. The Pearson correlation coefficient between WT and DO is 0.8 (*n* = 336), suggesting that surface DO was mainly controlled by WT for the lake.

3.3. Profiling Measurements

Profiles of WT, DO, CHLA and PC at site 4 (deepest area) for 1 January 2016–10 July 2020 are shown in Figure 4. The maximum measurement depth was 65 m at this site. WT (Figure 4A) profiles show a monomictic pattern of mixing with thermal stratification in summer and mixing in winter, although some periods and layers lacked measurements. The TDs were 9.2 m, 9.4 m, 12.1 m and 7.8 m in the summer (July–September) of 2016, 2017 and 2018 and July of 2019 (data not available in August and September). The TD in 2018 were greater than other years, which suggests that the stratification in the summer of 2018 was more intensive than in 2016, 2017 and 2019. Correspondingly, bottom hypoxia events were observed during the stratification of all years. Average bottom DO values (Figure 4B) during stratification were 7.0 mg L^{-1}, 7.7 mg L^{-1}, 8.2 mg L^{-1} and 7.7 mg L^{-1} in 2016, 2017, 2018 and 2019, respectively. The lowest DO values were 1.0 mg L^{-1}, 2.0 mg L^{-1}, 1.6 mg L^{-1} and 2.9 mg L^{-1}, observed on 4 April 2016, 9 December 2017, 1 January 2018 and 13 January 2019, respectively.

CHLA (Figure 4C) followed a similar pattern to WT, with higher values in summer than in winter, suggesting that the biomass of phytoplankton is mainly regulated by water temperature instead of nutrients. The phytoplankton was mostly distributed in the upper 15 m except for late 2018 and early 2019, when phytoplankton could still be found at a depth of 30 m. PC values (Figure 4D) were much smaller than CHLA at the same depth. It didn't have distinct seasonal variation, but was obviously stratified in the summer of 2018 with a higher concentration in the lower layer than the upper layer.

3.4. Real-Time Early Warning Information

The Ministry of Ecology and Environment of China (MEEC) issued state standards for surface water quality in 2002 [25] in order to better manage surface water in China. Lake Qiandao was required to meet Grade I (Table 4, the requirements for heavy metal were not shown in the table) since it provides drinking water for approximately half a million people in Chun'an County (located to the northeast of lake), and a total of 10 million people in both Hangzhou City and Jiaxing City, with water diverted through a tunnel of more than 110 km in length. DO and pH are the only two parameters which were measured by wireless sonde, deployed with monitoring buoys at the lake. The statistical analysis results for DO and pH are shown in Table 5. There were totally 275 and 130 samples of pH and DO out of their MMRs at site 5, accounting for 11.0% and 8.0% of all the valid samples, respectively. The percentages of pH (21.5%) and DO (11.3%) at site 6 were more than those at site 5. The observed maximum/minimum/average values of pH at site 5 and site 6 were 9.9/9.002/9.22 and 12.2/9.0002/9.62, respectively. Thus, the maximum pH at site 6 is much greater than that at site 5, showing that pH was more variable at site 6 than at site 5.

However, DO followed a reverse pattern, with more varied values at site 5 compared to site 6. Its lowest value was as low as 1.74 mg L^{-1}, observed on 3 April 2016.

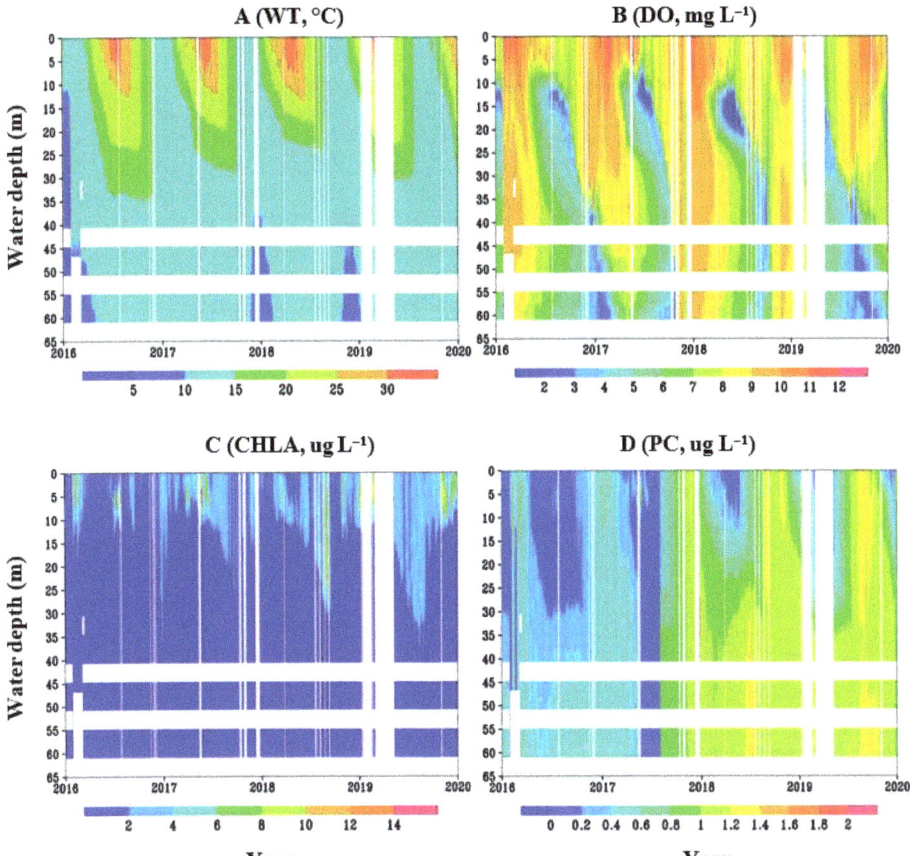

Figure 4. Daily (**A**)WT, (**B**) DO, (**C**) CHLA and (**D**) PC profiles at site 4 for 1 January 2016–10 July 2020. Y-axes for (**A–D**) represent water depth (m) and X-axes represent years from 2016 to 2020. White areas denote lack of measurements.

Table 4. Threshold of WQ parameters required for Lake Qiandao.

Parameter	pH	DO (mg L^{-1})	PI (mg L^{-1})	COD (mg L^{-1})	BOD5 (mg L^{-1})	NH$_4$-N (mg L^{-1})	TP (mg L^{-1})	TN (mg L^{-1})
Threshold for Grade I	6–9	≥7.5	≤2	≤15	≤3	≤0.15	≤0.01	≤0.2

Table 5. Statistical analysis results of DO and pH for early warning at site 5 and site 6.

Statistical Results	Site 5		Site 6	
	pH	DO	pH	DO
Number of all available data	1573	1618	1128	1128
Number of EWS data	275	130	242	128
Max	9.9	7.5	12.2	7.5
Min	9.002	1.74	9.0002	6.2
Avg.	9.22	6.92	9.62	6.99

4. Discussion

The EWS is now being used for giving real-time early warning signals by judging if the measurements of each sensor are out of the corresponding MMR and meet the required WQ grade. Unfortunately, DO and pH are the only two parameters which were directly measured by the buoy sensor and which can be used for the early warning of WQ at this lake. Therefore, it should be considered seriously whether sensors measuring nutrients (e.g., TN, TP, NH_4-N, COD and PI) should be added to the buoys, or whether a model (e.g., AEM3D) simulating these nutrients based on high-frequency monitoring data, needs to be developed for providing more satisfactory early warning signals.

The system is also providing data for horizontal interpolation, to produce an initial condition for AEM3D predicting HABs and WQ at a time scale of seven days. For a WQ and HAB prediction system, initial and boundary conditions are very important to improve prediction accuracy. The initial conditions include horizontal and vertical WQ distribution within the waterbody (e.g., TN, TP and CHLA concentrations at each grid location). A huge challenge with initial conditions derives from the limited number of monitoring buoys collecting high-frequency WQ data, due to economic considerations, which leads to an inaccurate spatial distribution of WQ (e.g., high spatial patchiness of cyanobacterial bloom). However, many advanced interpolation methods are now available to address this issue. Inaccurate spatial WQ distribution in model initialisation can lead to inaccurate WQ prediction at each grid location, and thus unconvincing algal aggregation caused by winds. However, temporal and spatial difficulties prevent conventional methods for water sampling and laboratory analysis to meet the requirements for producing model initial conditions. A comprehensive approach is required, integrating high-frequency buoy monitoring data, laboratory data, satellite images and other available resources to provide a satisfactory spatial WQ concentration for the initialization of predictive simulation systems.

WQ sonde (multi-sensor probes) measurements can efficiently provide data wirelessly at a high temporal resolution, but potential problems could include data distortion due to sensor faults, or data gaps because of failed data storage and/or transfer. Therefore, data QC procedures are necessary before using the sensor measurements. The first step for QC is typically to detect missing series and estimate missing values by relying on neighbouring observations, then to detect unreasonable values out of range between the upper and the lower limits for each parameter, ideally guided by experience for a specific water body and measurement type. Unreasonable values can be removed and substituted with interpolated values. If many successive measurements from the same sensor are of the same exact value, they should usually also be removed and interpolated with neighbouring values, or excluded from data analysis. The final step is to detect outliers (incorrect or out-of-range) measurements, which can be removed or assumed to be missing [26]. Outliers are typically those observations which represent abrupt increases or decreases compared to the neighbouring values. There are many methods [26–28] and pieces of software [29] available for data QC. In this paper, we adopted the outlier detection method for finding anomalous values, which were removed and generally replaced with interpolated values. The detections of change points and trends will further help to find abnormal values or sonde problems. In our system, interpolation was not implemented to reproduce actual missing values due to the high potential for erroneous measurement generation when interpolating over longer time periods. Therefore, interpolation methods or software need to be integrated into this system in order to produce data without measurement.

The whole buoy monitoring system was originally designed to provide essential WQ information, in order to meet Grade I at the required sites. Therefore, Buoy_profilers were deployed at site 1, site 3 and site 4, and Buoy_surface was deployed at site 5. However, the buoys can only monitor WT, DO, pH, CHLA, PC and TURB. It is very difficult and expensive to directly and accurately measure TN, TP, NH_4-N, PI, COD and BOD_5 at a high frequency and in near-real-time [30]. An optional solution in this early warning system, is to calculate these parameters based on their regressed relations with sonde-measured values. The calculated values from regressive equations can then provide vital WQ information at different zones of the lake, feeding the AEM3D (http://www.hydronumerics.com.au, accessed on 15 January 2022) model for WQ prediction.

The collected data may assist water environmental managers in identifying and predicting the impacts of climatic extreme events [31]. For example, at Lake Qiandao, rainstorms with high rainfall typically result in a large inflow of water and nutrient loading, including N, P, and organic/inorganic matters, leading to an abrupt increase in water level and a significant increase of regional N and P concentrations [22]. This remarkably alters the spectral absorption properties of chromophoric dissolved organic matter (CDOM) and particles at the northwestern, southwestern and northeastern areas [32]. HFMS can also provide a useful basis for theory and model developments, improving our understanding of lake (reservoir) responses to perturbations caused by human activities and climate change at different time scales (e.g., sub-hourly, hourly, daily, monthly, seasonally, annually and every decade).

Although HFMS has a wide range of applications, it is now still hampered by several factors. For example, we have a limited choice of water quality sensors that are robust, economic and low-maintenance. The accuracy of the sondes measuring chemical parameters (i.e., phosphorus, ammonium, ammonia and nitrite) and biological parameters (i.e., bacterial enumeration, cyanobacteria, biota and cyanotoxins) still needs to improve, although the fast spread of HFMS is encouraging sensor developers to improve technology as quickly as they can. For giving better early warning signals and real-time WQ assessment, in the future, it will be necessary to add sondes measuring chemical and biological parameters to the current HFMS.

Author Contributions: Conceptualization, L.L. and H.L.; methodology, L.L.; software, L.L.; validation, J.L. (Jia Lan), Y.W., Z.W. and C.M.; formal analysis, (J.L.) Jialong Li; investigation, Y.W.; resources, Y.W.; data curation, F.L., L.C. and G.W.; writing—original draft preparation, L.L.; writing—review and editing, C.M. and H.Z.; visualization, L.L., R.Z. and F.G.; supervision, L.L.; project administration, L.L. and H.L.; funding acquisition, L.L. All authors have read and agreed to the published version of the manuscript.

Funding: This research was funded by Yunnan University (C176220100043), Yunnan Provincial Department of Science and Technology (202001BB050078), National Natural Science Foundation of China (41671205, 42171034), and the Yunnan Provincial Government Leading Scientist Program (2015HA024).

Institutional Review Board Statement: Not applicable.

Informed Consent Statement: Not applicable.

Acknowledgments: The project was supported by Yunnan University (C176220100043), Yunnan Provincial Department of Science and Technology (202001BB050078), National Natural Science Foundation of China (41671205, 42171034), and the Yunnan Provincial Government Leading Scientist Program (2015HA024).

Conflicts of Interest: The authors declare no conflict of interest.

References

1. Greeson, P.E. Lake eutrophication—A natural process. *J. Am. Water Resour. Assoc.* **1969**, *5*, 16–30. [CrossRef]
2. Anderson, D.M.; Glibert, P.M.; Burkholder, J.M. Harmful algal blooms and eutrophication: Nutrient sources, composition, and consequences. *Estuaries* **2002**, *25*, 704–726. [CrossRef]
3. Smith, V.H.; Schindler, D.W. Eutrophication science: Where do we go from here? *Trends Ecol. Evol.* **2009**, *24*, 201–207. [CrossRef]
4. Conley, D.J.; Paerl, H.W.; Howarth, R.W.; Boesch, D.F.; Seitzinger, S.P.; Havens, K.E.; Lancelot, C.; Likens, G.E. Controlling eutrophication: Nitrogen and phosphorus. *Sciences* **2009**, *323*, 1014–1015. [CrossRef]
5. Lee, J.H.W.; Hodgkiss, I.J.; Wong, K.T.M.; Lam, I.H.Y. Real time observations of coastal algal blooms by an early warning system. *Estuar. Coast. Shelf Sci.* **2005**, *65*, 172–190. [CrossRef]
6. Wang, X.; Su, J.; Cai, Y.; Tan, Y.; Yang, Z.; Wang, F. An integrated approach for early warning of water stress in shallow lakes: A case study in Lake Baiyangdian, North China. *Lake Reserv. Manag.* **2013**, *29*, 285–302. [CrossRef]
7. Burchard-Levine, A.; Liu, S.; Vince, F.; Li, M.; Ostfeld, A. A hybrid evolutionary data driven model for river water quality early warning. *J. Environ. Manag.* **2014**, *143*, 8–16. [CrossRef]

8. Park, Y.; Cho, K.H.; Park, J.; Cha, S.M.; Kim, J.H. Development of early-warning protocol for predicting chlorophyll-a concentration using machine warning models in freshwater and estuarine reservoirs, Korea. *Sci. Total Environ.* **2015**, *502*, 31–41. [CrossRef]

9. Biddanda, B.A.; Weinke, A.D.; Kendall, S.T.; Gereaux, L.C.; Holcomb, T.M.; Snider, M.J.; Dila, D.K.; Long, S.A.; VandenBerg, C.; Knapp, K.; et al. Chronicles of hypoxia: Time-series buoy observations reveal annually recurring seasonal basin-wide hypoxia in Muskegon Lake—A Great Lakes estuary. *J. Great Lakes Res.* **2018**, *44*, 219–229. [CrossRef]

10. Delpla, I.; Florea, M.; Rodriguez, M.J. Drinking water source monitoring using early warning systems based on data mining techniques. *Water Resour. Manag.* **2019**, *33*, 129–140. [CrossRef]

11. Glasgow, H.B.; Burkholder, J.M.; Reed, R.E.; Lewitus, A.J.; Kleinman, J.E. Real-time remote monitoring of water quality: A review of current applications, and advancements in sensor, telemetry, and computing technologies. *J. Exp. Mar. Biol. Ecol.* **2004**, *300*, 409–448. [CrossRef]

12. Ungureanu, F.; Lupu, R.G.; Stan, A.; Craciun, I.; Teodosiu, C. Towards real time monitoring of water quality in river basins. *Environ. Eng. Manag. J.* **2010**, *9*, 1267–1274. [CrossRef]

13. Storey, M.V.; van der Gaag, B.; Burns, B.P. Advances in on-ling drinking water quality monitoring and early warning systems. *Water Res.* **2011**, *45*, 741–747. [CrossRef]

14. Qi, X.; Wang, S.; Jiang, Y.; Liu, P.; Li, Q.; Hao, W.; Han, J.; Zhou, Y.; Huang, X.; Liang, P. Artificial electrochemically active biofilm for improved sensing performance and quickly devising of water quality early warning biosensors. *Water Res.* **2021**, *198*, 117164. [CrossRef]

15. Zhang, C.; Pei, H.; Jia, Y.; Bi, Y.; Lei, G. Effects of air quality and vegetation on algal bloom early warning systems in large lakes in the middle–lower yangtze river basin. *Environ. Pollut.* **2021**, *285*, 117455. [CrossRef] [PubMed]

16. Kumar, T.; Naik, S.; Jujjavarappu, S.E. A critical review on early-warning electrochemical system on microbial fuel cell-based biosensor for on-site water quality monitoring. *Chmesphere* **2022**, *291*, 133098. [CrossRef] [PubMed]

17. Ungureanu, F.; Balasa, M.I. Sensor networks in waste water treatment monitoring. *Environ. Manag. J.* **2007**, *6*, 65–70. [CrossRef]

18. Tang, M.; Si, W.; Zhong, S. Development of automatic water quality monitoring network at Lake Taihu, Jiangsu Province. *Environ. Dev.* **2017**, *2*, 91–95. (In Chinese)

19. Wu, J.; Kong, L.; Yang, J.; Li, Y. Design and development of the integrated platform for environmental monitoring and information management for Lake Dianchi. *Technol. Innov. Appl.* **2018**, *15*, 79–80. (In Chinese)

20. Zhang, Y.; Wu, Z.; Liu, M.; He, J.; Shi, K.; Wang, M.; Yu, Z. Thermal structure and response to long-term climatic changes in Lake Qiandaohu, a deep subtropical reservoir in China. *Limnol. Oceanogr.* **2014**, *59*, 1193–1202. [CrossRef]

21. Durre, I.; Menne, M.J.; Vose, R. Strategies for evaluating quality assurance procedures. *J. Appl. Meteorol. Climatol.* **2008**, *47*, 1785–1791. [CrossRef]

22. Zhang, Y.; Lan, J.; Li, H.; Liu, F.; Luo, L.; Wu, Z.; Yu, Z.; Liu, M. Estimation of external nutrient loadings from the main tributary (Xin'anjiang) into Lake Qiandao, 2006–2016. *J. Lake Sci.* **2019**, *31*, 1534–1546. (In Chinese)

23. Xiang, C.; Yan, L.; Han, Y.; Wu, Z.; Yang, W. Valuation of ecosystem services of the Thousand-Island Lake, Zhejiang, China. *Chin. J. Appl. Ecol.* **2019**, *30*, 3875–3884. (In Chinese)

24. Pettitt, A.N. A nonparametric approach to the change-point problem. *J. R. Stat. Society. Ser. C* **1979**, *28*, 126–135.

25. Ministry of Ecology and Environment of China (MEEC). *State Standards for Surface Water Quality (GB 3838–2002)*; China Environmental Press: Beijing, China, 2002.

26. Gonzalez-Rouco, J.F.; Jimenez, J.L.; Quesada, V.; Valero, F. Quality control and homogeneity of precipitation data in the southwest of Europe. *J. Clim.* **2001**, *14*, 964–978. [CrossRef]

27. Li, X.; Li, F. Quality control of online oceanic and meteorological data. *Mar. Forecast.* **1997**, *14*, 71–78. (In Chinese)

28. Gokturk, O.M.; Bozkurt, D.; Sen, O.L.; Karaca, M. Quality control and homogeneity of Turkish precipitation data. *Hydrol. Processes* **2008**, *22*, 3210–3218. [CrossRef]

29. Araya-Lopez, J.L.; Kalyuzhnaya, A.V.; Kosukhin, S.S.; Ivanov, S.V. Data quality control for St. Petersburg flood warning system. *Procedia Comput. Sci.* **2016**, *80*, 2128–2140. [CrossRef]

30. Banna, M.H.; Imran, S.; Francisque, A.; Najjaran, H.; Sadiq, R.; Rodriguez, M.; Hoorfar, M. Online drinking water quality monitoring: Review on available and emerging technologies. *Crit. Rev. Environ. Sci. Technol.* **2014**, *44*, 1370–1421. [CrossRef]

31. Marcé, R.; George, G.; Buscarinu, P.; Deidda, M.; Dunalska, J.; de Eyto, E.; Flaim, G.; Grossart, H.-P.; Istvanovics, V.; Jennings, E.; et al. Automatic High Frequency Monitoring for Improved Lake and Reservoir Management. *Environ. Sci. Technol.* **2016**, *50*, 10780–10794. [CrossRef]

32. Shi, L.; Mao, Z.; Liu, M.; Zhang, Y. Effects of rainstorm on the spectral absorption properties of chromophoric dissolved organic matter and particles in Lake Qiandao. *J. Lake Sci.* **2018**, *30*, 358–374. (In Chinese)

Article

Remote Analysis of the Chlorophyll-a Concentration Using Sentinel-2 MSI Images in a Semiarid Environment in Northeastern Brazil

Thaís R. Benevides T. Aranha [1,*], Jean-Michel Martinez [2], Enio P. Souza [3], Mário U. G. Barros [4] and Eduardo Sávio P. R. Martins [5]

[1] Department of Hydraulic and Environmental Engineering, Federal University of Ceará, Campus do Pici, Block 713, Fortaleza 60440-970, Ceará, Brazil

[2] Géosciences Environnement Toulouse (GET), Unité Mixte de Recherche 5563, IRD/CNRS/Université, 31400 Toulouse, France; jean-michel.martinez@ird.fr

[3] Department of Atmospheric Sciences, Federal University of Campina Grande, Campina Grande 58429-140, Paraíba, Brazil; enio.souza@ufcg.edu.br

[4] Water Resources Management Company (COGERH), Fortaleza 60824-140, Ceará, Brazil; mario.barros@cogerh.com.br

[5] Research Institute of Meteorology and Water Resources (FUNCEME), Fortaleza 60115-221, Ceará, Brazil; espr.martins@gmail.com

* Correspondence: thais_benevides@hotmail.com

Citation: Aranha, T.R.B.T.; Martinez, J.-M.; Souza, E.P.; Barros, M.U.G.; Martins, E.S.P.R. Remote Analysis of the Chlorophyll-a Concentration Using Sentinel-2 MSI Images in a Semiarid Environment in Northeastern Brazil. *Water* **2022**, *14*, 451. https://doi.org/10.3390/w14030451

Academic Editors: Fei Zhang, Ngai Weng Chan, Xinguo Li and Xiaoping Wang

Received: 8 December 2021
Accepted: 29 January 2022
Published: 2 February 2022

Publisher's Note: MDPI stays neutral with regard to jurisdictional claims in published maps and institutional affiliations.

Abstract: In this paper, the authors use remote-sensing images to monitor the water quality of reservoirs located in the semiarid region of Northeast Brazil. Sentinel-2 MSI TOA Level 1C reflectance images were used to remotely estimate the concentration of chlorophyll-a (chl-a), the main indicator of the trophic state of aquatic environments, in five reservoirs in the state of Ceará, Brazil. A three-spectral band retrieval model was calibrated using 171 water samples, collected from November 2015 through July 2018 in 5 reservoirs. For validation, 71 additional samples, collected from August 2018 through December 2019, were used to ensure a robust accuracy assessment. The TOA Level 1C products performed very well, achieving a relative RMSE of 28% and $R^2 = 0.80$. Data on wind direction and speed, solar radiation and reservoir volume were used to generate a conceptual model to analyze the behavior of chl-a in the surface waters of the Castanhão reservoir. During 2019, the reservoir water quality showed strong variation, with concentration fluctuating from 30 to 95 µg/L We showed that the end of the dry season is marked by strong eutrophic conditions corresponding to very low water inflows into the reservoir. During the rainy season there is a large decrease in the chl-a concentration following the increase of the lake water storage. During the following dry season, satellite data show a progressive improvement of the trophic state controlled by wind intensity that promotes a better mixing of the reservoir waters and inhibiting the development of most phytoplankton.

Keywords: remote sensing; water quality; chlorophyll-a; reservoirs; semiarid

1. Introduction

The Brazilian semiarid portion covers an area of about 1 million km², which corresponds to 86% of the Northeast region of Brazil [1]. This region has irregular rainfall and high temperatures, with more than 2800 h of sunshine per year, in addition to high evaporation rates, around 2000 mm per year [2]. The soil is predominantly crystalline, and this is one of the factors responsible for the low levels of groundwater availability as well as the low quality of water accumulated in the aquifers, due to the high levels of salts from this type of rock [3].

Due to the strong interannual rainfall variability and the overall scarcity of rainfall in addition to unfavorable soil conditions, the construction of reservoirs such as weirs and dams to store water during the rainy season, and of water transposition channels to take this resource to the locations most affected by the drought, appeared as a reliable alternative to mitigate water scarcity in this semiarid region. Hence, construction of reservoirs is central to the storage and distribution of water in arid and semiarid regions. However, this intervention in the natural landscape, resulting from the transformation of lotic to lentic environments, entails a series of impacts on ecosystems, with changes in the natural behavior and water quality of engineered rivers and streams [4]. This anthropogenic interference generates a great ecological impact on aquatic ecosystems by reducing the flow of water, and increasing the sedimentation rate, water residence time, thermal stratification, and artificial enrichment of nutrients such as nitrogen and phosphorus (eutrophication) caused by human activities [5,6].

Eutrophication presents changes in the physicochemical and biological characteristics of these aquatic environments, arising from the use of fertilizers and pesticides in agriculture, the discharge of industrial and domestic sewage without adequate treatment, destruction of riparian vegetation in water sources, and high urbanization rate. This process produces changes in water quality, such as reduced dissolved oxygen, fish death, excessive proliferation of phytoplanktonic organisms and, consequently, increased incidence of potentially toxin-producing algae and cyanobacterial blooms [7,8]. Given this scenario, the concern with the degradation of these aquatic environments has grown in the scientific community, making it necessary to increase studies and expand knowledge on the subject.

Phytoplankton refers to the set of microscopic aquatic organisms that have photosynthetic capacity and that live dispersed floating in the water column. This group includes organisms traditionally considered algae. However, among these, there is a group of great public health importance, which is also classified as bacteria, namely the cyanobacteria [9,10]. The occurrence of cyanobacterial blooms in water bodies used for urban supply can represent a serious risk to the health of the population, due to the ability of these organisms to produce toxins (cyanotoxins), which can be lethal to mammals and other warm-blooded animals [11–13]. Blooms can also interfere with the balance of aquatic ecosystems, as when algae are in suspension they can modify the water's transparency, which can lead to deoxygenation of the water body and, consequently, fish mortality. In addition, they represent a serious problem for water treatment plants, as they can cause loss of filter load and change in the odor and taste of treated water [14].

Chlorophyll is the photosynthetic pigment present in all phytoplankton organisms. Chlorophyll-a (chl-a) is the most common of the chlorophylls (a, b, c, and d) and represents approximately 1 to 2% of the dry weight of the organic material in all algae. For this reason, knowledge of its concentration is used to detect algal blooms and understand its dynamics [15,16], and to estimate primary productivity [17].

Thus, chl-a is the main indicator of the trophic status of aquatic environments [18,19]. This statement is confirmed for the calculation of the Trophic State Index (TSI), which aims to classify water bodies in different degrees of trophy. The oligotrophic state is characterized by the lowest TSI values, with TSI values increasing through the mesotrophic, eutrophic, to the hypereutrophic, with the highest TSI values. In other words, the TSI aims to assess water quality in terms of nutrient enrichment and its effect related to the overgrowth of algae and cyanobacteria [20]. This index was adopted by [20] for temperate climates and was adapted for lentic environments of tropical climate by [21], using only two variables, namely chl-a and total phosphorus. In this index, the results corresponding to phosphorus should be understood as a measure of the eutrophication potential, since this nutrient acts as the causative agent of the process. The evaluation corresponding to chl-a, on the other hand, should be considered a measure of the response of the water body to the causative agent, adequately indicating the level of algae growth. Thus, the average index satisfactorily encompasses the cause and effect of the process.

As a result, there has been increasing interest in studies of this compound and its derivatives. On the one hand, the irregular nature of the distribution of phytoplankton, in particular the proliferation of cyanobacteria, is characterized by frequent migration

dynamics in the vertical layers. In addition to very fast replication rates, this dynamic makes it difficult to carry out a quantitative monitoring of the number of cells and spatiotemporal distribution, as surface blooms can appear and disappear quickly, usually within a few hours [22–24]. This poses an appreciable challenge to any research effort for larger water bodies. Normally, measurements for analyzing the quality of water in reservoirs are carried out by collecting water from strategic points, usually near the reservoir banks. These samples do not comprise the entire extent of the dams, and thus may not represent the real spatial distribution of water quality. In addition, these collections are carried out over a long-time interval, often without periodicity [25–27].

In this context, remote sensing (RS) reinforces the ability to monitor and understand the composition and dynamics of small and large reservoirs, especially with the advancement of this technology in recent decades. The latest generation of mid-resolution multispectral sensors, with free image availability, such as the Landsat-8 (L8) and Sentinel-2 (S2) satellites, offer advanced opportunities for synoptic view of the entire area of interest, fine-scale, and high-frequency monitoring [28]. These satellites were not specifically designed for water observation but are promising for a detailed analysis of water quality [15,29,30], thanks to their fine radiometric sensitivity [31,32]; 10 to 30-m spatial resolution; high revisit frequency (every 2–3 days combining L8 and S2 satellites) and the improved configuration of the spectral bands in the visible and near infrared range [33].

Orbital sensors are able to record the effects of the interaction of solar radiation with constituents in water [34], and the spectral information contained in satellite images is useful in the development of bio-optical models [35]. Optically active constituents (OACs) such as suspended solids, photosynthetic pigments (chlorophyll), and colored dissolved organic matter (CDOM) are used as indicators of water quality during the characterization of the aquatic environment [36]. The RS applied to the study and monitoring of aquatic environments is based on the processes of selective absorption and scattering of solar radiation by water and its OACs [37,38]. Thus, OACs concentrations can be estimated empirically or semi-analytically, based on RS data [39].

Phytoplankton is capable of synthesizing organic matter from photosynthesis, being mainly responsible for the primary production of cells, causing changes in the spectral behavior of water [40,41]. In general, photosynthetic pigments take advantage of radiation from the blue and red region for the photosynthesis process. Thus, waters with phytoplankton have two bands of maximum absorption in the electromagnetic spectrum, one in the blue region, around 440 nm, and the other in the red region, around 670 nm. The green band, around 560 nm, due to its low absorption coefficient, indicates a high chlorophyll reflectance. In addition, the presence of phytoplankton is also characterized by its peak reflectance at 700 nm, in the near infrared region [42–46].

Importantly, there is a consensus that harmful algal blooms are complex events, normally caused by several environmental factors that occur simultaneously [47]. In addition to the environmental conditions generated by anthropogenic activities, hydrological and climatic variables play an important role in the behavior of phytoplankton [48]. These organisms are sensitive to water-level fluctuations, regarding the abundance, composition and diversity of biomass [49–53], as this changes the physicochemical conditions of water such as the mixing regime, light availability, and nutrient concentrations [54–56].

Studies have shown that low water levels together with the high retention time in reservoirs in semiarid regions, resulting from irregularity in precipitation, are often associated with high algal biomass in freshwater ecosystems, due to the high availability of nutrients for primary producers [57–64].

These conditions combined with high irradiation and high temperatures, although they do not result in greater overall phytoplankton biomass, considerably favor the growth of cyanobacteria, which are extremely harmful to human health [65–68]. As temperatures approach and exceed 20 °C, growth rates of freshwater eukaryotic phytoplankton generally stabilize or decline, while growth rates of many cyanobacteria increase [69–72], providing a competitive advantage due to physiological factors such as faster growth and physical factors such as improved stratification [73–75].

On the other hand, the reduction of the water level can also increase the resuspension of sediments due to the turbulence caused by the wind and, consequently, the increase in inorganic turbidity. This reduces the availability of light and phytoplankton biomass, changing the structure of the phytoplankton community [76–79].

This study aims at the development of a water-quality analysis methodology based on remote sensing. It is applied to a semiarid environment, strongly influenced by an irregular hydrological regime. The objective is to reproduce trophic variations of water reservoirs of a semiarid region by applying a retrieval algorithm that estimates the concentration of chl-a from Sentinel-2 images. For this purpose, an empirical spectral model will be developed, in which statistical relationships are established between chl-a concentration data collected in situ and the reflectance of spectral bands extracted from Sentinel-2 images. An analysis with hydroclimatic data is also carried out in an attempt to better understand and characterize the eutrophication of the Castanhão reservoir. Section 2 presents the available data and the methodology while results are presented and discussed in Section 3; and, finally, conclusions are presented in Section 4.

2. Data and Methodology

2.1. Study Sites

This study was carried out in the state of Ceará, located in the semiarid region of northeastern Brazil, close to the Equator. The studied reservoirs were Gavião, Pacoti, Pacajús, Castanhão, and Orós. The first three reservoirs are closer to the metropolitan region of Fortaleza (capital of Ceará), and comprise the integrated Gavião system, responsible for the water supply of the metropolitan region. Castanhão and Orós are the two largest reservoirs in Ceará (Castanhão is the largest). Both are located in the Jaguaribe River Basin and play a key role in water security and in flood control of the Jaguaribe valley, in addition to transferring these accumulated waters to the reservoirs in the metropolitan region of Fortaleza. Thus, they are a strategic water reserve for the state. Despite their storage capacity, a severe drought, such as that which occurred between 2012 and 2018, could lead these large reservoirs to have very low volumes of stored water. In 2019 and 2020, Castanhão reached 2.8% and 2.4%, respectively, and Orós 5.2% and 4.7%, respectively, of its water volume capacity [80]. These conditions significantly affect the state's water supply management and these reservoirs' water quality.

The climate of this region is hot and dry, and the rainfall is irregular in space and time [81]. It is concentrated between the months of February and May [82], with maximum values in the months March and April [83]. Due to recurrent droughts and high annual average temperature, around 31 °C, it has a negative water balance for most of the year, with high potential evaporation rates, reaching over 2000 mm year^{-1}, and precipitation below 900 mm year^{-1} [84]. Moreover, the region is characterized by shallow soils on a crystalline basement. All this combined results in intermittent rivers [85,86].

In order to promote greater water security, the construction of surface reservoirs has been the most common and important strategy adopted [87]. These reservoirs, however, are subject to long periods of low, or even, zero inflows and long water residence time (more than 12 months), high solar radiation and high temperatures during most of the rainy season, in addition to intense anthropogenic activity in their basins. These factors contribute to an intense accumulation and concentration of nutrients, making these systems considerably more vulnerable to eutrophication [88]. In order to analyze different water-quality conditions, this study was carried out in reservoirs in which total water-storage capacity varies from 32.9 million to 6.7 billion m^3 and trophic state ranges from mesotrophic to hypereutrophic, as shown in Table 1.

Table 1. Reservoir studied (capacity = maximum water volume; average depth combined with standard deviation—sd). Source: Portal Hidrológico do Ceará (http://www.hidro.ce.gov.br/hidro-ce-zend/acude/eutrofizacao, accessed on 14 January 2022).

Reservoir	Maximum Depth (2019)		Average Depth (2019)		Capacity (m^3)	Trophic State (2015 a 2019)
	Rainy Season	Dry Season	Rainy Season (sd)	Dry Season (sd)		
Gavião	12.88	11.91 m	11.98 (±0.51) m	11.54 (±0.13) m	32.9 millions	eutrophic to hypereutrophic
Pacoti	21.95 m	21.8 m	18.1 (±3.28) m	19.93 (±1.21) m	380 millions	eutrophic to hypereutrophic
Pacajús	14.36 m	14.19 m	11.78 (±2.37) m	13.15 (±0.63) m	240 millions	eutrophic to mesotrophic
Castanhão	33.03 m	32.93 m	30.58 (±1.60) m	30.58 (±1.67) m	6.7 billion	eutrophic to hypereutrophic
Orós	20.86 m	20.52 m	18.97 (±1.52) m	19.13 (±0.95) m	1.94 billion	mesotrophic to eutrophic

2.2. Data

Over decades of satellite water monitoring, remote monitoring of small reservoirs has been greatly hampered by the lack of appropriate satellite sensors [89]. Ocean color sensors such as the MODIS (MODerate Resolution Imaging Spectroradiometer) and the MERIS (Medium Resolution Imaging Spectrometer) have a spatial resolution of 250–1200 m, making them suitable only for very large reservoirs. The first Landsat series satellites (Landsat 1–7) had good spatial resolution (30–79 m) but restricted radiometric resolution (6–8 bits), which made them a limited tool for mapping water-quality parameters [90]. The radiometric resolution of Landsat 8 is 12 bits, and this makes it suitable for remote analysis even in dark lakes (rich in CDOM). However, its revisit time is quite long (16 days), limiting its use in systematic water quality monitoring [91].

In 2015, the launch of the Multispectral Imager (MSI), aboard satellite Sentinel-2, opened up a new potential for the use of RS for reservoir monitoring [92]. The images have a spatial resolution of 10 m, 20 m, and 60 m, which means that even small reservoirs can be studied. Data are acquired in 13 spectral bands, distributed along the visible and infrared regions. This includes narrow bands that capture phytoplankton spectral characteristics, such as the chl-a absorption maximum, around 670 nm, and the reflectance peak near 700 nm associated with phytoplankton backscatter. The radiometric resolution of the sensor is 12 bits. Sentinel-2 comprises a constellation of two identical satellites that are part of the European Commission Copernicus program and operated by the European Space Agency (ESA), located in Paris, France. They operate simultaneously in the same sun-synchronized polar orbit and, in opposite 180° positions, are designed to provide a high 5-day revisit frequency, which makes them suitable for routine monitoring.

In this study, WGS84 UTM zone 24 South images of the Sentinel-2 satellite with a resolution of 10 m were used. The passage date was the closest to the dates of chl-a sample collection in the reservoir, in the period from November 2015 to December 2019. The S2 MSI Level 1C (L1C) images were obtained from the Copernicus Open Access Hub (https://scihub.copernicus.eu/, accessed on 9 September 2021). The Sentinel Application Platform (SNAP) version 7.0 on Windows 10 (64-bit) was used to process the images and the Band Maths function to generate the chl-a concentration maps from the chl-a recovery algorithm.

To calibrate and validate the empirical models, data from in situ chl-a collection were used, provided by the Ceará Water Resources Management Company (COGERH). The APHA 10200 H spectrophotometric [93] was the laboratory method used to analyze the concentration of collected chl-a. The collections were carried out in a dark flask, free of interferences, and the pigments were cold extracted with 90% acetone. Considering the 5 reservoirs analyzed, 171 collection data were used to calibrate the spectral model from satellite images, over 122 campaigns. Calibration data covered the period from

November 2015 to July 2018. To validate the spectral model, 71 collection data were used, over 61 campaigns, covering the period from August 2018 to December 2019. Regarding the validation period, turbidity data were also used to correlate with the chl-a data. The locations of the in situ data collection points are shown in Figure 1.

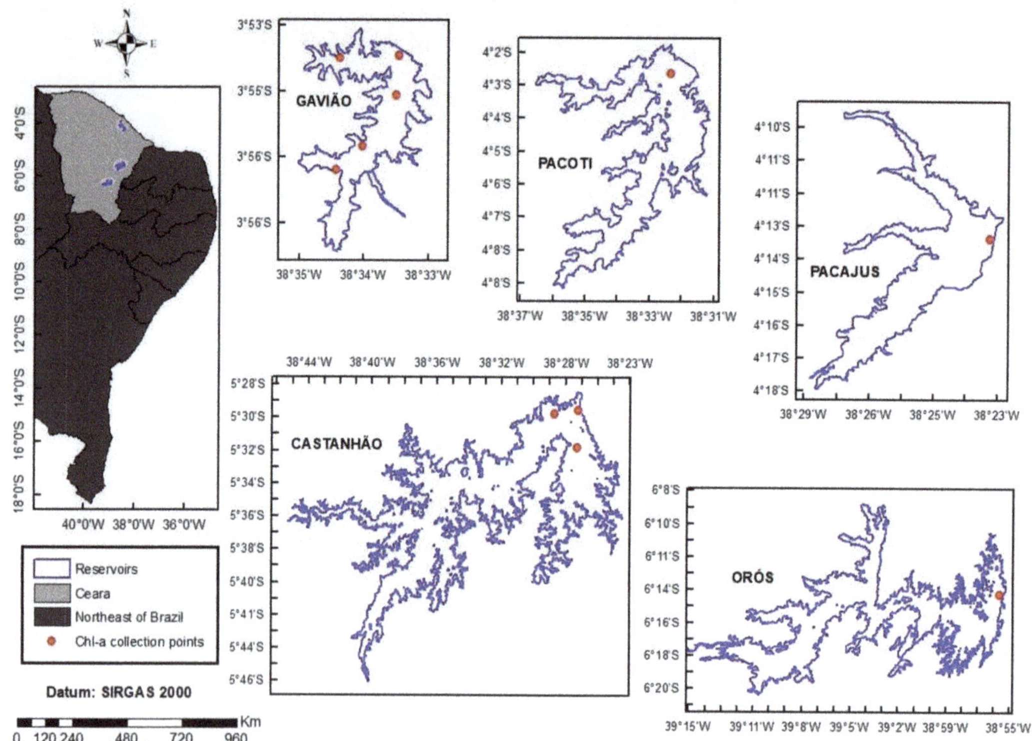

Figure 1. Location map of the five analyzed reservoirs and of the chlorophyll-a (chl-a) collection points collected and made available by Ceará Water Resources Management Company (COGERH).

Among the reservoirs studied, Castanhão is very important for Ceará due to its storage capacity of around 6 billion, 700 million cubic meters, which gives it a residence time of 400 days, on average, similar to a closed system. In this way, a large part of the sediments and dissolved elements that are transported and carried by the Jaguaribe river are retained, around 98%, according to [94]. Pollution sources, agricultural practices, average depth, conductivity, and TN/TP (nitrogen over phosphorus) ratio, associated with the low dilution capacity of the region and the static nature of the reservoir, explain its state of eutrophication. However, this is not enough to explain variations in the trophic state over time. According to [95], eutrophication and, consequently, the development of phytoplankton, is also linked to climatic and hydrological factors.

Thus, in this study, daily data were used, for the months of 2019, of wind (speed and direction) and solar radiation, available from the National Institute of Meteorology (INMET), on the portal https://portal.inmet.gov.br/ (accessed on 26 February 2020), and data on the volume of the Castanhão reservoir, available from the Ceará Meteorology and Water Resources Foundation (FUNCEME), through the website http://www.hidro.ce.gov.br/ (accessed on 31 January 2022), to analyze the interference of these factors in the distribution of phytoplankton on the surface of the Castanhão reservoir. The data provided by INMET were extracted from the conventional meteorological station Morada Nova, located about 42 km north of the Castanhão reservoir, taking into account the direction of the wind, which comes from the North and Northeast.

Estimated chl-a data for the Castanhão reservoir were also used, through the MODIS satellite, in a study developed within the scope of the MEG-HIBAM project, from the partnership between the National Water Agency (ANA) and the Institut de Recherche pour le Development (IRD). These data are available on the website (http://hidrosat.ana.gov.br/, accessed on 12 July 2021), by the brazilian National Water Agency (ANA).

2.3. Image Processing

Atmospheric corrections are widely used successfully in oceanic waters, as they assume the color of the ocean to be black (complete absorption of incident radiation) in the near-infrared (NIR) spectral region. However, this assumption does not apply to optically complex waters (type case 2), where chlorophyll, suspended sediments, and bottom reflectance lead to a non-zero brightness in the NIR [96,97]. Thus, extracting reflectance in turbid water through remote sensing (RS) by satellites has been hampered by the lack of an atmospheric correction that does not assume zero leaving irradiation of water in the NIR. This can lead to an overestimation of atmospheric radiation across the visible spectrum, with increasing severity at shorter wavelengths. This can result in significant errors in algorithms developed to estimate the concentration of chl-a [98].

Thus, for a broad assessment of the trophic state, some authors such as [46,92,99,100] concluded that only a simplified atmospheric correction procedure that normalizes the top of the atmosphere (TOA) signal for Rayleigh effects is possible, avoiding more complex atmospheric corrections for aerosols, given the large uncertainties associated with these corrections, as they are typically prone to errors in turbid water and high biomass. Moreover, improved processing time and simpler implementation for operational monitoring systems are additional advantages of using TOA-type data. Therefore, it was preferred to use Level 1C radiance images of the TOA type, from the Sentinel-2 satellite, to extract the chl-a reflectance.

2.4. Model Derivation

Remote sensing, when used to estimate concentrations of water constituents, is based on the relationship between the reflectance, $R(\lambda)$, and the inherent optical properties, namely, the total absorption and backscatter coefficients of water. To retrieve chl-a concentrations from spectral reflectance, it is necessary to isolate the chl-a absorption coefficient. For this, an empirical model was used, in which statistical relationships are established between observed concentrations of optically active constituents and some spectral index. As a spectral index, one can use reflectance at a single wavelength or an operation (ratio, difference or mixed operations) involving reflectance at two or more wavelengths.

In this work, an empirical three-band reflectance model developed by [101] was adopted. It has been widely used in the literature for inland waters by authors such as [45,102–110]. The three-band spectral index is written as

$$3BSI = \left[R^{-1}(\lambda_1) - R^{-1}(\lambda_2) \right] \cdot R(\lambda_3) \tag{1}$$

where 3BSI is three-band spectral index, and $R(\lambda_i)$ is the reflectance at wavelength λ_i, i = 1, 2 and 3. The choices of wavelengths (λ_i) are explained below. This index was chosen because it was created specifically for inland waters, has obtained excellent results in its applications, and its adaptation to Sentinel-2 bands showed good statistical correlations [102].

Ref. [101] developed an algorithm with the objective of isolating the chl-a absorption coefficient. For this, it is necessary to minimize the effect of the absorption of dissolved organic matter (CDOM) and the set of non-algal suspended organic and inorganic particles (Tripton or NAP or TSS), both commonly found in interior turbid waters, as well as to normalize the backscattering effect by all particular matter. This algorithm considers that $R(\lambda_1)$ must be a reflectance at a wavelength with a maximum chl-a absorption (in the range

660–690 nm), although it is still affected by CDOM, tripton, and backscatter absorption. $R(\lambda_2)$ is a reflectance at a near wavelength that is minimally sensitive to chl-a absorption (700–730 nm), but still affected by the other water constituents. If both $R(\lambda_1)$ and $R(\lambda_2)$ are similarly affected by the presence of tripton and CDOM, it is possible to remove the effects of absorption by these constituents by subtracting these wavelengths. However, this difference remains affected by the tripton particles backscattering. As a solution, a third spectral band $R(\lambda_3)$ in the 740–760 nm spectral range was introduced, where the reflectance is minimally affected by the absorption of water constituents (chl-a, CDOM and tripton) and is basically controlled by pure water absorption and tripton backscattering. Finally, the division of $[R^{-1}(\lambda_1) - R^{-1}(\lambda_2)]$ by $R(\lambda_3)$ will remove the effects of tripton backscattering variability, turning this 3-band combination as a robust estimator of the chl-a absorption coefficient [102,111,112]. For Sentinel-2 data, $R(\lambda_1)$ is equivalent to band 4 (665 nm); $R(\lambda_2)$ matches band 5 (705 nm); and $R(\lambda_3)$ corresponds to band 6 (740 nm).

To extract the reflectance of Sentinel-2 bands 4, 5 and 6, an area of 100 m^2 free from clouds was selected in the image, around the collection points of the chl-a samples, and the average reflectance of the pixels of the selected area was extracted. Images with passage dates close to the collection dates were selected, most of them ranging from 1 to 3 days, with some images with an interval up to a week. According to [113], images with a 3-day interval of collection dates are as good as exact match, a time difference of 1 week reduces correlation, but a month time difference is too much to even estimate a parameter as relatively stable as CDOM. On the other hand, [114] found that even larger time differences are still reliable.

To calibrate and validate the three-band model, a relationship was obtained by using statistical regression, considering the coefficient of determination (R^2) and the square root mean square error (RMSE) as metrics to identify the best fit between the spectral index generated from Equation (1) and the observed chl-a.

The regression model was calibrated using the Bootstrap method developed by [115], which consists of a random resampling procedure with replacement and each calculated estimate. Based on the replications, it is possible to determine confidence intervals (e.g., 95%) and hypothesis test on the generated estimators [116].

2.5. Hydroclimatic Data

Data on wind direction and wind speed, solar radiation and reservoir volume were used to generate a conceptual model to analyze the behavior of chl-a in surface waters of the Castanhão reservoir. This reservoir was chosen because, among the five reservoirs studied in this work, it presented a continuous monthly image sequence to estimate the concentration of chl-a using Sentinel-2 without cloud interference that could harm the understanding of the eutrophication cycle during a whole year, that is, for 2019.

For the wind data, a chart of the wind rose was generated, with the average of the values of the 4 days preceding the date of passage of the satellite, and at times of greatest gust (17:00, 18:00 and 19:00 LT; LT = local time = UTC-3:00). These peaks can be important because they control water column vertical mixing and resuspension of bottom sediment.

A graph was generated with the values of solar radiation in the days and times of the passage of the Sentinel-2 satellite (10:00 LT), to analyze the behavior of the concentration of chl-a against the possible thermal stratification of the column of water, generated by solar radiation incidence.

Daily variation of the water volume storage in the Castanhão reservoir was also retrieved, for the whole year 2019, to verify how water storage can interfere in the dilution or concentration of chl-a levels.

3. Results and Discussion

3.1. Chlorophyll-a Algorithm Definition and Performance

Table 2 shows the values of the chl-a samples collected in the 5 reservoirs considered in this work, from 2015 to 2018, used for calibration of the 3-band spectral model (Equation (1)). The datasets encompassed varying concentrations conditions. There was a strong variability in chl-a concentration from one reservoir to another, ranging from 0.2 to close to 90 µg L^{-1}, and strong seasonality was also detected. The Pacajus reservoir showed a highly differentiated eutrophication pattern with low to intermediate concentration of chl-a [0.2–15.6 µg L^{-1}].

Table 2. Descriptive statistics (average combined with standard deviation—sd) of the chlorophyll-a (chl-a) and turbidity (Turb) parameters collected in the five reservoirs considered in this study. The chl-a data were used to calibrate the reflectance model to estimate chl-a from the Sentinel-2 images.

Reservoir	Samples	Time Period of Collection	[Range] Measured chl-a (µg.L^{-1})	Average (sd) chl-a	Median chl-a	[Range] Measured Turb (NTU)	Average (sd) Turb	Median Turb
Gavião	51	4 November 2015–3 July 2018	8.4–79.6	51.8 (±15.6)	53.4	6.59–13.5	9.5 (±2.1)	8.7
Pacoti	34	10 November 2015–10 July 2018	7.9–89.2	56.9 (±21.3)	66.2	3.95–9.83	7.6 (±1.6)	7.3
Pacajús	33	11 November 2015–5 July 2018	0.2–15.6	7.2 (±3.5)	6.8	4.27–25.5	11.7 (±6.8)	11.7
Castanhão	34	2 December 2015–11 July 2018	12.8–56.1	38.3 (±15.1)	42.3	9.82–36.7	21.3 (±12.4)	17.6
Orós	9	30 November 2015–21 February 2018	26.0–66.6	41.5 (±16.2)	33.6	7.21–20	13.1 (±4.8)	13.3

The relationship between the chl-a concentration in the calibration dataset (Table 2) and the 3-band spectral index (Equation (1)) considering the spectral bands defined for Sentinel-2, $R(\lambda_1)$ = 665 nm; $R(\lambda_2)$ = 705 nm; and $R(\lambda_3)$ = 740 nm, obtained fine results with a a relative RMSE of 28% and R^2 = 0.80. The calibrated chl-a model is:

$$\text{Chl-a}\left(\mu g\, L^{-1}\right) = 279.95(3\text{BSI}) + 38.06 \tag{2}$$

From this chl-a retrieval algorithm, one can estimate the concentration of chl-a using Sentinel-2 Level 1C images.

Figure 2 shows the scatter plot of the values generated from the 3BSI for Sentinel-2 Level 1C images against the measured in situ calibration dataset.

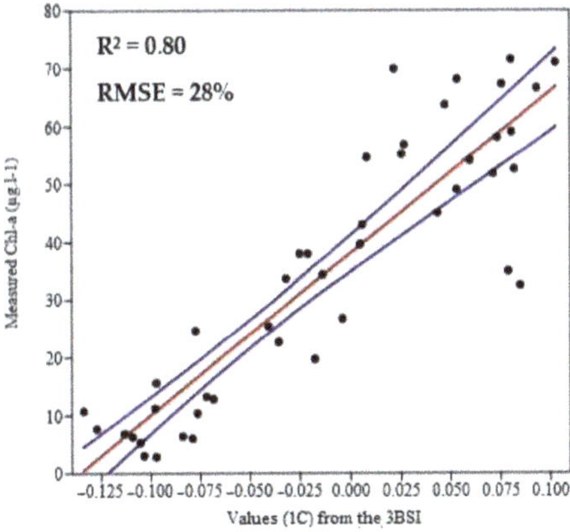

Figure 2. Relationship between the measured in situ chl-a, for the samples considered in the calibration dataset, and the values generated by the three-band spectral index (3BSI) for the Sentinel-2 Level 1-C image datasets. The red line represents the regression model calculated using the bootstrap resampling technique. Blue lines indicate two RMSE estimates of chl-a.

The independent validation dataset was collected from August 2018 to December 2019. The estimated chl-a concentrations were compared to the observed chl-a concentrations (Figure 3). The correlation between the validation data, considering the 5 reservoirs, presented a coefficient of determination $R^2 = 0.77$. Overall, the retrieved chl-a concentration followed well the temporal behavior of the field data except for the Pacajus reservoir, which presented the lowest chl-a concentration levels observed.

Figure 3. Comparison between the chl-a estimated by the three-band algorithm (Equation (2)) and the chl-a measured for the Castanhão, Orós, Pacajús, Pacoti and Gavião reservoirs, referring to the validation data covering the period from August 2018 to December 2019.

The difference in the performance of the algorithm for estimating low concentrations chl-a is probably related to the use of reflectance at λ_3 in the three-band model because λ_3 is practically insensitive to the absorption of water constituents (chl-a, CDOM and tripton), but is susceptible to suspended particle scattering (TSS). For low concentrations of chl-a, the correlation between backscatter and TSS increases considerably for longer wavelengths (λ_3). This indicates an increased effect of the concentration of suspended particles. Thus, backscattering effects may not be fully removed using (λ_3) and may introduce uncertainties in the concentrations chl-a estimation if the three-band model is applied. This can cause significant changes to the model output. Therefore, although the model has been calibrated for a wide range of biophysical and optical water quality parameters, the accuracy of the retrieval for a low chl-a concentration level (i.e., <10 µg L^{-1}) may be considered with caution.

Figure 4 shows that chl-a concentration and turbidity (the latter is a proxy of scattering by phytoplankton and inorganic suspended matter) are not correlated ($R^2 < 0.09$), denoting that that reservoir water presents complex behavior belonging clearly to case 2 waters.

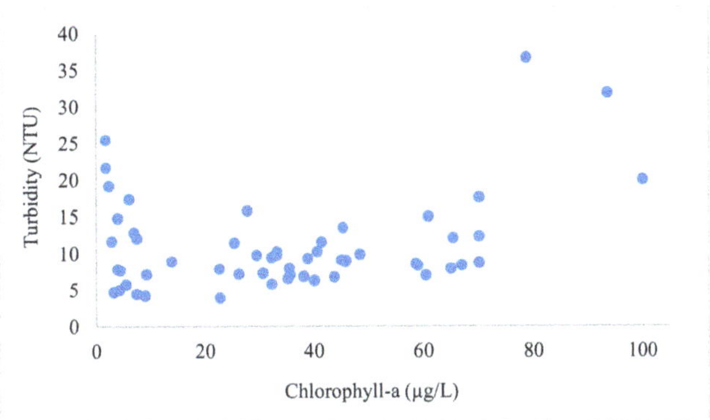

Figure 4. Chlorophyll-a concentration vs. turbidity of samples collected in situ for validation data, in all reservoirs of this study. Determination coefficient for linear relationship (R^2) < 0.09. Chl-a and turbidity are practically independent, indicating these reservoirs belong to case 2 waters.

3.2. Chl-a Retrieval Comparison between Sentinel-2 and MODIS Sensors

We compared the chl-a time series retrieved with Sentinel-2 MSI sensor with an independent dataset based on MODIS (Terrra and Aqua satellites) available through the Hidrosat database (http://hidrosat.ana.gov.br/, accessed on 31 January 2022), developed by the Brazilian Water Agency (ANA) [117]. MODIS data offer higher sampling frequency than Sentinel 2 data with near daily acquisition rate. The Hidrosat database was calibrated using other field data and can be considered a totally independent dataset. Figure 5A presents the comparison between MODIS-derived and MSI-derived chl-a time series for 2019 over the Castanhão reservoir jointly to the field measurement already presented above. It can be seen that all datasets show close agreement, but some differences can be noticed during January, April, and October. It is worthwhile to note that the chl-a values assessed using Sentinel-2 are systematically closer to the field chl-adata, when compared to the values estimated by MODIS. The coarse resolution of MODIS data (i.e., 500 m) in relation to the 20-m Sentinel-2 data used for this study may partially explain this lower accuracy of the MODIS-derived estimates. Sentinel-2 images allow for greater refinement of the characteristics of the water body, reinforcing the gain in monitoring capacity and understanding of water quality processes in small and large reservoirs.

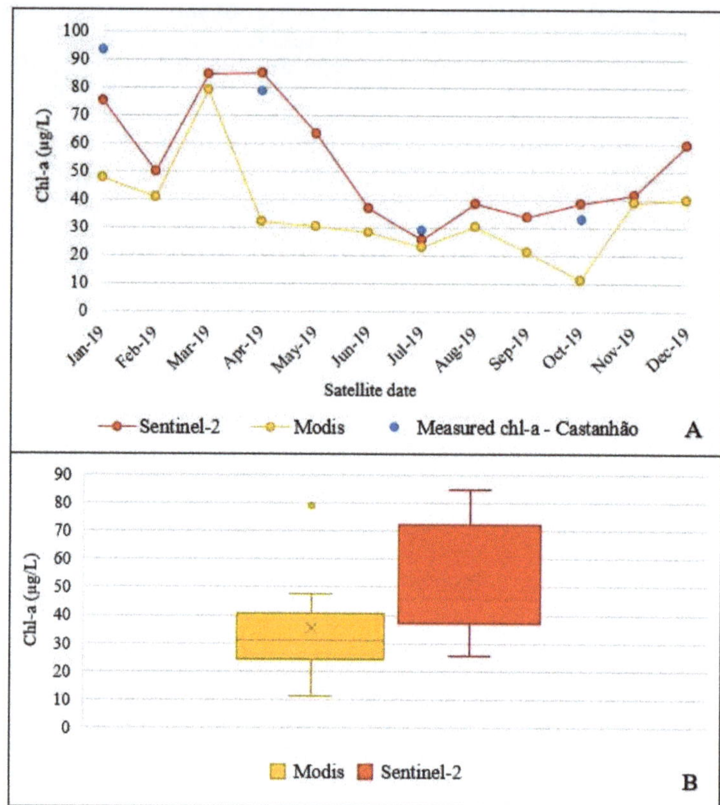

Figure 5. Comparison between the chl-a estimated by the Sentinel-2 satellite and by MODIS satellite, for the Castanhão reservoir, for the months of 2019. (**A**) shows the temporal variation jointly with the quarterly field measurements. (**B**) displays the boxplot with the distribution of chl-a values estimated respectively by Sentinel-2 and MODIS.

Figure 5B shows the distribution of both MODIS and Sentinel-2 derived chl-a data in a boxplot format. It is observed that the chl-a estimated by MODIS shows lower concentration values when compared to Sentinel-2 data. It is noteworthy that, in 2019, the Castanhão had its trophic status varying between hypereutrophic and eutrophic, as shown in Figure 6, which suggests a high chl-a concentration in that graph. In addition, Sentinel-2 has a greater range of concentration values throughout the year. That is more consistent with the field data collected in the reservoir, which showed greater chl-a concentration in the first months of the year, and a drop of concentration as the year goes by.

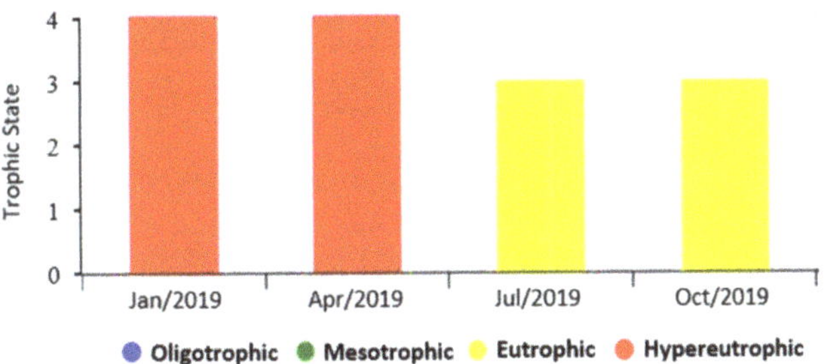

Figure 6. Trophic state of the Castanhão reservoir retrieved from quarterly measurements for the year 2019. Adapted from COGERH and made available at the electronic address http://www.hidro.ce.gov.br/hidro-ce-zend/acude/eutrofizacao (accessed on 17 January 2022). In 2019, no oligotrophic and mesotrophic conditions were found in the Castanhão as represented by the Trophic State Index (TSI): 1—Oligotrophic ($24 \leq$ TSI ≤ 44); 2—Mesotrophic ($44 <$ TSI ≤ 54); 3—Eutrophic ($54 <$ TSI ≤ 74); 4—Hypereutrophic (TSI > 74) [3].

3.3. Analysis of the Trophic State Evolution in the Castanhão Reservoir

The trophic state graph, provided by COGERH, is presented (Figure 6) to compare and confirm the water quality results presented in the chl-a concentration maps (Figure 7) for the Castanhão reservoir during 2019.

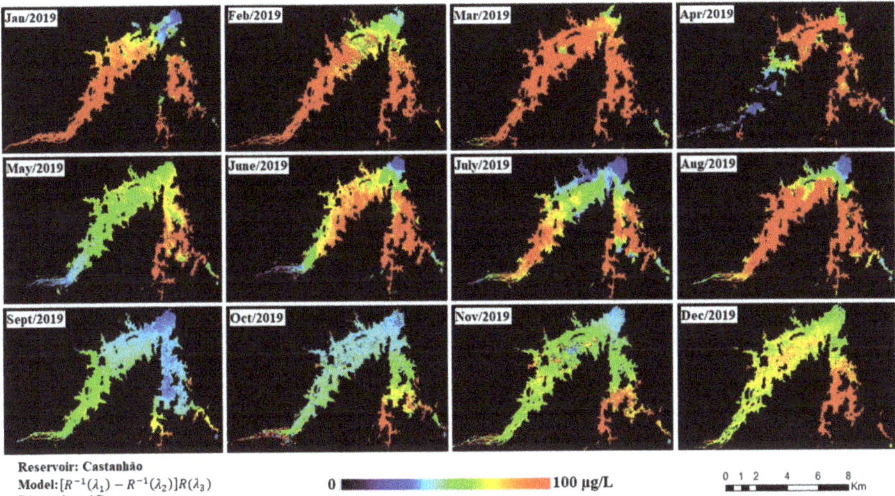

Figure 7. Monthly map of the concentration of chl-a in the Castanhão reservoir, during 2019, produced from the 1C level TOA products and the three-band model (Equation (2)).

The trophic state index used by COGERH to qualify the degree of trophy of the Ceará reservoirs is based on the proposal of [20] adapted by [21] to lentic environments of tropical climate. This index is based on chlorophyll-a and total phosphorus data. In other words, the greater the concentration of these parameters, the worse the water quality of aquatic environments will be.

For [21], in general, it is accepted that the trophic level of a water body can be evaluated from the chl-a concentration level alone.

Figure 7 shows the time series of the chl-a maps, generated for the Castanhão reservoir for every month of 2019 and prepared from Level 1C products, using Equation (2).

The chl-a concentration maps are consistent with the trophic state graph based on the in situ measurements (Figure 6) provided by COGERH. Figure 7 shows that between January and April the concentration of chl-a is high (i.e., >80 µg L^{-1}) in a substantial part of the water body. In the following months, there is a decreasing chl-a concentration trend, indicating an improvement in the reservoir's water quality, as shown in the trophic state graph (Figure 6), in which the Castanhão becomes eutrophic in the second half of the year. This can be explained by 1. Inflow: most of the inflow in 2019 concentrated in the months of April and May; 2. Reservoir volume: reduction in reservoir volume during the second semester since there is no inflow; 3. Wind speed: stronger winds starting in September, which promotes a better mixing of the reservoir waters. (As will be shown below.)

It is worth noting in the Sentinel-2 derived chl-a concentration maps (Figure 7) that the minimum chl-a concentration appears in the vicinity of the Castanhão dam, as shown in Figure 8, matching the deepest portion of the reservoir.

Figure 8. Hydrological flow in the Castanhão reservoir and location of the reservoir dam.

Figures 8–10 show the variability of some hydroclimatic variables, namely reservoir volume, wind speed and direction, and solar radiation, respectively, to detect any relation with satellite-retrieved chl-a concentrations in the Castanhão reservoir, as shown in Figure 7. Temperature was not included in the analysis, as it varies little throughout the year in the study region, since it is close to the equator line, as well as relative humidity, which is mainly related to evaporation, and should have no effect on algal biomass development.

In the analyzed time scale, from a hydrodynamic point of view, the concentration of chl-a increases with the beginning of the rainy season in Ceará (March to May). This is probably because there are still no appreciable river water inputs from the upstream catchment, but there is a high input of nutrients by leaching and runoff from the baresoils areas around the dam. When the river starts to flow, a dilution effect is observed, with an increase in the volume of the reservoir and a simultaneous decrease in the concentration of chlorophyll. Thus, as shown in the estimated chl-a maps for the year 2019 (Figure 7), we can see a biomass dilution effect upstream of the reservoir (Figure 8), from April onwards, with the arrival of Jaguaribe River waters. In May there is a dilution of the biomass in most of the water mirror, due to the continuous increase in the volume of water in the reservoir, as a result of the rainy season in Ceará State. In June, as shown in Figure 9, the water volume starts to decrease as a result of the end of the rainy season, evaporation and multiple water uses. This decrease in the water volume matches an increase in the concentration of chl-a at the water surface. The analysis of the sequence of Sentinel-2-derived maps make it possible to monitor the rapid changes induced by the hydrological and meteorological conditions, confirming the strong control of these factors on the phytoplankton biomass in the Castanhão reservoir.

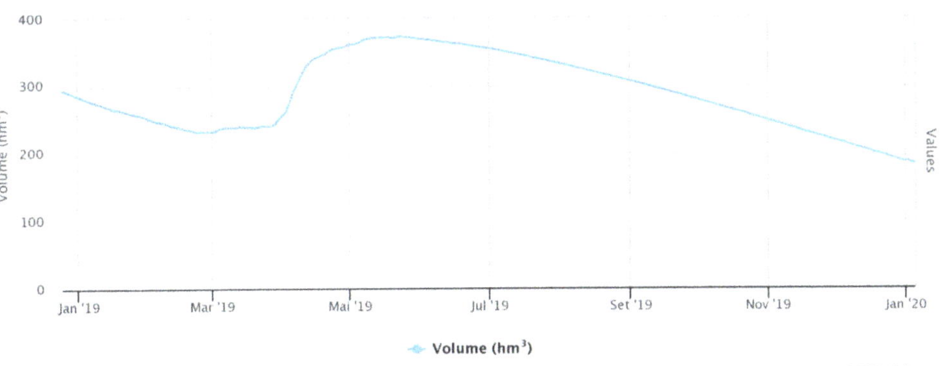

Figure 9. Variation in the daily water volume in the Castanhão reservoir during 2019. Release discharge: 7476 m^3/s. Available by COGERH at the website http://www.hidro.ce.gov.br/hidro-ce-zend/acude/nivel-diario (accessed on 17 January 2022).

Figure 10 displays both wind intensity and mean direction for each month during 2019. It can be seen that there is a net increase in wind intensity from September to December. The analysis of the chl-a maps (Figure 7) shows a strong decrease in the values of chl-a in the entire surface of the Castanhão reservoir for the same period. This overall chl-a decrease at the lake surface may be directly related to both the wind speed increase and predominantly northeast direction from September to December 2019 (Figure 10) that generates a longitudinal turbulence in the reservoir mirror. It is worth highlighting the chl-a concentration in the month of August (Figure 7), as this month presents a more accentuated concentration compared to previous and following months. This concentration maximum may correspond to a combination of unique conditions with decreasing water storage at the beginning of the dry season, which corresponds to increasing water residence time and mild wind conditions.

From a meteorological point of view, the chlorophyll peak can occur simultaneously in the period of low wind speed. Wind can be an inducer of water mass movements, both horizontaly and spatially, and most phytoplankton species will not be able to develop. Furthermore, the action of the wind can cause sediment resuspension from the lake bottom, especially in shallow conditions when the reservoir faces low storage rate. This generates a lower availability of the light in the water column and may lead to inhibition of phytoplanktonic growth, resulting in lower chl-a concentrations values [77,79].

In this study, the radiation variability over the year did not seem to have a clear impact on chl-a variability at the surface level (Figure 11), since Castanhão is located close to the Equator and the solar irradiance is quite constant. Looking at finer time scales, both wind speed and radiation are closely related to the diurnal cycle of water quality: stratification of the reservoir during the peak of radiation and mixing of its water during the peak of the wind speed [118]. However, the study of the diurnal cycles remains out of the reach of the satellite date due to their lower time-revisit frequency.

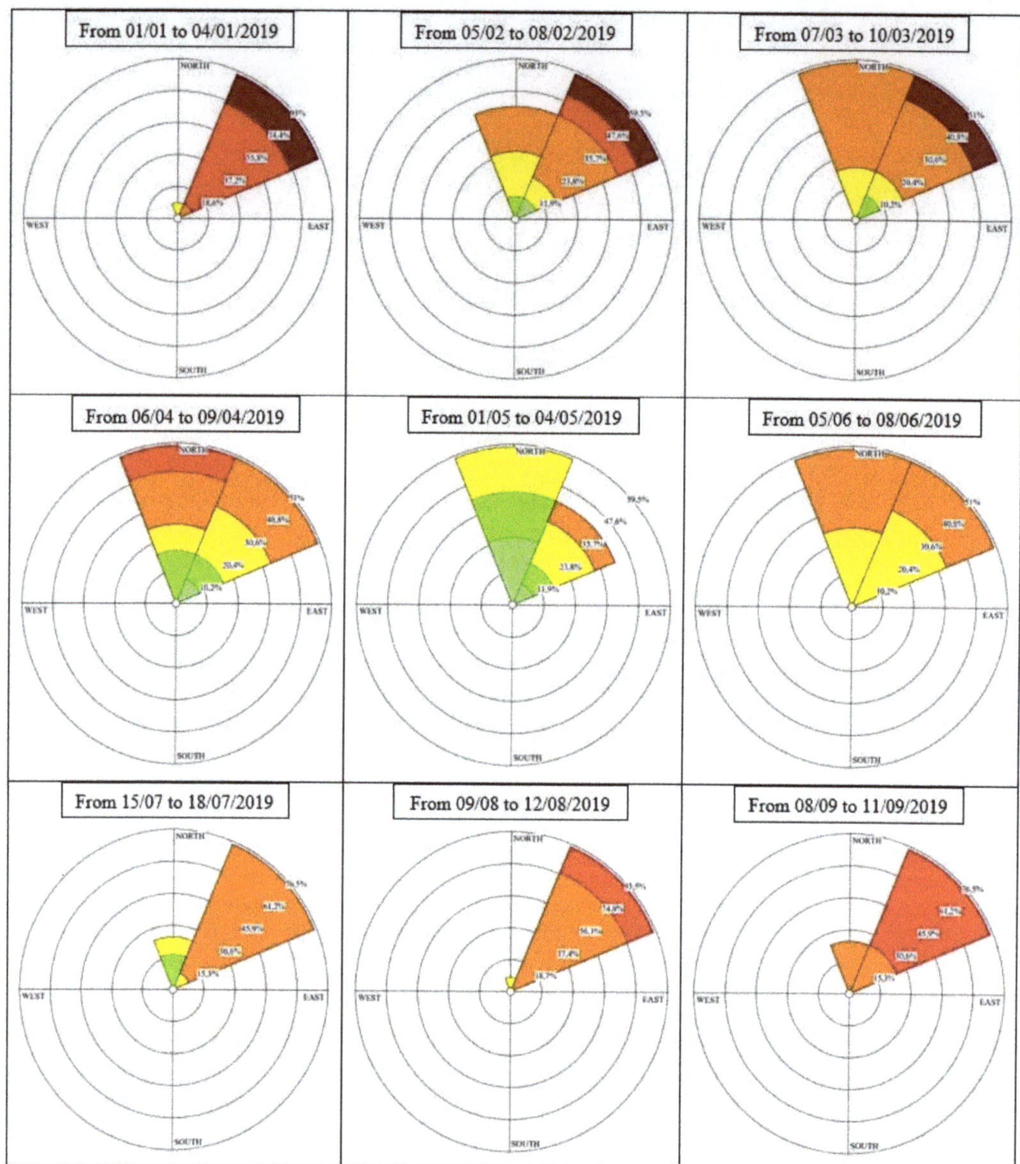

Figure 10. *Cont.*

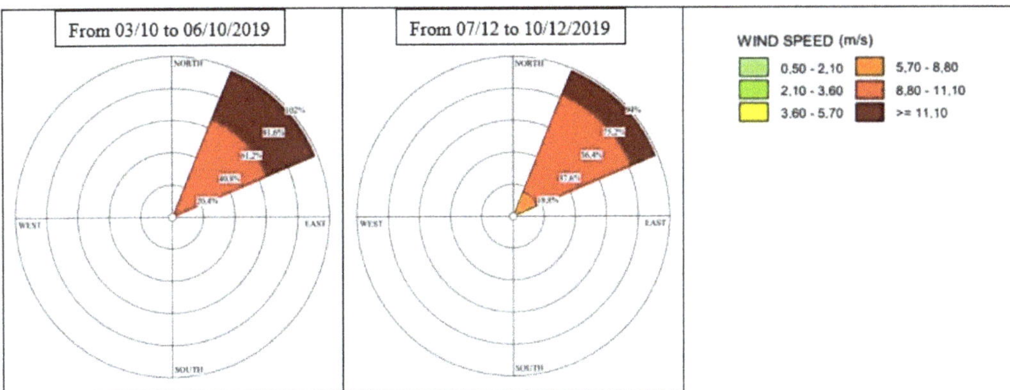

Figure 10. Compass rose containing wind speed (gust) and direction over the months of 2019. It was not possible to generate the graph for the month of November due to lack of data. The data were made available by INMET, on the portal https://portal.inmet.gov.br/ (accessed on 26 February 2020), and extracted from the conventional meteorological station Morada Nova/CE.

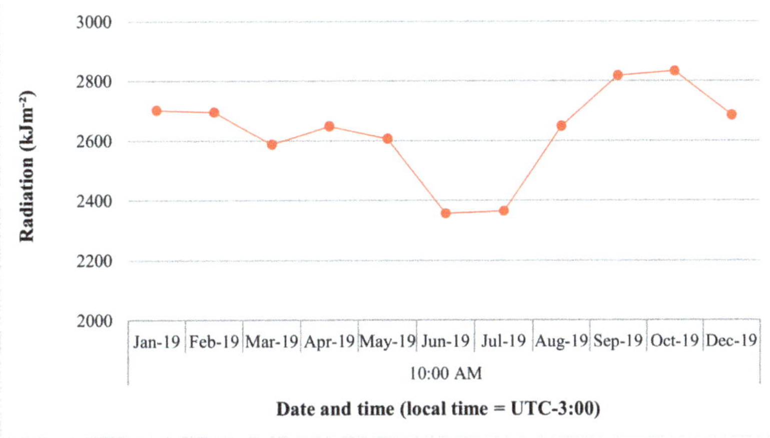

Figure 11. Graph of solar radiation extracted on the days and times of the Sentinel-2 satellite passage, referring to the data used to generate the annual map of chl-a concentration in Castanhão reservoir (Figure 7). November radiation was not presented due to lack of data. The data were made available by INMET, at https://portal.inmet.gov.br/ (accessed on 26 February 2020), and extracted from the Morada Nova/CE conventional weather station.

4. Concluding Remarks

This work showed that the use of Sentinel-2 MSI remote-sensing data to monitor the quality of complex case 2 waters in semiarid environments has strong potential.

The defined 3-band spectral model for Sentinel-2 Level 1C, where $R(\lambda_1) = 665$ nm; $R(\lambda_2) = 705$ nm; and $R(\lambda_3) = 740$ nm presented acceptable statistical performance to estimate the concentration of chl-a, with statistical regression parameters reaching a relative RMSE of 28% and $R^2 = 0.80$.

The hydroclimatic data (wind direction and speed and reservoir volume) used to analyze the behavior of chl-a in the surface waters of the Castanhão reservoir was important, making it possible to interpret the temporal and spatial distribution of satellite-retrieved chl-a maps over the entire surface of the reservoir.

Thus, this study contributes to the development of continuous monitoring of water quality, based on remote analysis by satellite, in small and large reservoirs, a strategic resource to promote greater water security in the Brazilian semiarid region, which is heavily impacted by strong interannual rainfall variability and an overall scarcity of rainfall.

Author Contributions: Conceptualization, T.R.B.T.A.; Data curation, M.U.G.B.; Funding acquisition, E.S.P.R.M.; Supervision, J.-M.M.; Writing—review & editing, E.P.S. All authors have read and agreed to the published version of the manuscript.

Funding: The present work was carried out with the support of the Coordination for the Improvement of Higher Education Personnel—Brazil (CAPES)—Financing Code 001 in the context of the projects CNPq/490688/2010-1 and AQUASENSE/FINEP.

Data Availability Statement: Sentinel-2 satellite images are available on the website (https://scihub.copernicus.eu/, accessed on 14 January 2022); The estimated chl-a data for the Castanhão reservoir, through the MODIS satellite, are available on the website (http://hidrosat.ana.gov.br/, accessed on 14 January 2022); Data on chl-a collected in-situ are available at the Ceará Hydrological Portal (http://www.hidro.ce.gov.br/hidro-ce-zend/acude/eutrofizacao, accessed on 14 January 2022); Data on the volume of water in the Castanhão reservoir are also available on the Ceará Hydrological Portal (http://www.hidro.ce.gov.br/hidro-ce-zend/acude/nivel-diario, accessed on 14 January 2022); Brazilian National Institute of Meteorology data on wind direction and speed, and solar radiation are available at "https://portal.inmet.gov.br", (accessed on 14 January 2022).

Acknowledgments: The first author is grateful to FUNCEME and CAPES for supporting her study.

Conflicts of Interest: The authors declare no conflict of interest.

References

1. Teixeira, M.N. O Sertão Semiárido. Uma Relação de Sociedade e Natureza Numa Dinâmica de Organização Social Do Espaço. *Soc. E Estado* **2016**, *31*, 769–780. [CrossRef]
2. de Araújo, J.C.; Güntner, A.; Bronstert, A. Loss of Reservoir Volume by Sediment Deposition and Its Impact on Water Availability in Semiarid Brazil. *Hydrol. Sci. J.* **2006**, *51*, 157–170. [CrossRef]
3. de Araújo, J.C.; Döll, P.; Güntner, A.; Krol, M.; Abreu, C.B.R.; Hauschild, M.; Mendiondo, E.M. Water Scarcity under Scenarios for Global Climate Change and Regional Development in Semiarid Northeastern Brazil. *Water Int.* **2004**, *29*, 209–220. [CrossRef]
4. Bucci, M.H.; De Oliveira, L.F.C. Índices de Qualidade Da Água e de Estado Trófico Na Represa Dr. João Penido (Juiz de Fora, MG). *Ambiente E Agua—Interdiscip. J. Appl. Sci.* **2014**, *9*, 243–251. [CrossRef]
5. de Lucena Barbosa, J.E.; Medeiros, E.S.F.; Brasil, J.; da Silva Cordeiro, R.; Crispim, M.C.B.; da Silva, G.H.G. Aquatic Systems in Semi-Arid Brazil: Limnology and Management. *Acta Limnol. Bras.* **2012**, *24*, 103–118. [CrossRef]
6. Pimenta, A.M.; Albertoni, E.F.; Palma-Silva, C. Characterization of Water Quality in a Small Hydropower Plant Reservoir in Southern Brazil. *Lakes Reserv. Res. Manag.* **2012**, *17*, 243–251. [CrossRef]
7. Padisák, J.; Reynolds, C.S. Selection of Phytoplankton Associations in Lake Balaton, Hungary, in Response to Eutrophication and Restoration Measures, with Special Reference to the Cyanoprokaryotes. *Hydrobiologia* **1998**, *384*, 41–53. [CrossRef]
8. Codd, G.A. Cyanobacterial Toxins, the Perception of Water Quality, and the Prioritisation of Eutrophication Control. *Ecol. Eng.* **2000**, *16*, 51–60. [CrossRef]
9. Christoffersen, K.; Kaas, H. Toxic Cyanobacteria in Water. A Guide to Their Public Health Consequences, Monitoring, and Management. *Limnol. Oceanogr.* **2000**, *45*, 1212. [CrossRef]
10. Sivonen, K. Cyanobacterial Toxins. In *Encyclopedia of Microbiology*; Elsevier: Amsterdam, The Netherlands, 2009; pp. 290–307.
11. CARMICHAEL, W. The Cyanotoxins. In *Incorporating in Plant Pathology—Classic Papers*; Elsevier: Amsterdam, The Netherlands, 1997; Volume 27, pp. 211–256. ISBN 0-12-005927-4.
12. Codd, G.A.; Ward, C.J.; Bell, S.G. Cyanobacterial Toxins: Occurrence, Modes of Action, Health Effects and Exposure Routes. *Arch. Toxicol. Suppl. Arch. Für Toxikol. Suppl.* **1997**, *19*, 399–410. [CrossRef]
13. Hunter, P.R. Cyanobacterial Toxins and Human Health. *J. Appl. Microbiol. Symp. Suppl.* **1998**, *84*, 35–40. [CrossRef]
14. Falconer, I.R. *Health Problems from Exposure to Cyanobacteria and Proposed Safety Guidelines for Drinking and Recreational Water*; The Royal Society of Chemistry: London, UK, 1994; p. 10.
15. Kutser, T. Quantitative Detection of Chlorophyll in Cyanobacterial Blooms by Satellite Remote Sensing. *Limnol. Oceanogr.* **2004**, *49*, 2179–2189. [CrossRef]
16. Richardson, L.L. Remote Sensing of Algal Bloom Dynamics. *BioScience* **1996**, *46*, 492–501. [CrossRef]
17. Ross, M.R.V.; Topp, S.N.; Appling, A.P.; Yang, X.; Kuhn, C.; Butman, D.; Simard, M.; Pavelsky, T.M. AquaSat: A Data Set to Enable Remote Sensing of Water Quality for Inland Waters. *Water Resour. Res.* **2019**, *55*, 10012–10025. [CrossRef]

18. Ha, N.T.T.; Koike, K.; Nhuan, M.T. Improved Accuracy of Chlorophyll-a Concentration Estimates from MODIS Imagery Using a Two-Band Ratio Algorithm and Geostatistics: As Applied to the Monitoring of Eutrophication Processes over Tien Yen Bay (Northern Vietnam). *Remote Sens.* **2013**, *6*, 421–442. [CrossRef]

19. Markogianni, V.; Kalivas, D.; Petropoulos, G.P.; Dimitriou, E. Estimating Chlorophyll-a of Inland Water Bodies in Greece Based on Landsat Data. *Remote Sens.* **2020**, *12*, 2087. [CrossRef]

20. Carlson, R.E. A Trophic State Index for Lakes. *Limnol. Oceanogr.* **1977**, *22*, 361–369. [CrossRef]

21. Toledo Júnior, A.P.; Talarico, M.; Chinez, S.J.; Agudo, E.G.A. A Aplicação de Modelos Simplificados Para a Avaliação de Processo Da Eutrofização Em Lagos e Reservatórios Tropicais. In Proceedings of the Congresso Brasileiro de Engenharia Sanitária, Camboriú, Brazil, 20–25 November 1983.

22. Halstvedt, C.B.; Rohrlack, T.; Andersen, T.; Skulberg, O.; Edvardsen, B. Seasonal Dynamics and Depth Distribution of Planktothrix Spp. in Lake Steinsfjorden (Norway) Related to Environmental Factors. *J. Plankton Res.* **2007**, *29*, 471–482. [CrossRef]

23. Reynolds, C.S.; Oliver, R.L.; Walsby, A.E. Cyanobacterial Dominance: The Role of Buoyancy Regulation in Dynamic Lake Environments. *N. Z. J. Mar. Freshw. Res.* **1987**, *21*, 379–390. [CrossRef]

24. Walsby, A.E.; Hayes, P.K.; Boje, R.; Stal, L.J. The Selective Advantage of Buoyancy Provided by Gas Vesicles for Planktonic Cyanobacteria in the Baltic Sea. *New Phytol.* **1997**, *136*, 407–417. [CrossRef]

25. Pitois, S.; Jackson, M.H.; Wood, B.J.B. Problems Associated with the Presence of Cyanobacteria in Recreational and Drinking Waters. *Int. J. Environ. Health Res.* **2000**, *10*, 203–218. [CrossRef]

26. Backer, L.C. Cyanobacterial Harmful Algal Blooms (CyanoHABs): Developing a Public Health Response. *Lake Reserv. Manag.* **2002**, *18*, 20–31. [CrossRef]

27. Johansen, R.; Beck, R.; Nowosad, J.; Nietch, C.; Xu, M.; Shu, S.; Yang, B.; Liu, H.; Emery, E.; Reif, M.; et al. Evaluating the Portability of Satellite Derived Chlorophyll-a Algorithms for Temperate Inland Lakes Using Airborne Hyperspectral Imagery and Dense Surface Observations. *Harmful Algae* **2018**, *76*, 35–46. [CrossRef] [PubMed]

28. Neil, C.; Spyrakos, E.; Hunter, P.D.; Tyler, A.N. A Global Approach for Chlorophyll-a Retrieval across Optically Complex Inland Waters Based on Optical Water Types. *Remote Sens. Environ.* **2019**, *229*, 159–178. [CrossRef]

29. Pahlevan, N.; Lee, Z.; Wei, J.; Schaaf, C.B.; Schott, J.R.; Berk, A. On-Orbit Radiometric Characterization of OLI (Landsat-8) for Applications in Aquatic Remote Sensing. *Remote Sens. Environ.* **2014**, *154*, 272–284. [CrossRef]

30. Tavares, M.H.; Lins, R.C.; Harmel, T.; Fragoso, C.R.; Martínez, J.M.; Motta-Marques, D. Atmospheric and Sunglint Correction for Retrieving Chlorophyll-a in a Productive Tropical Estuarine-Lagoon System Using Sentinel-2 MSI Imagery. *ISPRS J. Photogramm. Remote Sens.* **2021**, *174*, 215–236. [CrossRef]

31. Dörnhöfer, K.; Oppelt, N. Remote Sensing for Lake Research and Monitoring—Recent Advances. *Ecol. Indic.* **2016**, *64*, 105–122. [CrossRef]

32. Hedley, J.; Roelfsema, C.; Koetz, B.; Phinn, S. Capability of the Sentinel 2 Mission for Tropical Coral Reef Mapping and Coral Bleaching Detection. *Remote Sens. Environ.* **2012**, *120*, 145–155. [CrossRef]

33. Bresciani, M.; Cazzaniga, I.; Austoni, M.; Sforzi, T.; Buzzi, F.; Morabito, G.; Giardino, C. Mapping Phytoplankton Blooms in Deep Subalpine Lakes from Sentinel-2A and Landsat-8. *Hydrobiologia* **2018**, *824*, 197–214. [CrossRef]

34. Dekker, A.G.; Brando, V.E.; Anstee, J.M.; Pinnel, N.; Kutser, T.; Hoogenboom, E.J.; Peters, S.; Pasterkamp, R.; Vos, R.; Olbert, C.; et al. Imaging Spectrometry of Water. In *Imaging Spectrometry*; Springer: Berlin/Heidelberg, Germany, 2002; pp. 307–359. [CrossRef]

35. Gholizadeh, M.H.; Melesse, A.M.; Reddi, L. A Comprehensive Review on Water Quality Parameters Estimation Using Remote Sensing Techniques. *Sensors* **2016**, *16*, 1298. [CrossRef]

36. Odermatt, D.; Gitelson, A.; Brando, V.E.; Schaepman, M. Review of Constituent Retrieval in Optically Deep and Complex Waters from Satellite Imagery. *Remote Sens. Environ.* **2012**, *118*, 116–126. [CrossRef]

37. Kupssinskü, L.S.; Guimarães, T.T.; De Souza, E.M.; Zanotta, D.C.; Veronez, M.R.; Gonzaga, L.; Mauad, F.F. A Method for Chlorophyll-a and Suspended Solids Prediction through Remote Sensing and Machine Learning. *Sensors* **2020**, *20*, 2125. [CrossRef]

38. Sun, D.; Li, Y.; Wang, Q.; Le, C.; Lv, H.; Huang, C.; Gong, S. Specific Inherent Optical Quantities of Complex Turbid Inland Waters, from the Perspective of Water Classification. *Photochem. Photobiol. Sci.* **2012**, *11*, 1299–1312. [CrossRef] [PubMed]

39. Ma, J.; Song, K.; Wen, Z.; Zhao, Y.; Shang, Y.; Fang, C.; Du, J. Spatial Distribution of Diffuse Attenuation of Photosynthetic Active Radiation and Its Main Regulating Factors in Inland Waters of Northeast China. *Remote Sens.* **2016**, *8*, 964. [CrossRef]

40. Lisboa, F.; Brotas, V.; Santos, F.D.; Kuikka, S.; Kaikkonen, L.; Maeda, E.E. Spatial Variability and Detection Levels for Chlorophyll-a Estimates in High Latitude Lakes Using Landsat Imagery. *Remote Sens.* **2020**, *12*, 2898. [CrossRef]

41. Soomets, T.; Uudeberg, K.; Kangro, K.; Jakovels, D.; Brauns, A.; Toming, K.; Zagars, M.; Kutser, T. Spatio-Temporal Variability of Phytoplankton Primary Production in Baltic Lakes Using Sentinel-3 OLCI Data. *Remote Sens.* **2020**, *12*, 2415. [CrossRef]

42. Gitelson, A. The Peak near 700 Nm on Radiance Spectra of Algae and Water: Relationships of Its Magnitude and Position with Chlorophyll Concentration. *Int. J. Remote Sens.* **1992**, *13*, 3367–3373. [CrossRef]

43. Gitelson, A.; Mayo, M.; Yacobi, Y.Z.; Parparov, A.; Berman, T. The Use of High-Spectral-Resolution Radiometer Data for Detection of Low Chlorophyll Concentrations in Lake Kinneret. *J. Plankton Res.* **1994**, *16*, 993–1002. [CrossRef]

44. Jensen, J.R. *Sensoriamento Remoto Do Ambiente: Uma Perspectiva Em Recursos Terrestres*; Tradução 2; Epiphanio, J.C.N., Formaggio, A.R., Santos, A.R., Rudorff, B.F.T., Almeida, C.M., Galvão, L.S., Eds.; Parêntese: São José dos Campos, Brazil, 2009; p. 672. ISBN 978-85-60507-06-1.

45. Le, C.; Li, Y.; Zha, Y.; Wang, Q.; Zhang, H.; Yin, B. Remote Sensing of Phycocyanin Pigment in Highly Turbid Inland Waters in Lake Taihu, China. *Int. J. Remote Sens.* **2011**, *32*, 8253–8269. [CrossRef]
46. Matthews, M.W.; Bernard, S.; Robertson, L. An Algorithm for Detecting Trophic Status (Chlorophyll-a), Cyanobacterial-Dominance, Surface Scums and Floating Vegetation in Inland and Coastal Waters. *Remote Sens. Environ.* **2012**, *124*, 637–652. [CrossRef]
47. Heisler, J.; Glibert, P.M.; Burkholder, J.M.; Anderson, D.M.; Cochlan, W.; Dennison, W.C.; Dortch, Q.; Gobler, C.J.; Heil, C.A.; Humphries, E.; et al. Eutrophication and Harmful Algal Blooms: A Scientific Consensus. *Harmful Algae* **2008**, *8*, 3–13. [CrossRef] [PubMed]
48. de Castro Medeiros, L.; Mattos, A.; Lürling, M.; Becker, V. Is the Future Blue-Green or Brown? The Effects of Extreme Events on Phytoplankton Dynamics in a Semi-Arid Man-Made Lake. *Aquat. Ecol.* **2015**, *49*, 293–307. [CrossRef]
49. Kangur, K.; Möls, T.; Milius, A.; Laugaste, R. Phytoplankton Response to Changed Nutrient Level in Lake Peipsi (Estonia) in 1992-2001. *Hydrobiologia* **2003**, *506–509*, 265–272. [CrossRef]
50. Naselli-Flores, L.; Barone, R. Water-Level Fluctuations in Mediterranean Reservoirs: Setting a Dewatering Threshold as a Management Tool to Improve Water Quality. *Hydrobiologia* **2005**, *548*, 85–99. [CrossRef]
51. Leira, M.; Cantonati, M. Effects of Water-Level Fluctuations on Lakes: An Annotated Bibliography. *Hydrobiologia* **2008**, *613*, 171–184. [CrossRef]
52. Geraldes, A.M.; Boavida, M.J. Zooplankton Assemblages in Two Reservoirs: One Subjected to Accentuated Water Level Fluctuations, the Other with More Stable Water Levels. *Aquat. Ecol.* **2007**, *41*, 273–284. [CrossRef]
53. Naselli-Flores, L. Morphological Analysis of Phytoplankton as a Tool to Assess Ecological State of Aquatic Ecosystems: The Case of Lake Arancio, Sicily, Italy. *Inland Waters* **2014**, *4*, 15–26. [CrossRef]
54. Reynolds, C.S. *The Ecology of Phytoplankton (Ecology, Biodiversity and Conservation)*; Cambridge University Press: Cambridge, UK, 2006.
55. Soares, M.C.S.; Marinho, M.M.; Azevedo, S.M.O.F.; Branco, C.W.C.; Huszar, V.L.M. Eutrophication and Retention Time Affecting Spatial Heterogeneity in a Tropical Reservoir. *Limnologica* **2012**, *42*, 197–203. [CrossRef]
56. da Costa, M.R.A.; Attayde, J.L.; Becker, V. Effects of Water Level Reduction on the Dynamics of Phytoplankton Functional Groups in Tropical Semi-Arid Shallow Lakes. *Hydrobiologia* **2016**, *778*, 75–89. [CrossRef]
57. Naselli-flores, L. Man-Made Lakes in Mediterranean Semi-Arid Climate: The Strange Case of Dr Deep Lake and Mr Shallow Lake. *Hydrobiologia* **2003**, *506*, 13–21. [CrossRef]
58. Geraldes, A.M.; Boavida, M.J. Seasonal Water Level Fluctuations: Implications for Reservoir Limnology and Management. *Lakes Reserv. Res. Manag.* **2005**, *10*, 59–69. [CrossRef]
59. Bond, N.R.; Lake, P.S.; Arthington, A.H. The Impacts of Drought on Freshwater Ecosystems: An Australian Perspective. *Hydrobiologia* **2008**, *600*, 3–16. [CrossRef]
60. Özen, A.; Karapinar, B.; Kucuk, I.; Jeppesen, E.; Beklioglu, M. Drought-Induced Changes in Nutrient Concentrations and Retention in Two Shallow Mediterranean Lakes Subjected to Different Degrees of Management. *Hydrobiologia* **2010**, *646*, 61–72. [CrossRef]
61. Teferi, M.; Declerck, S.; Lemmens, P.; Gebrekidan, A. Strong Effects of Occasional Drying on Subsequent Water Clarity and Cyanobacterial Blooms in Cool Tropical Reservoirs. *Freshw. Biol.* **2014**, *59*, 870–884. [CrossRef]
62. Braga, G.G.; Becker, V.; Neuciano Pinheiro de Oliveira, J.; Rodrigues de Mendonça Junior, J.; Felipe de Medeiros Bezerra, A.; Macêdo Torres, L.; Marília Freitas Galvão, Â.; Mattos, A. Influence of Extended Drought on Water Quality in Tropical Reservoirs in a Semiarid Region. *Acta Limnol. Bras.* **2015**, *27*, 15–23. [CrossRef]
63. Jeppesen, E.; Brucet, S.; Naselli-Flores, L.; Papastergiadou, E.; Stefanidis, K.; Nõges, T.; Nõges, P.; Attayde, J.L.; Zohary, T.; Coppens, J.; et al. Ecological Impacts of Global Warming and Water Abstraction on Lakes and Reservoirs Due to Changes in Water Level and Related Changes in Salinity. *Hydrobiologia* **2015**, *750*, 201–227. [CrossRef]
64. do Vale Figueiredo, A.; Becker, V. Influence of Extreme Hydrological Events in the Quality of Water Reservoirs in the Semi-Arid Tropical Region. *RBRH* **2018**, *23*, 1–8. [CrossRef]
65. Kosten, S.; Huszar, V.L.M.; Bécares, E.; Costa, L.S.; van Donk, E.; Hansson, L.A.; Jeppesen, E.; Kruk, C.; Lacerot, G.; Mazzeo, N.; et al. Warmer Climates Boost Cyanobacterial Dominance in Shallow Lakes. *Glob. Change Biol.* **2012**, *18*, 118–126. [CrossRef]
66. Romo, S.; Soria, J.; Fernández, F.; Ouahid, Y.; Barón-Solá, Á. Water Residence Time and the Dynamics of Toxic Cyanobacteria. *Freshw. Biol.* **2013**, *58*, 513–522. [CrossRef]
67. Wagner, C.; Adrian, R. Cyanobacteria Dominance: Quantifying the Effects of Climate Change. *Limnol. Oceanogr.* **2009**, *54*, 2460–2468. [CrossRef]
68. Willis, A.; Chuang, A.W.; Orr, P.T.; Beardall, J.; Burford, M.A. Subtropical Freshwater Phytoplankton Show a Greater Response to Increased Temperature than to Increased PCO2. *Harmful Algae* **2019**, *90*, 101705. [CrossRef] [PubMed]
69. Accoroni, S.; Ceci, M.; Tartaglione, L.; Romagnoli, T.; Campanelli, A.; Marini, M.; Giulietti, S.; Dell'Aversano, C.; Totti, C. Role of Temperature and Nutrients on the Growth and Toxin Production of Prorocentrum Hoffmannianum (Dinophyceae) from the Florida Keys. *Harmful Algae* **2018**, *80*, 140–148. [CrossRef]
70. Gobler, C.J. Climate Change and Harmful Algal Blooms: Insights and Perspective. *Harmful Algae* **2020**, *91*, 101731. [CrossRef] [PubMed]

71. Liu, X.; Lu, X.; Chen, Y. The Effects of Temperature and Nutrient Ratios on Microcystis Blooms in Lake Taihu, China: An 11-Year Investigation. *Harmful Algae* **2011**, *10*, 337–343. [CrossRef]

72. Paerl, H.W.; Huisman, J. Climate: Blooms like It Hot. *Science* **2008**, *320*, 57–58. [CrossRef]

73. Peperzak, L. Climate Change and Harmful Algal Blooms in the North Sea. *Acta Oecologica* **2003**, *24*, S139–S144. [CrossRef]

74. Paerl, H.W.; Huisman, J. Climate Change: A Catalyst for Global Expansion of Harmful Cyanobacterial Blooms. *Environ. Microbiol. Rep.* **2009**, *1*, 27–37. [CrossRef]

75. O'Neil, J.M.; Davis, T.W.; Burford, M.A.; Gobler, C.J. The Rise of Harmful Cyanobacteria Blooms: The Potential Roles of Eutrophication and Climate Change. *Harmful Algae* **2012**, *14*, 313–334. [CrossRef]

76. Allende, L.; Tell, G.; Zagarese, H.; Torremorell, A.; Pérez, G.; Bustingorry, J.; Escaray, R.; Izaguirre, I. Phytoplankton and Primary Production in Clear-Vegetated, Inorganic-Turbid, and Algal-Turbid Shallow Lakes from the Pampa Plain (Argentina). *Hydrobiologia* **2009**, *624*, 45–60. [CrossRef]

77. Feng, L.; Chen, B.; Hayat, T.; Alsaedi, A.; Ahmad, B. Modelling the Influence of Thermal Discharge under Wind on Algae. *Phys. Chem. Earth* **2015**, *79–82*, 108–114. [CrossRef]

78. Xia, M.; Jiang, L. Influence of Wind and River Discharge on the Hypoxia in a Shallow Bay. *Ocean Dyn.* **2015**, *65*, 665–678. [CrossRef]

79. Zhang, Y.; Loiselle, S.; Shi, K.; Han, T.; Zhang, M.; Hu, M.; Jing, Y.; Lai, L.; Zhan, P. Wind Effects for Floating Algae Dynamics in Eutrophic Lakes. *Remote Sens.* **2021**, *13*, 800. [CrossRef]

80. Governo do Estado do Ceará. Portal Hidrológico Do Ceará. Tabela Detalhamento Açudes. Available online: http://www.hidro.ce.gov.br/ (accessed on 8 January 2022).

81. Molisani, M.M.; Becker, H.; Barroso, H.S.; Hijo, C.A.G.; Monte, T.M.; Vasconcellos, G.H.; Lacerda, L.D. The Influence of Castanhão Reservoir on Nutrient and Suspended Matter Transport during Rainy Season in the Ephemeral Jaguaribe River (CE, Brazil). *Braz. J. Biol.* **2013**, *73*, 115–123. [CrossRef] [PubMed]

82. Alves, J.M.B.; Campos, J.N.B.; Servain, J. Reservoir Management Using Coupled Atmospheric and Hydrological Models: The Brazilian Semi-Arid Case. *Water Resour. Manag.* **2012**, *26*, 1365–1385. [CrossRef]

83. dos Santos, J.C.N.; de Andrade, E.M.; de Araújo Neto, J.R.; Meireles, A.C.M.; de Queiroz Palácio, H.A. Land Use and Trophic State Dynamics in a Tropical Semi-Arid Reservoir. *Rev. Cienc. Agron.* **2014**, *45*, 35–44. [CrossRef]

84. Medeiros, P.H.A.; de Araújo, J.C.; Bronstert, A. Medidas de Interceptação e Avaliação Do Desempenho Do Modelo de Gash Para Uma Região Semi-Árida. *Rev. Cienc. Agron.* **2009**, *40*, 165–174.

85. Lima Neto, I.E.; Wiegand, M.C.; Carlos de Araújo, J. Redistribution Des Sédiments Due à Un Réseau Dense de Réservoirs Dans Un Grand Bassin Versant Semi-Aride Du Brésil. *Hydrol. Sci. J.* **2011**, *56*, 319–333. [CrossRef]

86. Campos, J.N.B.; Souza Filho, F.A.; Lima, H.V.C. Risques et Incertitudes de Rendement de Réservoir Dans Des Riviéres Intermittentes Trés Variables: Cas Du Réservoir Castanhão Dans Le Brésil Semi-Aride. *Hydrol. Sci. J.* **2014**, *59*, 1184–1195. [CrossRef]

87. Pereira, B.; Medeiros, P.; Francke, T.; Ramalho, G.; Foerster, S.; De Araújo, J.C. Assessment of the Geometry and Volumes of Small Surface Water Reservoirs by Remote Sensing in a Semi-Arid Region with High Reservoir Density. *Hydrol. Sci. J.* **2019**, *64*, 66–79. [CrossRef]

88. Barros, M.U.G.; Wilson, A.E.; Leitão, J.I.R.; Pereira, S.P.; Buley, R.P.; Fernandez-Figueroa, E.G.; Capelo-Neto, J. Environmental Factors Associated with Toxic Cyanobacterial Blooms across 20 Drinking Water Reservoirs in a Semi-Arid Region of Brazil. *Harmful Algae* **2019**, *86*, 128–137. [CrossRef]

89. Palmer, S.C.J.; Kutser, T.; Hunter, P.D. Remote Sensing of Inland Waters: Challenges, Progress and Future Directions. *Remote Sens. Environ.* **2015**, *157*, 1–8. [CrossRef]

90. Kallio, K.; Attila, J.; Härmä, P.; Koponen, S.; Pulliainen, J.; Hyytiäinen, U.M.; Pyhälahti, T. Landsat ETM+ Images in the Estimation of Seasonal Lake Water Quality in Boreal River Basins. *Environ. Manage.* **2008**, *42*, 511–522. [CrossRef] [PubMed]

91. Kutser, T.; Paavel, B.; Verpoorter, C.; Ligi, M.; Soomets, T.; Toming, K.; Casal, G. Remote Sensing of Black Lakes and Using 810 Nm Reflectance Peak for Retrieving Water Quality Parameters of Optically Complex Waters. *Remote Sens.* **2016**, *8*, 497. [CrossRef]

92. Toming, K.; Kutser, T.; Laas, A.; Sepp, M.; Paavel, B.; Nõges, T. First Experiences in Mapping Lakewater Quality Parameters with Sentinel-2 MSI Imagery. *Remote Sens.* **2016**, *8*, 640. [CrossRef]

93. Rice, E.W.; Bridgewater, L. *Standard Methods for the Examination of Water and Wastewater*; Rice, E.W., Ed.; American Public Health Association, American Water Works Association, Water Environment Federation: Washington, DC, USA, 2012; ISBN 978-0-87553-013-0.

94. Hijo, C.A.G. Quantificação Do Efeito Do Açude Castanhão Sobre o Fluxo Fluvial de Material Particulado Em Suspensão e Nutrientes Para o Estuário Do Rio Jaguaribe, Ceará—Brasil. Master's Thesis, Universidade Federal do Ceará, Fortaleza, Brazil, 2009.

95. Bouvy, M.; Molica, R.; De Oliveira, S.; Marinho, M.; Beker, B. Dynamics of a Toxic Cyanobacterial Bloom (Cylindrospermopsis Raciborskii) in a Shallow Reservoir in the Semi-Arid Region of Northeast Brazil. *Aquat. Microb. Ecol.* **1999**, *20*, 285–297. [CrossRef]

96. Schroeder, T.; Behnert, I.; Schaale, M.; Fischer, J.; Doerffer, R. Atmospheric Correction Algorithm for MERIS above Case-2 Waters. *Int. J. Remote Sens.* **2007**, *28*, 1469–1486. [CrossRef]

97. Topp, S.N.; Pavelsky, T.M.; Jensen, D.; Simard, M.; Ross, M.R.V. Research Trends in the Use of Remote Sensing for Inland Water Quality Science: Moving towards Multidisciplinary Applications. *Water* **2020**, *12*, 169. [CrossRef]

98. Jamet, C.; Loisel, H.; Kuchinke, C.P.; Ruddick, K.; Zibordi, G.; Feng, H. Comparison of Three SeaWiFS Atmospheric Correction Algorithms for Turbid Waters Using AERONET-OC Measurements. *Remote Sens. Environ.* **2011**, *115*, 1955–1965. [CrossRef]

99. Matthews, M.W.; Odermatt, D. Improved Algorithm for Routine Monitoring of Cyanobacteria and Eutrophication in Inland and Near-Coastal Waters. *Remote Sens. Environ.* **2015**, *156*, 374–382. [CrossRef]

100. Soriano-González, J.; Angelats, E.; Fernández-Tejedor, M.; Diogene, J.; Alcaraz, C. First Results of Phytoplankton Spatial Dynamics in Two NW-Mediterranean Bays from Chlorophyll-A Estimates Using Sentinel 2: Potential Implications for Aquaculture. *Remote Sens.* **2019**, *11*, 1756. [CrossRef]

101. Dall'Olmo, G.; Gitelson, A.A. Effect of Bio-Optical Parameter Variability on the Remote Estimation of Chlorophyll-a Concentration in Turbid Productive Waters: Experimental Results. *Appl. Opt.* **2005**, *44*, 412. [CrossRef] [PubMed]

102. Gitelson, A.A.; Dall'Olmo, G.; Moses, W.; Rundquist, D.C.; Barrow, T.; Fisher, T.R.; Gurlin, D.; Holz, J. A Simple Semi-Analytical Model for Remote Estimation of Chlorophyll-a in Turbid Waters: Validation. *Remote Sens. Environ.* **2008**, *112*, 3582–3593. [CrossRef]

103. Sun, D.; Li, Y.; Wang, Q. A Unified Model for Remotely Estimating Chlorophyll a in Lake Taihu, China, Based on SVM and in Situ Hyperspectral Data. *IEEE Trans. Geosci. Remote Sens.* **2009**, *47*, 2957–2965. [CrossRef]

104. Hunter, P.D.; Tyler, A.N.; Carvalho, L.; Codd, G.A.; Maberly, S.C. Hyperspectral Remote Sensing of Cyanobacterial Pigments as Indicators for Cell Populations and Toxins in Eutrophic Lakes. *Remote Sens. Environ.* **2010**, *114*, 2705–2718. [CrossRef]

105. Mishra, S.; Mishra, D.R. Normalized Difference Chlorophyll Index: A Novel Model for Remote Estimation of Chlorophyll-a Concentration in Turbid Productive Waters. *Remote Sens. Environ.* **2012**, *117*, 394–406. [CrossRef]

106. Lyu, H.; Li, X.; Wang, Y.; Jin, Q.; Cao, K.; Wang, Q.; Li, Y. Evaluation of Chlorophyll-a Retrieval Algorithms Based on MERIS Bands for Optically Varying Eutrophic Inland Lakes. *Sci. Total Environ.* **2015**, *530–531*, 373–382. [CrossRef]

107. Shanmugam, P.; He, X.; Singh, R.K.; Varunan, T. A Modern Robust Approach to Remotely Estimate Chlorophyll in Coastal and Inland Zones. *Adv. Space Res.* **2018**, *61*, 2491–2509. [CrossRef]

108. Moses, W.J.; Saprygin, V.; Gerasyuk, V.; Povazhnyy, V.; Berdnikov, S.; Gitelson, A.A. OLCI-Based NIR-Red Models for Estimating Chlorophyll- a Concentration in Productive Coastal Waters—a Preliminary Evaluation. *Environ. Res. Commun.* **2019**, *1*, 011002. [CrossRef]

109. Van Nguyen, M.; Lin, C.H.; Chu, H.J.; Jaelani, L.M.; Syariz, M.A. Spectral Feature Selection Optimization for Water Quality Estimation. *Int. J. Environ. Res. Public. Health* **2020**, *17*, 272. [CrossRef]

110. Ogashawara, I.; Kiel, C.; Jechow, A.; Kohnert, K.; Ruhtz, T.; Grossart, H.P.; Hölker, F.; Nejstgaard, J.C.; Berger, S.A.; Wollrab, S. The Use of Sentinel-2 for Chlorophyll-A Spatial Dynamics Assessment: A Comparative Study on Different Lakes in Northern Germany. *Remote Sens.* **2021**, *13*, 1542. [CrossRef]

111. Zimba, P.V.; Gitelson, A. Remote Estimation of Chlorophyll Concentration in Hyper-Eutrophic Aquatic Systems: Model Tuning and Accuracy Optimization. *Aquaculture* **2006**, *256*, 272–286. [CrossRef]

112. Gurlin, D.; Gitelson, A.A.; Moses, W.J. Remote Estimation of Chl-a Concentration in Turbid Productive Waters—Return to a Simple Two-Band NIR-Red Model? *Remote Sens. Environ.* **2011**, *115*, 3479–3490. [CrossRef]

113. Kutser, T. The Possibility of Using the Landsat Image Archive for Monitoring Long Time Trends in Coloured Dissolved Organic Matter Concentration in Lake Waters. *Remote Sens. Environ.* **2012**, *123*, 334–338. [CrossRef]

114. Cardille, J.A.; Leguet, J.B.; del Giorgio, P. Remote Sensing of Lake CDOM Using Noncontemporaneous Field Data. *Can. J. Remote Sens.* **2013**, *39*, 118–126. [CrossRef]

115. Efron, B. Bootstrap Methods: Another Look at the Jackknife. *Ann. Stat.* **1979**, *7*, 1–26. [CrossRef]

116. de Almeida, R.M.V.R.; Infantosi, A.F.C.; Gismondi, R.C. Replicação Bootstrap e Análise de Sensibilidade Em Redes Neurais Artificiais. In Proceedings of the Anais do 5. Congresso Brasileiro de Redes Neurais, Rio de Janeiro, Brazil, 2–5 April 2001; pp. 295–300.

117. Martinez, J.M.; Ventura, D.; Cochonneau, G.; De Oliveira, E.; Cerqueira Piscoya, R.; Santos Guimarães, V. Monitoring of Water Quality and Water Level of Rivers and Lakes in Brazil: Towards a Remote Sensing-Based Operational Monitoring Application at the Brazilian National Water Agency. In *Applications of Satellite Earth Observations: Serving Society, Science & Industry*; CEOS: Washington, DC, USA, 2015; Available online: https://www.vista-geo.de/wp-content/uploads/CEOS_20151013_Satellite_Observations_2015.pdf (accessed on 14 January 2022).

118. Filho, S.; de Assis de Martins, F.; Rodrigues, E.S.P. O Processo de Mistura Em Reservatórios Do Semi-Árido e Sua Implicação Na Qualidade Da Água. *Rev. Bras. Recur. Hídricos* **2006**, *11*, 109–119. [CrossRef]

Article

Dynamics of Mid-Channel Bar during Different Impoundment Periods of the Three Gorges Reservoir Area in China

Qingqing Tang [1,2,3], Daming Tan [4], Yongyue Ji [1,2], Lingyun Yan [1,2], Sidong Zeng [1], Qiao Chen [1,2], Shengjun Wu [1,2] and Jilong Chen [1,2,*]

[1] Chongqing Institute of Green and Intelligent Technology, Chinese Academy of Sciences, Chongqing 400714, China; tangqingqing830@163.com (Q.T.); jiyongyue_happy@163.com (Y.J.); yanlingyun20@mails.ucas.ac.cn (L.Y.); zengsidong@cigit.ac.cn (S.Z.); chenqiao@cigit.ac.cn (Q.C.); chenjilong@cigit.ac.cn (S.W.)
[2] Key Laboratory of Reservoir Aquatic Environment, Chinese Academy of Sciences, Chongqing 400714, China
[3] School of Architecture and Urban Planning, Chongqing Jiaotong University, Chongqing 400074, China
[4] Institute of Agricultural Resources and Environment, Tibet Academy of Agricultural and Animal, Husbandry Science, Lhasa 850000, China; tdmxz@126.com
* Correspondence: cjl47168@163.com; Tel.: +86-023-6593-5878

Abstract: The dynamics of the mid-channel bars (MCBs) in the Three Gorges Reservoir (TGR) were substantially impacted by the large water-level changes due to the impoundments of the TGR. However, it is still not clear how the morphology of the MCBs changed under the influence of water level and hydrological regime changes induced by the impoundments and operation of the TGR. In this work, the MCBs in the TGR were retrieved using Landsat remote sensing images from 1989 to 2019, and the spatio-temporal variations in the number, area, morphology and location of the MCBs during different impoundment periods were investigated. The results showed that the number and area of MCBs changed dramatically with water-level changes, and the changes were dominated by MCBs with an area less than 0.03 km^2 and larger than 1 km^2. The area of MCBs decreased progressively with the rising water level, and the number generally showed a decreasing trend, with the minimum number occurring at the third stage when the water level reached 139 m, resulting in the maximum average area at this period. The ratio of length to width of the MCBs generally decreased with the changes in hydrological and sediment regimes, leading to a shape adjustment from narrow–long to relatively short–round with the rising of the water level. The water impoundments of the TGR led to the migration of the dominant area from the upper section to the middle section of the TGR and resulted in a more even distribution of MCBs in the TGR. The results improve our understanding of the mechanisms of the development of MCBs in the TGR under the influence of water impoundment coupled with the annually cyclic hydrological regime and longer periods of inundation and exposure.

Keywords: mid-channel bars; morphological change; spatial distribution; different impoundment periods; Three Gorges Reservoir

Citation: Tang, Q.; Tan, D.; Ji, Y.; Yan, L.; Zeng, S.; Chen, Q.; Wu, S.; Chen, J. Dynamics of Mid-Channel Bar during Different Impoundment Periods of the Three Gorges Reservoir Area in China. *Water* **2021**, *13*, 3427. https://doi.org/10.3390/w13233427

Academic Editors: Fei Zhang, Ngai Weng Chan, Xinguo Li and Xiaoping Wang

Received: 2 November 2021
Accepted: 1 December 2021
Published: 3 December 2021

Publisher's Note: MDPI stays neutral with regard to jurisdictional claims in published maps and institutional affiliations.

1. Introduction

The mid-channel bar (MCB) is formed by favorable hydrological conditions in the river. It is a stable island above the river's water level, formed by the gradual development and shaping of the river siltation over a long period [1,2]. The development of MCB is influenced by exogenous materials such as sediment, the transport capacity of flowing water and the sediment concentration, as well as by dam construction and reservoir regulations [3–6]. The dynamics of MCBs were significantly impacted by the water-level changes due to the impoundment of the TGR. [7,8]. On the one hand, some original MCBs in the Yangtze River were submerged, while some new MCBs were formed from the inundation of low-lying lands and point bars by the reservoir, due to the rising of the water level [9,10]. On the other hand, some MCBs were periodically exposed and submerged due to the annually cyclic hydrological regime induced by the TGR operation. Therefore, the morphology of the MCBs in the TGR changed significantly under the influence of these hydrological regime changes, which are expected to have an important impact on channel stability, water–land interactions and biological diversity [11–13].

Over the past century, great efforts have been made to investigate the formation and development processes of MCBs in bifurcated channel stretches, using field observations [14], remote sensing [5], theoretical generalized models and mathematical models [15–17]. Many experiments have also been conducted to explore the morphological dynamics of MCBs [18–20]. However, most of the studies are based on ideal environmental conditions, including constant flow, slope, etc. Furthermore, many other environmental factors affecting the development of MCBs were not considered [21,22]. In recent years, the rapid development of numerical simulation technology, remote sensing and spatial analysis in geographic information science has provided an opportunity to monitor and model the dynamics of MCBs at multi-spatial and multi-temporal scales [23,24]. Schuurman et al. [15] generated datasets of water depth, flow and sediment transport of MCBs based on physical models, and further developed a conceptual network model describing the interactions of MCBs, sub-branches and river channels. Liu et al. [25] investigated the proportion of riverine sand partitioning when MCBs reached their stable equilibrium form, using an analytical hydrodynamics method. Rasbold et al. [26] identified the development signatures of MCBs based on the theory of sedimentology. Adami et al. [22] used wavelength, migration rate and height to investigate the spatio-temporal variations of the morphological dynamics of the MCBs in the Alpine Rhine over the last 30 years.

As the longest river in China, the Yangtze River has an important strategic position and a role in boosting the development of the cities along its length [27]. The morphological development of MCBs in the Yangtze River is of great significance in maintaining the stability of the river and enhancing the function of the "golden channel" [28]. However, the construction of the Three Gorges Dam (TGD) has significantly changed the hydrological and sediment regimes downstream of the TGD over the last 30 years, altering the hydrological conditions for the development of the MCBs [28–30]. Based on long-term observations, multi-temporal remote sensing data and model simulations, many studies have been conducted to monitor the changes in the MCBs in the middle and lower reaches of the Yangtze River [31,32]. The results showed significant morphological changes in the MCBs after the TGD operation [5,33], and revealed the process [28] and mechanism for the development of MCBs [33] downstream of the TGD. In contrast, under the influence of the annually cyclic hydrological regime and of longer inundation and exposure periods induced by the TGR operation, the morphological development process of MCBs in the TGR and their response to hydrological and sediment regime changes differs greatly from those downstream of the TGD [34]. However, due to the lack of relevant studies, it is still not clear how the morphology of the MCBs changes under the influence of water level and hydrological regime changes induced by the impoundments and operation of the TGR.

Therefore, this study was carried out to fill the knowledge gap. The main objectives of this study were to: (1) retrieve the MCBs from Landsat images and construct datasets of morphological changes of MCBs in the TGR; (2) investigate the spatio-temporal variations in the numbers, areas, morphology and locations of the MCBs during different impoundment periods. The study helps to reveal the mechanisms for the development of MCBs in the TGR; it also offers a scientific basis for the planning, optimal utilization and ecological restoration of the MCBs in the TGR.

2. Materials and Methods

2.1. Study Area

The Three Gorges Reservoir (TGR) is located in the lower section of the main waterway in the upper reaches of the Yangtze River, which is a typical mountainous river. It extends from Jiangjin District in Chongqing to Yichang City in Hubei Province, from west to east

(Figure 1). The topography of the Three Gorges Reservoir Area (TGRA) is dominated by mountains and hills. The TGRA has a subtropical monsoon climate with an average annual temperature of 17–19° and annual precipitation of 1000–1800 mm [35]. After the official operation of the TGR, it formed a narrow-valley reservoir with a total length of 660 km and a surface area of 1084 km^2. The geographical location of the TGR is between 28°56′ and 31°44′ east (longitude) and between 106°16′ and 111°28′ north (latitude), which includes 25 districts and counties in Hubei and Chongqing municipalities.

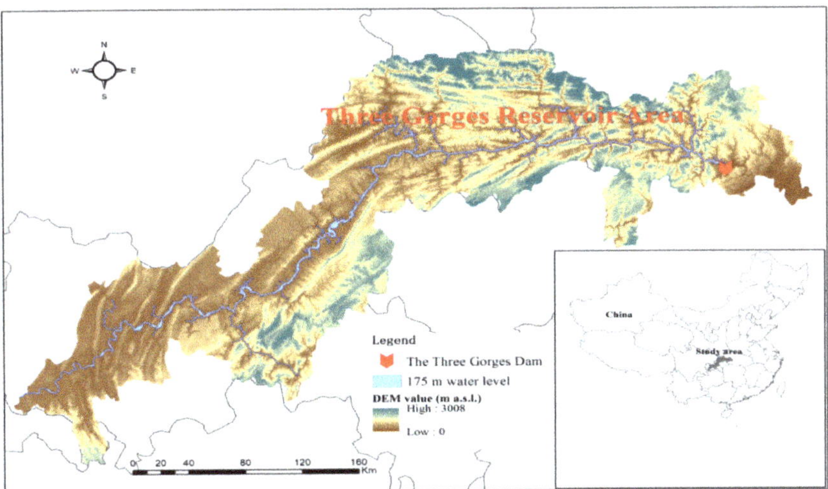

Figure 1. Geographic location of the study area.

2.2. Division of Impoundment Periods

The upper reaches of the Yangtze River were successfully intercepted by the TGD in 1997, raising the reservoir water level to approximately 66 m above sea level. The TGR began storing water in steps from 139 m in June 2003, to 156 m in October 2006 and 175 m in November 2009. It was officially operated after one year of experimental water storage, in 2009 [36]. Thus, based on the construction phase and the changes in the water levels of the TGD (Figure 2), five stages were identified to investigate the morphological changes in the MCBs.

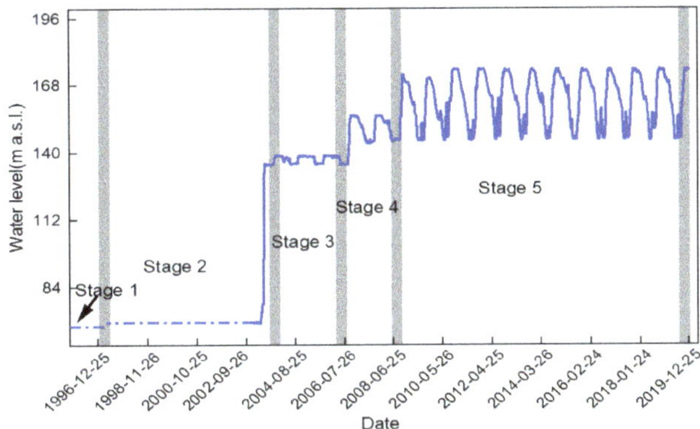

Figure 2. The water-level changes of TGR and division of impoundment period.

2.3. Data Collection

The spatial information of the MCBs was retrieved using multi-temporal Landsat images. Since the TGR had been in operation for more than twenty years from the interception of the Yangtze River, three criteria were employed to ensure data consistency regarding the spatial-temporal resolution and retrieval accuracy for the MCBs. Firstly, images with less cloud cover were used where possible [5,23]. Secondly, images acquired during the dry season (i.e., November to March) were used, to reduce the difference in water levels during the different stages. Lastly, due to the influence of the 16-day-long revisit cycle of the Landsat satellite, it was difficult to obtain images with the same water levels for different strips. Therefore, images with similar water levels were obtained as far as possible.

A total of 36 images were collected, with an spatial resolution of 30 m, following these criteria. These images, spanning the 39th–42nd strips of the Landsat satellite were obtained from the Chinese Geospatial Data Cloud (http://www.gscloud.cn/). The data pre-processing, including single-band extraction, false color synthesis and geometric correction, etc., was carried out using ENVI 5.3 software. Many previous studies proved that these images can be used to monitor the landscape dynamics with reasonable accuracy [5,23,24]. Detailed information on the collected image data is shown in Table 1. The data on the water level, sediment and siltation of the TGR were obtained from the Yangtze River Sediment Bulletin (http://www.cjw.gov.cn/zwzc/bmgb/) and Yangtze River Three Gorges Group (https://www.ctg.com./sxjt/sqqk/index.html).

Table 1. Detailed information on the collected remote sensing images.

Impoundment Period	Sensor	Acquisition Date	Number	Resolution	Data Sources
Stage 1	Landsat5 TM	24 January 1993, 29 January 1994, 1 November 1995, 17 November 1995, 26 December 1995, 5 February 1996	6	30 m	Geospatial Data Cloud
Stage 2	Landsat5 TM	5 November 2000, 17 January 2001, 10 February 2001, 12 March 2001, 27 December 2001, 8 January 2002, 28 November 2002	7	30 m	Geospatial Data Cloud
Stage 3	Landsat5 TM	24 January 2004, 7 December 2004, 6 January 2005, 2 February 2005, 7 December 2005, 2 January 2006, 4 February 2006	7	30 m	Geospatial Data Cloud
Stage 4	Landsat5 TM	3 February 2007, 1 March 2007, 22 February 2008, 23 March 2008, 15 November 2008, 21 November 2008 2 December 2008, 3 February 2009	8	30 m	Geospatial Data Cloud
Stage 5	Landsat8 OLI	22 January 2015, 17 December 2015, 25 January 2016, 15 November 2016, 22 December 2017, 31 December 2017, 14 January 2018, 12 February 2018	8	30 m	Geospatial Data Cloud

2.4. Retrieval of MCBs from Landsat Images

MCBs were retrieved from Landsat images using auto-classification coupled with manual inspection and digitization. They were initially auto-retrieved using the modified normalized difference water index (MNDWI) developed by Xu [37]. This index was derived from the normalized difference water index (NDWI), which highlighted the water information in the image by normalizing the spectral difference between the green band and the mid-infrared band [38]. The MNDWI has been proved to be an effective method for retrieval of the MCBs with reasonable accuracy [5,23]. The MNDWI was calculated as:

$$\text{MNDWI} = \frac{\rho_{Green} - \rho_{MIR}}{\rho_{Green} + \rho_{MIR}} \tag{1}$$

where ρ_{Green} and ρ_{MIR} are the reflectances of the green band and mid-infrared band [37], respectively.

Due to differences in the spectral features among different images, some MCBs were misclassified as other landscapes, while some other landscapes were misclassified as MCBs since they had similar spectral features. Thus, the automatically retrieved vector data of MCBs need to be manually modified and verified by combing them with the observed data such as water levels, hydrological data and land use data. Such modification can significantly improve the retrieval accuracy of MCBs from Landsat images.

2.5. Classification of MCBs

A field survey showed that the areas of MCBs varied greatly in the TGR, and previous studies found that MCBs with different sizes had different responses to the changes in the hydrological and sediment regimes [5]; thus, all the MCBs were reclassified into 4 types based on different areas: small MCBs with an area less than 0.03 km^2 (SMB), medium MCBs with an area less than 0.1 km^2 (MMB), medium–large MCBs with an area less than 1 km^2 (MLMB) and large MCBs with an area greater than 1 km^2 (LMB).

2.6. Analysis Method for the Dynamics of MCBs
2.6.1. Index of Area and Shape

The area and perimeter of a single MCB can be directly calculated from the vector data of MCBs using ArcGIS software. The length and width changes of MCBs can reflect their adjustment to the changes in hydrological and sediment regimes [39]. The ratio of length to width (LWR) was used as a comprehensive index to investigate the morphological characteristics of, and changes in, MCBs [5,23]. This was calculated as:

$$\text{LWR} = \frac{L}{W} \tag{2}$$

where L and W are the length and width of the MCB, respectively.

2.6.2. The Coefficient of Variation (CV)

The CV was used to measure the spatial variability of the morphological characteristics of MCBs. It has been proved to be a useful indicator for investigating the variations in spatial features and is widely used in landscape ecology [5]. It was calculated as:

$$\text{CV} = \frac{1}{x}\sqrt{\frac{1}{n}\sum_{i=1}^{n}(x_i - \overline{x})^2} \tag{3}$$

where x_i, x and n are the LWR of each MCB, average LWR and number of MCBs, respectively.

2.6.3. The Gravity Center Shifting Model

The gravity center shifting model was used for investigating the spatial change trends of MCBs during different impoundment stages [40].

$$X_s = \sum_{i=1}^{n}(A_{si}x_i) \Big/ \sum_{i=1}^{n} A_{si} \tag{4}$$

$$Y_s = \sum_{i=1}^{n}(A_{si}y_i) \Big/ \sum_{i=1}^{n} A_{si} \tag{5}$$

where X_s and Y_s are the latitude and longitude of the gravity center of all the MCBs at stage s, respectively, A_{si} is the area of the ith MCB at stage s, x_i and y_i are the latitude and longitude of the geometric center of the ith MCB, respectively, and n is the total number of MCBs.

The following equation was used for calculating the shifting distance of the gravity center:

$$D_{s'-s} = \sqrt{(Y_{s'} - Y_S)^2 + (X_{s'} - X_S)^2}$$

(6)

where $D_{s'-s}$ is the shifting distance of the gravity center, $X_{s'}$ and $Y_{s'}$ are the latitude and longitude of the gravity center of all the MCBs at stage s', respectively, and X_s and Y_s are the latitude and longitude of the gravity center of all the MCBs at stage s, respectively.

3. Results

3.1. Variations in Area and Number of MCBs

Retrieved results from Landsat images showed significant variations in the area and number of MCBs in the TGR (Figure 3). The area and number presented different trends with respect to the changes in water level during different impoundment periods. The number of MCBs ranged between 89 and 150, with an average of 113 in the TGR; 90 MCBs were located in the main stream and accounted for 79.7% of the total number, and the remaining MCBs were located in tributaries (Figure 3a).

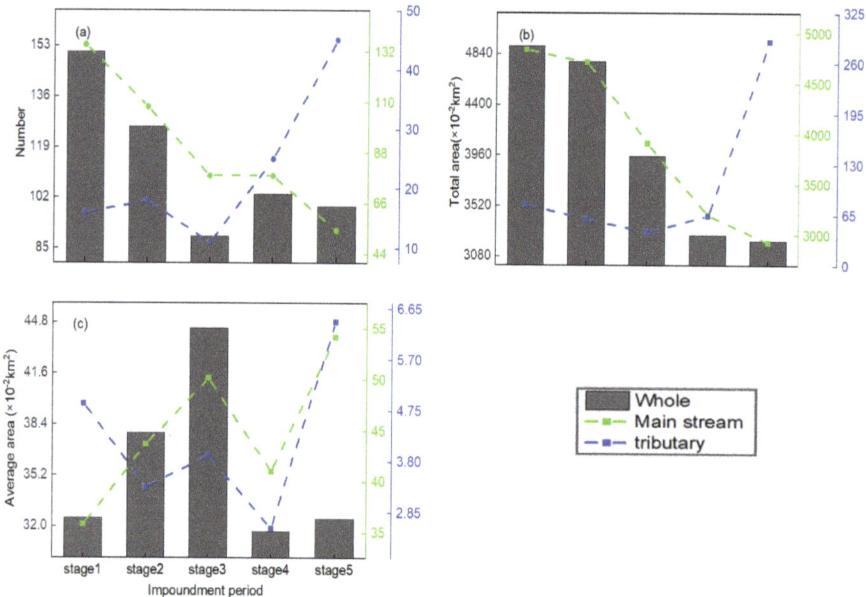

Figure 3. Number and area changes for MCBs during different impoundment periods: (**a**) number, (**b**) total area and (**c**) average area.

The number and area of MCBs changed dramatically with water-level changes (Figure 3). The maximum number of MCBs occurred at stage 1 under the natural hydrological regime. The number decreased sharply to 89 at stage 3 when the water level reached 139 m, while it increased to 103 at stage 4 when the water level rose to 156 m and slightly declined to 99 at stage 5 when the water level increased to 175 m. The number trend for MCBs in the main stream differed greatly from that in tributaries. It decreased progressively from 139 at stage 1 to 54 at stage 5, in the main stream. The number slightly increased from stage 1 to stage 2 and then slightly decreased in stage 3 in tributaries, while it sharply increased from a minimum of 11 at stage 3 to a maximum of 45 at stage 5.

The area of MCBs in the TGR varied greatly from 0.2 to 1134 ($\times 10^{-2}$ km^2), with the average area ranging between 31.69 ($\times 10^{-2}$ km^2) at stage 4 and 44.42 ($\times 10^{-2}$ km^2) at stage 3. The average area in the main stream was much higher than that in tributaries, with maximum values of 54.19 ($\times 10^{-2}$ km^2) and 6.4 ($\times 10^{-2}$ km^2), respectively (Figure 3c). The total area decreased progressively from a maximum of 4910.56 ($\times 10^{-2}$ km^2) at stage 1 to a minimum of 3214.6 ($\times 10^{-2}$ km^2) at stage 5, with the rising of the water level, and the most obvious changes occurred between stage 2 and stage 4 (Figure 3b). The total area in the main stream presented a similar trend to that in the whole reservoir. The total area changed slightly from stage 1 to stage 4 in tributaries, and then increased sharply to a maximum of 288.07 ($\times 10^{-2}$ km^2) at stage 5.

The area and number changes in MCBs for the different classes are presented in Figure 4. In terms of the number, the MCBs in the TGR were dominated by SMBs followed by MMBs, accounting for 55.6% and 20.9% of the total number on average, respectively. In contrast, the MCBs were dominated by LMBs followed by LMMBs in terms of area, accounting for 80.2% and 14.8% of the total area on average, respectively. The number and area variations in the MCBs differed greatly for different sizes, with the most obvious changes in number and area appearing in SMBs (Figure 4a) and LMBs (Figure 4b), respectively. The number of SMBs dramatically decreased from stage 1 to stage 3, then increased sharply to stage 5, which was the main reason for the sharp change in the total number in tributaries from stage 4 to stage 5. The numbers of MMBs, LMMBs and LMBs generally showed a similar decreasing trend as the water level rose. The trend for area changes in MCBs in the TGR was determined by that of LMBs generally, due to the dominant role of LMBs in the total area. MMBs presented a similar trend to LMMBs regarding area, with an increase from stage 1 to stage 2 and a significantly decreasing trend from stage 2 to stage 5.

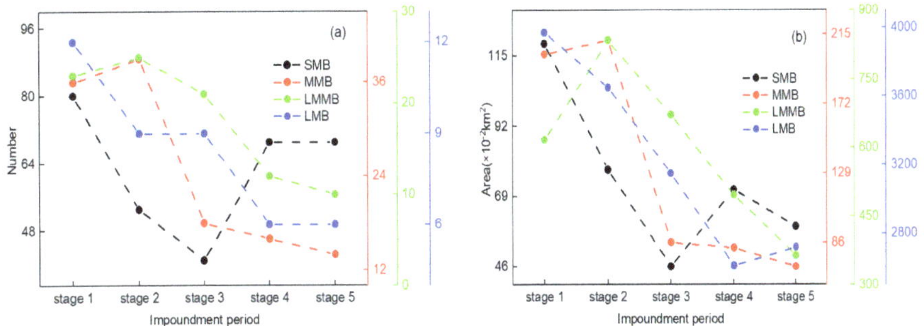

Figure 4. Number and area changes in MCBs with different sizes: (**a**) number, (**b**) area.

3.2. Morphological Changes of MCBs

The temporal variation of the LWR is presented in Figure 5. The LWR of MCBs in the TGR ranged between 2.09 and 3.05, with an average of 2.56. The changing LWR generally indicates a morphological adjustment of MCBs following the changes in hydrological and sediment regimes induced by the operation of the TGR. The LWR increased from stage 1 to a peak at stage 2, and then decreased from stage 2 to a minimum at stage 5. This suggests that the morphology of MCBs tended to change from a narrow–long shape to a short–round shape with the rising of the water level. On average, LMMBs had the highest LWR followed by MMBs, while the lowest LWR was observed for SMBs, generally indicating that the LMMBs and MMBs tended to be a narrow–long shape, while the SMBs tended to be a short–round shape. The LWR variation of MCBs differed greatly among the different classes, with the most obvious change occurring in LMBs, suggesting that the effect of impoundment on the morphology of LMBs was more pronounced than the effect on SMBs, MMBs and LMMBs. The effect for LMBs was relatively small from stage 1 to stage 2, resulting in only slight changes in LWR, while a significant effect appeared at stage 3 when the water level rose to 139 m. MMBs showed similar trend to LMMBs regarding the LWR, with an increase from stage 1 to stage 2 and a decreasing trend from stage 3 to stage 5. Compared to LMBs, MMBs and LMMBs, SMBs had a more stable LWR with lower fluctuations probably due to the fact that SMBs often had a shorter development time and a short–round morphology.

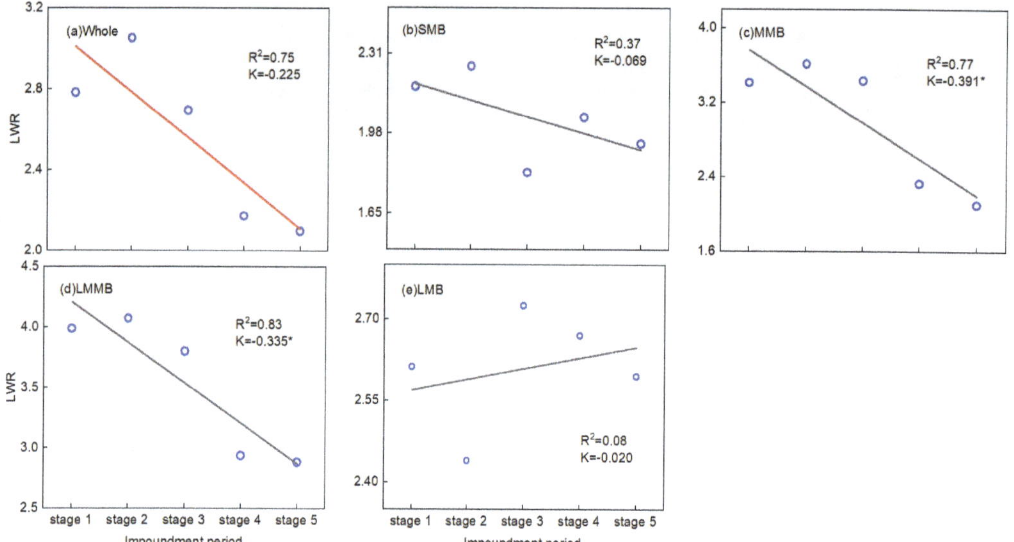

Figure 5. LWR changes in MCBs with different sizes: (**a**) whole, (**b**) SMBs, (**c**) MMBs, (**d**) LMMBs and (**e**) LMBs.

The LWR stability of the MCBs is shown in Figure 6. Generally, the CV decreased with an increase in the area of the MCB, indicating that larger MCBs tended to have a more stable morphology, and vice versa. The MCBs in the river, with natural hydrological and sediment regimes, often showed large morphological changes due to the large variations in erosion and siltation, resulting in a higher CV at stage 1. However, the conversion from natural river to man-made reservoir resulted in a rise in the water level, which significantly weakens the hydrodynamic conditions, leading to the CV changing from being scattered to being more clustered, as the water level rose from stage 1 to stage 5, as can been seen in Figure 6. This change was more evident for the LMBs.

3.3. Spatial Distribution of MCBs in TGR

The spatial distributions of the number and area of MCBs in the TGR are shown in Figures 7 and 8, respectively. It can be seen that the MCBs were unevenly distributed spatially. They were mostly distributed in the upper section of the TGR at 300 km from the TGD, accounting for 96.7% and 99.4% of the total number (Figure 7a) and area (Figure 8a) at stage 1. The distributions of the MCBs at the second stage (Figures 7b and 8b) were generally similar to those at stage 1. The number and area of MCBs in the upper section of the TGR were reduced mainly due to the influence of human activities such as sand mining, which contributed mostly to the number decrease in MCBs from stage 1 to stage 2. A total of 41 MCBs with a total area of 1068.03 ($\times 10^{-2}$ km^2) disappeared in the section of TGR from 250 to 500 km from the TGD (Figure 7c), due to the inundation occurring when the water level rose to 139 m at stage 3. The number of MCBs changed dramatically in the whole reservoir as the water level increased from 139 m to 156 m at stage 4 (Figure 7d). In the section from the dam to 400 km from the dam, 40 new MCBs formed following the inundation of low-lying mountain tops by the reservoir, while the area changed slightly (Figure 8d), and meanwhile 26 MCBs disappeared in the section from 400 to 660 km from the dam due to the inundation when the water level rose, leading to a drastic decline in the area. After the official operation of the TGR in 2009, 26 new MCBs formed in the section from 200 to 350 km from the TGD (Figure 7e), resulting in the number proportion increasing from 6.3% at stage 3 to 48.5% at stage 5, and the area proportion increasing to 44.1% (Figure 8e). However, 21 MCBs with a total area up to 1287.11 ($\times 10^{-2}$ km^2) disappeared in the section from 500 to 660 km from the TGD due to the inundation when the water level rose to 175 m. The spatial variation of MCBs during the different stages was caused mainly by the rising water level and the expansion of the inundation area created by the water impoundments of the TGR. The results suggest that water impoundments at the TGR had led to the migration of the dominant area from the upper to the middle section of the TGR, resulting in a more even distribution of MCBs in the TGR.

3.4. Gravity Center Migration of MCBs

The weighted gravity center by area of the MCBs was compared during different impoundment periods, and the migration routes are presented in Figure 9. Generally, the spatial location of the gravity center of the MCBs varied obviously with the water level changes during different water impoundment periods. The gravity center migrated 2.04 km south-westwards after the interception of the Yangtze River at stage 2. It continued to move south-westwards with the rising of the water level, with migration distances of 25.68 km and 9.39 km when the water level reached 139 m at stage 3 and 156 m at stage 4, respectively. The migration direction of MCBs was consistent with the tail direction of the TGR, which was significantly associated with the expansion of the inundation area towards the tail direction of the TGR induced by the rising water level. However, many new MCBs formed in the middle section of the TGR, owing to the inundation of low-lying mountain tops by the reservoir, induced by the water level rising to 175 m at stage 5. This resulted in a notable migration of the gravity center to the middle section of the TGR. As can been seen from Figure 8, the gravity center moved 70.63 km north-eastwards from stage 4 to stage 5, which was opposite to the migrations from stage 1 to stage 4.

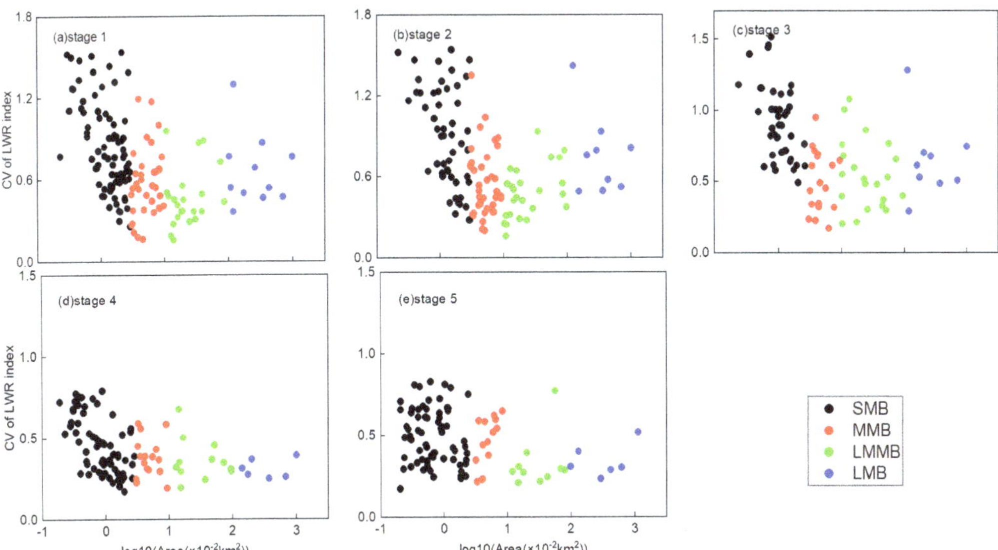

Figure 6. Relationship between CV of LWR and area of MCBs with different sizes in different impoundment periods: (**a**) stage 1, (**b**) stage 2, (**c**) stage 3, (**d**) stage 4 and (**e**) stage 5.

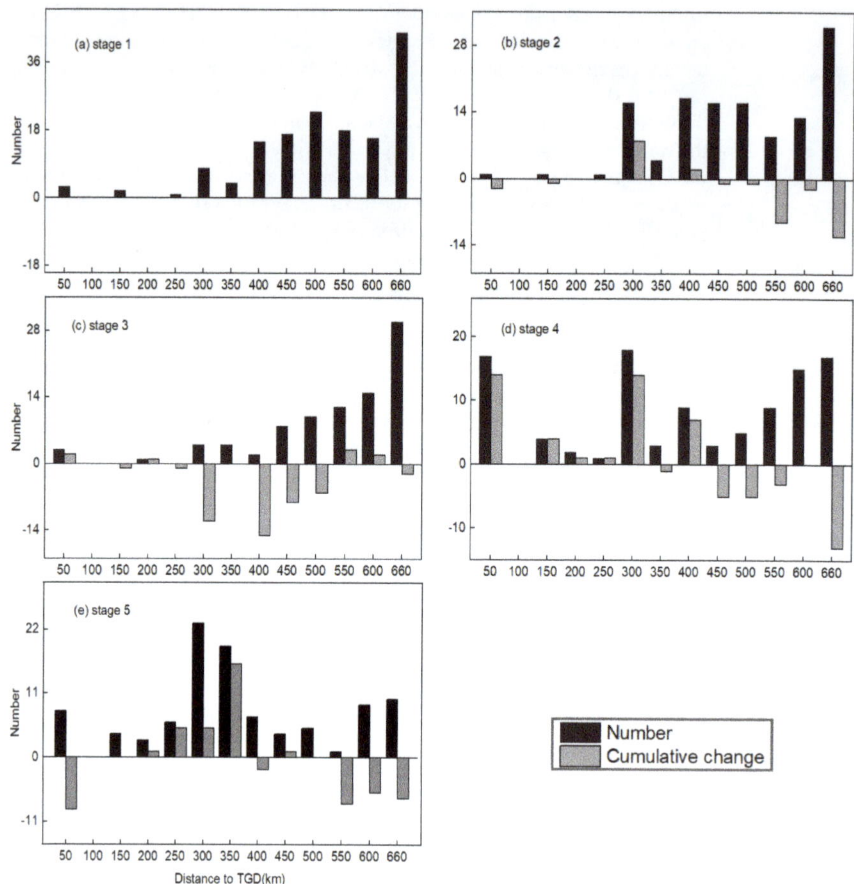

Figure 7. Spatial distribution and changes in the number of MCBs in different impoundment periods: (**a**) stage 1, (**b**) stage 2, (**c**) stage 3, (**d**) stage 4 and (**e**) stage 5.

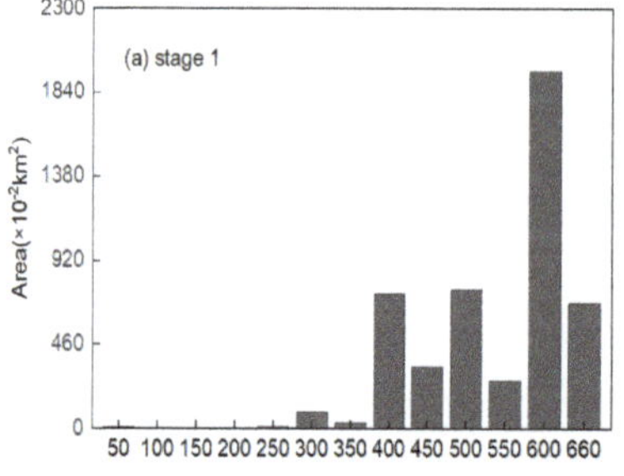

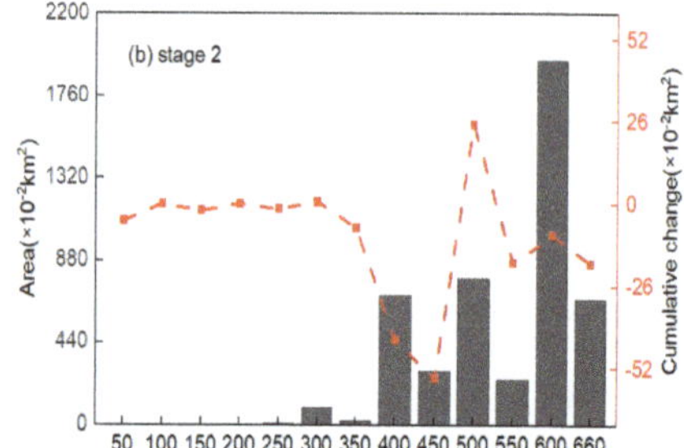

Figure 8. *Cont.*

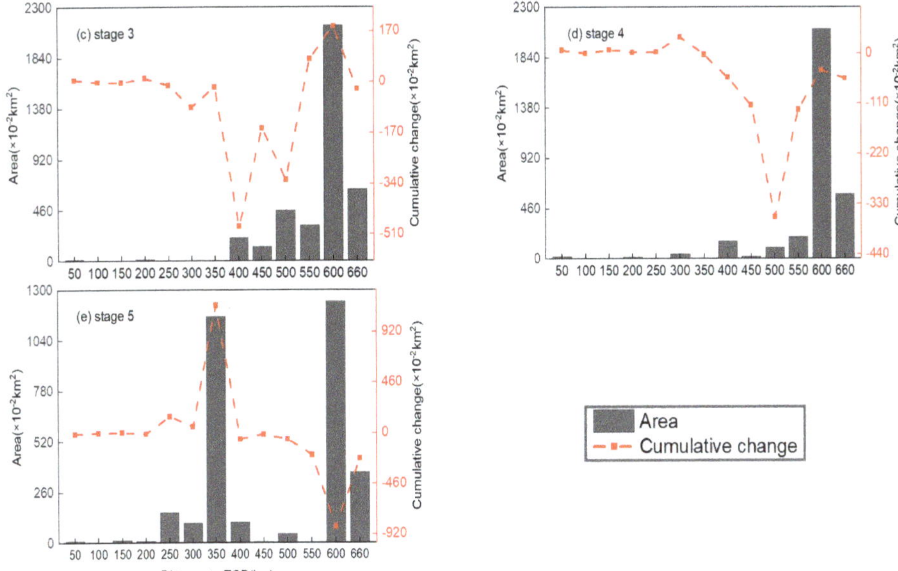

Figure 8. Spatial distribution and changes in the area of MCBs in different impoundment periods: (**a**) stage 1, (**b**) stage 2, (**c**) stage 3, (**d**) stage 4 and (**e**) stage 5.

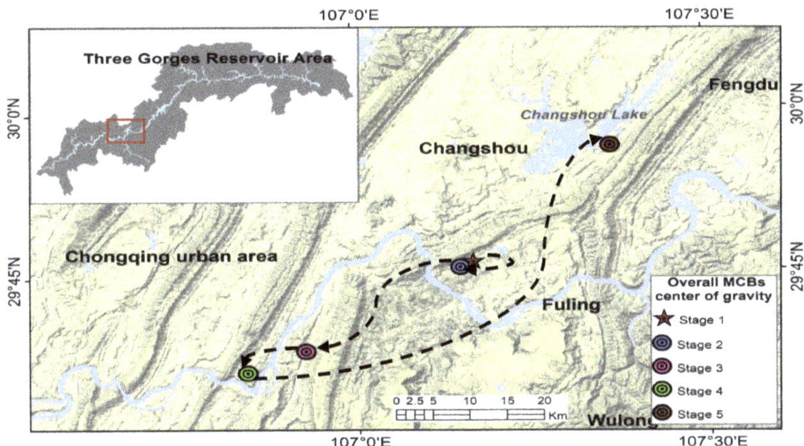

Figure 9. Gravity center migration of MCBs during different impoundment periods.

4. Discussion

The Yangtze River had a natural hydrological and sediment regime before the construction of the TGD, where the MCBs developed with a relatively stable balance between erosion and siltation [37]. However, the construction of the TGD has significantly changed the hydrological and sediment regimes of the Yangtze River over the past decades, which has seriously disrupted this balance. Many previous observations and studies found tremendous riverbed erosion in the middle and lower reaches of the Yangtze River, mainly due to the intercepting of sediment and discharging of clear water since the initial impoundment of the TGD [30,32,41]. The degree of erosion became weaker as the distance from the TGD increased [5,23,28]. The MCBs in the TGR varied dramatically under the influence of notable changes in the hydrological and sediment regimes of the TGR induced by the weakened hydrodynamic condition and rising water levels from stage 1 to stage 5 [36,42]. The surface area of the TGR has expanded as the water level rose since the initial impoundment of the TGD [42,43]. This resulted in a large portion of the previous MCBs being submerged, contributing to a subsequent reduction in the areas of the exposed MCBs. Meanwhile, many new MCBs formed from the inundation of point bars and low-lying mountain tops by the reservoir (Figure 10). A large number of new MCBs appeared, especially in the area of the TGR with the largest fluctuations in water level, in Kaizhou county (Figure 11).

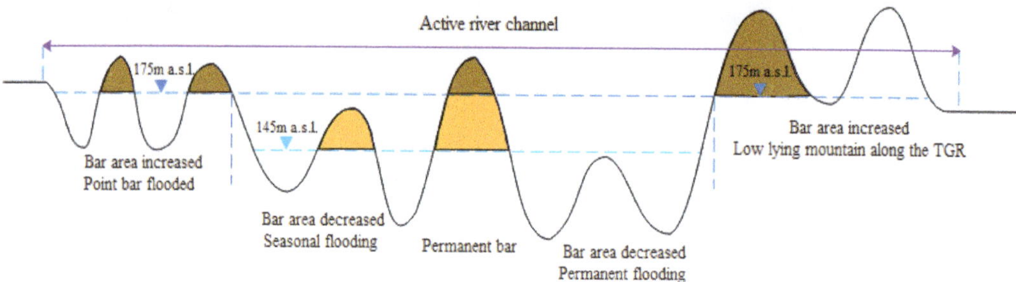

Figure 10. Schematic diagram of impacts of water-level rising on MCBs of the TGR.

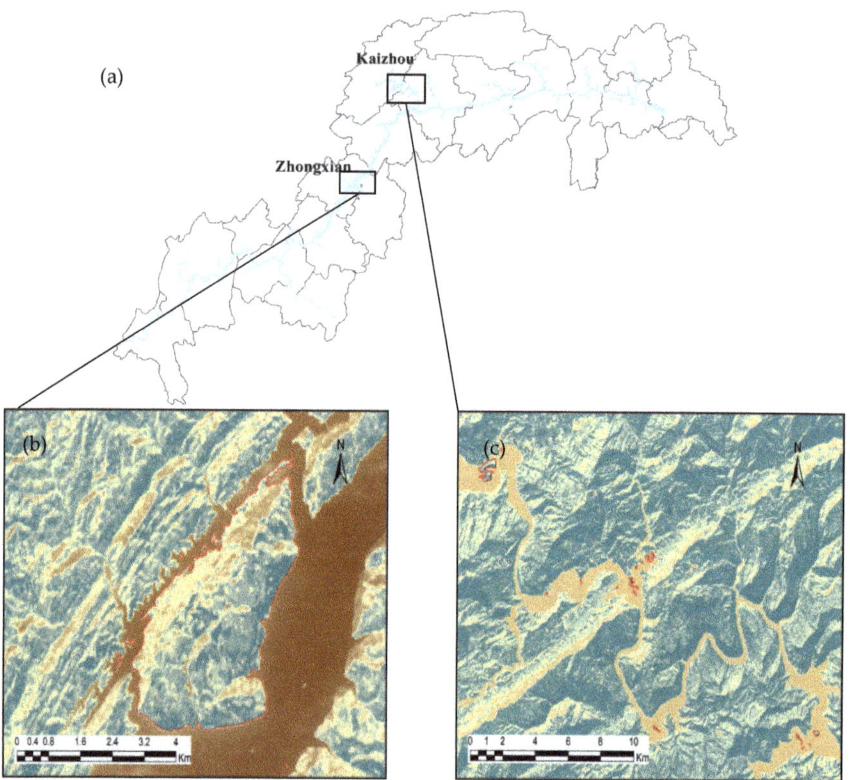

Figure 11. MCBs formed by the flooding of low-lying mountain tops: (**a**) Three Gorges Reservoir Area, (**b**) in Zhongxian and (**c**) in Kaizhou.

Human activities such as sand mining caused an unsaturated sediment transportation capacity of the flow of the TGR (Figure 12) before stage 3, which led to poor stability of the riverbed and MCBs in the TGR. In addition, the conversion from river to reservoir weakened the hydrodynamic condition when the water level reached 139 m at stage 3. The flow rate slowed and a great deal of sediment was trapped in the reservoir, and the water level was the dominant influence on the MCBs [43]. The morphology of the naturally developed MCBs often took typical forms such as oval, bamboo-leaf and sickle-shaped forms. However, the morphology of MCBs changed greatly as the water level rose, regulated by the man-made dam [6,7]. Due to the coupled effect of natural hydrological and sediment regimes and water-level changes regulated by the TGD, the MCBs in the fluctuating backwater zone of the TGR were more affected by siltation than those in the perennial backwater zones. Thus, more attention and protection should be paid to MCBs in fluctuating backwater zones.

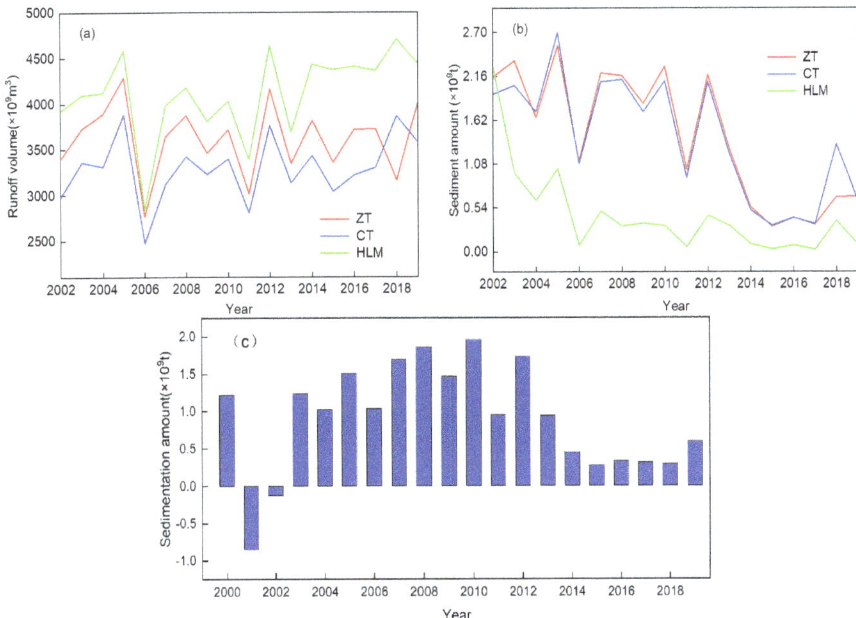

Figure 12. Water discharge and sediment amounts at the three key gauging stations in the different impoundment periods. (**a**) Annual runoff volume from 2002 to 2019, (**b**) annual sediment amount from 2002 to 2019 and (**c**) annual sedimentation amount from 2000 to 2019.

However, limitations and uncertainties associated with the remote sensing image data may exist and need to be improved in future work. In the data acquisition stage, we did our best to collect Landsat images with similar times and water levels to reduce the potential effect on the morphology, numbers and areas of the retrieved MCBs. However, it is very difficult to obtain ideal image data, mainly due to the influence of cloud cover, the difference in spectral features and the revisit cycle of the Landsat satellite, leading to a certain degree of uncertainty in the analysis and results. Thus, more remote sensing images need to be employed and collected from other platforms to reduce the influence of acquisition times, spectral differences and water levels. In the context of the "storing clean water and discharging sandy water" operation schedule [7,8], the influence of the annually cyclic hydrological regime and the longer inundation and exposure induced by the TGR operation, the water level of the TGR fluctuated dramatically between 145 and 175 m, with increase periods from September to January in the following year, and decrease periods from January to September. The present study only investigated the dynamics of MCBs from October to January. It will be necessary to further investigate the changes in MCBs during the year.

Due to the coupled effect of natural hydrological and sediment regimes and water-level changes regulated by the TGD, the development mechanism of MCBs in the fluctuating backwater zones differed from that in the perennial backwater zones, and this was mainly influenced by water-level changes regulated by the TGD. This was not explored in this study. Nevertheless, the scientific findings help to reveal the mechanisms of the development of MCBs in the TGR and can also offer a scientific basis for planning, optimal utilization and ecological restoration of the MCBs in the TGR.

5. Conclusions

This work investigated the spatio-temporal variations in the number, area, morphology and location of MCBs in the TGR during different impoundment periods, using Landsat images. The results showed that the number of MCBs ranged between 89 and 150 with an average of 113, and the area varied greatly from 0.2 to 1134 ($\times 10^{-2}$ km^2) with an average of 31.69. The number and area of MCBs changed dramatically with the water-level changes induced by the impoundments and operation of the TGR. The total area of MCBs decreased progressively from stage 1 to stage 5, with the most significant changes occurring between stage 2 and stage 4. Although the number showed a decreasing trend, the minimum number appeared at stage 3, which was dominated by the change in the number of SMBs. The number and area variations of MCBs differed greatly among MCBs with different sizes, with the most obvious changes appearing for SMBs and LMBs, respectively. The LWR of MCBs in the TGR ranged between 2.09 and 3.05 with an average of 2.56. It generally decreased as the water level rose, suggesting that the morphology of MCBs tended to change from a narrow–long shape to a short–round shape. The LWR variation of MCBs differed greatly among different sizes, with the most obvious changes occurring in SMBs, suggesting that the effect of impoundment on the morphology of SMBs was more pronounced than the effect on MMBs, LMMBs and LMBs. The MCBs were unevenly distributed spatially. They were mostly distributed in the upper section of the TGR at stage 1, under a natural hydrological regime. The water impoundments of the TGR led to the migration of the dominant area from the upper to the middle section of the TGR, resulting in a more even distribution of MCBs in the TGR and the migration of the gravity center of MCBs from the upper to the middle section of the TGR.

This study showed the enormous impacts of the operation of the TGD on the morphological dynamics of MCBs. While the mechanisms of the development of MCBs in the TGR are complex, it will be necessary to investigate these changes further in the future.

Author Contributions: Q.T.: conceptualization, data curation and writing—original draft. D.T.: visualization, formal analysis and funding acquisition. Y.J.: software and methodology. L.Y.: investigation. S.Z.: funding acquisition. Q.C.: writing—review. S.W.: project administration. J.C.: resources, reviewing and supervision. All authors have read and agreed to the published version of the manuscript.

Funding: This research was funded by the Youth Innovation Promotion Association, grant number 2018417, 2021385, the National Natural Science Foundation of China, grant number 41901130, 42061015, the Strategic Priority Research Program of the Chinese Academy of Sciences, grant number XDA23040500, Chongqing Science and Technology project, grant number cstc2019jscx-dxwtBX0007, cstc2021jcyj-msxmX0187, Chongqing Municipal Bureau of Water Resources, grant number 5000002021BF40001 and the Institute of Agricultural Resources and Environment, Tibet Academy of Agricultural and Animal Husbandry Science, grant number 2020ZZKT-01.

Institutional Review Board Statement: Not applicable.

Informed Consent Statement: Not applicable.

Data Availability Statement: Landsat TM, Landsat OLI and the water level, sediment and siltation data for the TGR used in this paper can be downloaded from the Chinese Geospatial Data Cloud, Yangtze River Sediment Bulletin and the Yangtze River Three Gorges Group website.

Acknowledgments: We thank the funding supporters of this work, and the scientific research foundation provided by the Key Laboratory of Reservoir Aquatic Environment, Chinese Academy of Sciences.

Conflicts of Interest: The authors declare no conflict of interest.

References

1. Słowik, M. Sedimentary record of point bar formation in laterally migrating anabranching and single-channel meandering rivers (The Obra Valley, Poland). *Z. Geomorphol.* **2016**, *60*, 259–279. [CrossRef]
2. Afzalimehr, H.; Maddahi, M.R.; Sui, J. Bedform characteristics in a gravel-bed river. *J. Hydrol. Hydromech.* **2017**, *65*, 366–377. [CrossRef]
3. Wang, P.; Fu, K.D.; Huang, J.; Duan, X.; Yang, Z. Morphological changes in the lower Lancang River due to extensive human activities. *PeerJ* **2020**, *8*, e9471. [CrossRef] [PubMed]
4. Long, J.; Li, H.; Wang, Z.; Wang, B.; Xu, Y. Three decadal morphodynamic evolution of a large channel bar in the middle Yangtze River: Influence of natural and anthropogenic interferences. *Catena* **2021**, *199*, 105128. [CrossRef]
5. Wen, Z.; Yang, H.; Zhang, C.; Shao, G.; Wu, S. Remotely Sensed Mid-Channel Bar Dynamics in Downstream of the Three Gorges Dam, China. *Remote Sens.* **2020**, *12*, 409. [CrossRef]
6. Jaballah, M.; Camenen, B.; Pénard, L.; Paquier, A. Alternate bar development in an alpine river following engineering works. *Adv. Water Resour.* **2015**, *81*, 103–113. [CrossRef]
7. Li, W.; Yang, S.; Xiao, Y.; Fu, X.; Hu, J.; Wang, T. Rate and Distribution of Sedimentation in the Three Gorges Reservoir, Upper Yangtze River. *J. Hydraul. Eng.* **2018**, *144*, 05018006. [CrossRef]
8. Zhang, W.; Yuan, J.; Han, J.; Huang, C.; Li, M. Impact of the Three Gorges Dam on sediment deposition and erosion in the middle Yangtze River: A case study of the Shashi Reach. *Hydrol. Res.* **2016**, *47*, 175–186. [CrossRef]
9. Wintenberger, C.L.; Rodrigues, S.; Claude, N.; Jugé, P.; Bréhéret, J.; Villar, M. Dynamics of nonmigrating mid-channel bar and superimposed dunes in a sandy-gravelly river (Loire River, France). *Geomorphology* **2015**, *248*, 185–204. [CrossRef]
10. Klösch, M.; Blamauer, B.; Habersack, H. Intra-event scale bar-bank interactions and their role in channel widening. *Earth Surf. Process. Landf.* **2015**, *40*, 1506–1523. [CrossRef]
11. Hooke, J.M.; Yorke, L. Channel bar dynamics on multi-decadal timescales in an active meandering river. *Earth Surf. Process. Landf.* **2011**, *36*, 1910–1928. [CrossRef]
12. Wang, J.; Dai, Z.; Mei, X.; Lou, Y.; Wei, W.; Ge, Z. Immediately downstream effects of Three Gorges Dam on channel sandbars morphodynamics between Yichang-Chenglingji Reach of the Changjiang River, China. *J. Geogr. Sci.* **2018**, *28*, 629–646. [CrossRef]
13. Li, D.F.; Lu, X.X.; Chen, L.; Wasson, R. Downstream geomorphic impact of the Three Gorges Dam: With special reference to the channel bars in the Middle Yangtze River. *Earth Surf. Process. Landf.* **2019**, *44*, 2660–2670. [CrossRef]
14. Adami, L. Long Term Morphodynamics Ofalternate Bars in Straightenedrivers: A multiple Perspective. Ph.D. Thesis, The University of Trento, Trento, Italy, April 2016.
15. Schuurman, F.; Kleinhans, M.G. Bar dynamics and bifurcation evolution in a modelled braided sand-bed river. *Earth Surf. Process. Landf.* **2015**, *40*, 1318–1333. [CrossRef]
16. Yang, H. Numerical investigation of avulsions in gravel-bed braided rivers. *Hydrol. Process.* **2020**, *34*, 3702–3717. [CrossRef]
17. Wang, D.; Ma, Y.; Liu, X.; Huang, H.Q.; Huang, L.; Deng, C. Meandering-anabranching river channel change in response to flow-sediment regulation: Data analysis and model validation. *J. Hydrol.* **2019**, *579*, 124209. [CrossRef]
18. Crosato, A.; Mosselman, E.; Desta, F.B.; Uijttewaal, W.S. Experimental and numerical evidence for intrinsic nonmigrating bars in alluvial channels. *Water Resour. Res.* **2011**, *47*, 77–79. [CrossRef]
19. Yang, H.; Lin, B.; Sun, J.; Huang, G. Simulating Laboratory Braided Rivers with Bed-Load Sediment Transport. *Water* **2017**, *9*, 686. [CrossRef]
20. Redolfi, M.; Welber, M.; Carlin, M.; Tubino, M.; Bertoldi, W. Morphometric properties of alternate bars and water discharge: A laboratory investigation. *Earth Surf. Dynam.* **2020**, *8*, 789–808. [CrossRef]
21. Tubino, M.; Repetto, R.; Zolezzi, G. Free bars in rivers. *J. Hydraul. Res.* **1999**, *37*, 759–775.
22. Adami, L.; Bertoldi, W.; Zolezzi, G. Multidecadal dynamics of alternate bars in the Alpine Rhine River. *Water Resour. Res.* **2016**, *52*, 8921–8938. [CrossRef]
23. Lyu, Y.; Fagherazzi, S.; Tan, G.; Zheng, S.; Feng, Z.; Han, S.; Shu, C. Hydrodynamic and geomorphic adjustments of channel bars in the Yichang-Chenglingji Reach of the Middle Yangtze River in response to the Three Gorges Dam operation. *Catena* **2020**, *193*, 104628. [CrossRef]
24. Cyples, N.; Ielpi, A.; Dirszowsky, R.W. Planform and stratigraphic signature of proximal braided streams: Remote-sensing and ground-penetrating-radar analysis of the Kicking Horse River, Canadian Rocky Mountains. *J. Sediment. Res.* **2020**, *90*, 131–149. [CrossRef]

25. Liu, X.; Huang, H.; Ddeng, C. A theoretical investigation of the hydrodynamic conditions for equilibrium island morphology in anabranching rivers. *Adv. Water Sci.* **2014**, *25*, 477–483.
26. Rasbold, G.G.; Stevaux, J.C.; Parolin, M.; Leli, I.T.; Luz, L.D.; Guerreiro, R.L.; Brito, H.D. Sponge spicules as indicators of paleoenvironmental changes in island deposits-Upper Paraná River, Brazil. *Palaeogeogr. Palaeoclimatol. Palaeoecol.* **2019**, *536*, 109391. [CrossRef]
27. Fu, B.-J.; Wu, B.-F.; Lue, Y.-H.; Xu, Z.-H.; Cao, J.-H.; Niu, D.; Yang, G.-S.; Zhou, Y.-M. Three Gorges Project: Efforts and challenges for the environment. *Prog. Phys. Geogr. Earth Environ.* **2010**, *34*, 741–754. [CrossRef]
28. Lou, Y.; Mei, X.; Dai, Z.; Wang, J.; Wei, W. Evolution of the mid-channel bars in the middle and lower reaches of the Changjiang (Yangtze) River from 1989 to 2014 based on the Landsat satellite images: Impact of the Three Gorges Dam. *Environ. Earth Sci.* **2018**, *77*, 394. [CrossRef]
29. Edmonds, R.L. The Sanxia (3 Gorges) Project—The environmental argument surrounding China super dam. *Glob. Ecol. Biogeogr. Lett.* **1992**, *2*, 105–125. [CrossRef]
30. Gao, B.; Yang, D.; Yang, H. Impact of the Three Gorges Dam on flow regime in the middle and lower Yangtze River. *Quatern. Int.* **2013**, *304*, 43–50. [CrossRef]
31. Gao, C.; Chen, S.; Yu, J. River islands' change and impacting factors in the lower reaches of the Yangtze River based on remote sensing. *Quatern. Int.* **2013**, *304*, 13–21. [CrossRef]
32. Yang, S.L.; Milliman, J.D.; Li, P.; Xu, K. 50,000 dams later: Erosion of the Yangtze River and its delta. *Glob. Planet. Chang.* **2011**, *75*, 14–20. [CrossRef]
33. Wang, Z.; Li, H.; Cai, X. Remotely Sensed Analysis of Channel Bar Morphodynamics in the Middle Yangtze River in Response to a Major Monsoon Flood in 2002. *Remote Sens.* **2018**, *10*, 1165. [CrossRef]
34. Ren, S.; Zhang, B.; Wang, W.; Yuan, Y.; Guo, C. Sedimentation and its response to management strategies of the Three Gorges Reservoir, Yangtze River, China. *Catena* **2021**, *199*, 105096. [CrossRef]
35. Chen, J.; Yang, H.; Lv, M.; Xiao, Z.; Wu, S. Estimation of monthly pan evaporation using support vector machine in Three Gorges Reservoir Area, China. *Theor. Appl. Climatol.* **2019**, *138*, 1095–1107. [CrossRef]
36. Xiang, R.; Wang, L.; Li, H.; Tian, Z.; Zheng, B. Temporal and spatial variation in water quality in the Three Gorges Reservoir from 1998 to 2018. *Sci. Total Environ.* **2021**, *768*, 144866. [CrossRef]
37. Xu, H. Modification of normalised difference water index (NDWI) to enhance open water features in remotely sensed imagery. *Int. J. Remote Sens.* **2006**, *27*, 3025–3033. [CrossRef]
38. McFeeters, S.K. The use of the Normalized Difference Water Index (NDWI) in the delineation of open water features. *Int. J. Remote Sens.* **1996**, *17*, 1425–1432. [CrossRef]
39. Li, Z.; Wang, Z.; Zhang, K. Relationship between morphology of typical sand bars and river channels. *J. Sediment. Res.* **2012**, *1*, 68–73. (In Chinese)
40. Hu, L.; Jiang, C.; Li, Z.B.; Zhang, X.; Li, P.; Wang, Q.; Zhang, W.J. Evolution Path Analysis of Economic Gravity Center and Air Pollutants Gravity Center in Shaanxi Province. *Adv. Mater. Res.* **2012**, *361–363*, 1359–1363. [CrossRef]
41. Huang, H.; Deng, C.; Nanson, G.C.; Fan, B.; Liu, X.; Liu, T.; Ma, Y. A test of equilibrium theory and a demonstration of its practical application for predicting the morphodynamics of the Yangtze River. *Earth Surf. Process. Landf.* **2014**, *39*, 669–675. [CrossRef]
42. Ji, Y.; Tang, Q.; Yan, L.; Wu, S.; Yan, L.; Tan, D.; Chen, J.; Chen, Q. Spatiotemporal Variations and Influencing Factors of Terrestrial Evapotranspiration and Its Components during Different Impoundment Periods in the Three Gorges Reservoir Area. *Water* **2021**, *13*, 2111. [CrossRef]
43. Tang, Q.; Bao, Y.; He, X.; Fu, B.; Collins, A.L.; Zhang, X. Flow regulation manipulates contemporary seasonal sedimentary dynamics in the reservoir fluctuation zone of the Three Gorges Reservoir, China. *Sci. Total Environ.* **2016**, *548*, 410–420. [CrossRef] [PubMed]

Article

Water Quality Index (WQI) as a Potential Proxy for Remote Sensing Evaluation of Water Quality in Arid Areas

Fei Zhang [1,*], Ngai Weng Chan [2], Changjiang Liu [1], Xiaoping Wang [1], Jingchao Shi [3], Hsiang-Te Kung [3], Xinguo Li [4], Tao Guo [5], Weiwei Wang [1] and Naixin Cao [1]

[1] Key Laboratory of Wisdom City and Environment Modeling of Resources and Environmental Science College, Xinjiang University, Urumqi 830046, China; liuchangjiang3s@163.com (C.L.); wxp4911@163.com (X.W.); wangweiwei0820@163.com (W.W.); cnx19971214@163.com (N.C.)
[2] School of Humanities, Universiti Sains Malaysia, George Town 11800, Malaysia; nwchan1@gmail.com
[3] Departments of Earth Sciences, The University of Memphis, Memphis, TN 38152, USA; jshi@memphis.edu (J.S.); hkung@memphis.edu (H.-T.K.)
[4] School of Geography and Tourism, Xinjiang Normal University, Urumqi 830046, China; onlinelxg@163.com
[5] Institute of Remote Sensing and Digital Agriculture, Sichuan Academy of Agricultural Sciences, Chengdu 610066, China; guotao3s@163.com
* Correspondence: zhangfei3s@xju.edu.cn or zhangfei3s@163.com

Citation: Zhang, F.; Chan, N.W.; Liu, C.; Wang, X.; Shi, J.; Kung, H.-T.; Li, X.; Guo, T.; Wang, W.; Cao, N. Water Quality Index (WQI) as a Potential Proxy for Remote Sensing Evaluation of Water Quality in Arid Areas. *Water* **2021**, *13*, 3250. https://doi.org/10.3390/w13223250

Academic Editor: Dimitrios E. Alexakis

Received: 25 September 2021
Accepted: 14 November 2021
Published: 17 November 2021

Publisher's Note: MDPI stays neutral with regard to jurisdictional claims in published maps and institutional affiliations.

Abstract: Water Resource Sustainability Management plays a vitally important role in ensuring sustainable development, especially in water-stressed arid regions throughout the world. In order to achieve sustainable development, it is necessary to study and monitor the water quality in the arid region of Central Asia, an area that is increasingly affected by climate change. In recent decades, the rapid deterioration of water quality in the Ebinur Lake basin in Xinjiang (China) has severely threatened sustainable economic development. This study selected the Ebinur Lake basin as the study target, with the purpose of revealing the response between the water quality index and water body reflectivity, and to describe the relationship between the water quality index and water reflectivity. The methodology employed remote sensing techniques that establish a water quality index monitoring model to monitor water quality. The results of our study include: (1) the Water Quality Index (WQI) that was used to evaluate the water environment in Ebinur Lake indicates a lower water quality of Ebinur Lake, with a WQI value as high as 4000; (2) an introduction of the spectral derivative method that realizes the extraction of spectral information from a water body to better mine the information of spectral data through remote sensing, and the results also prove that the spectral derivative method can improve the relationship between the water body spectral and WQI, whereby R^2 is 0.6 at the most sensitive wavelengths; (3) the correlation between the spectral sensitivity index and WQI was greater than 0.6 at the significance level of 0.01 when multi-source spectral data were integrated with the spectral index (DI, RI and NDI) and fluorescence baseline; and (4) the distribution map of WQI in Ebinur Lake was obtained by the optimal model, which was constructed based on the third derivative data of Sentinel 2 data. We concluded that the water quality in the northwest of Ebinur Lake was the lowest in the region. In conclusion, we found that remote sensing techniques were highly effective and laid a foundation for water quality detection in arid areas.

Keywords: Water Quality Index (WQI); Ebinur Lake; remote sensing

1. Introduction

Water problems can be a great barrier to economic development in any corner of the world [1–3], especially in such arid regions as Xinjiang, China, where water shortages (and other water issues) aggravate ecological environment deterioration. Therefore, studying and monitoring water quality is very important to reduce the potential negative impacts on the ecological environment in Xinjiang. However, traditional water quality monitoring methods are time-consuming, cumbersome, and limited to a small scale. Therefore, they can no longer meet the needs of water quality monitoring in terms of speed, large areas, or a long time series. In order to have more accurate estimates, new data sources and new methods need to be introduced in the monitoring of comprehensive water quality indicators [4,5]. The development of multi-source observation and the monitoring of remote sensing technology increasingly brings huge opportunities for speedy, high-precision water environment monitoring and evaluation over large areas.

Satellite remote sensing technology has developed very quickly since 1970. Consequently, more water resource researchers started to apply remote sensing technology in their research, and the water quality within remote sensing monitoring mechanisms has also gradually improved. In recent years, remote sensing satellites have been widely used to observe pollutants in rivers and lakes. As a result, the detectable types of pollutants retrieved from satellite images have greatly increased, and the inversion accuracy has been further improved, as well [6].

Remote sensing applied to water quality monitoring is mainly used to map the water quality indexes of rivers and lakes through the relationship between water quality indexes and spectral data with satellite image data such as Landsat, MODIS, ENVISAT, and SPOT data [6–8]. However, the spatial resolution of the above remote sensing data is greater than 10 m, which makes it difficult to meet monitoring requirements. Only a few water quality parameters can be monitored by these remote sensing data, such as Chlorophy ll, SS, NTU, and CDOM. [4,9,10]. Other chemical indicators of water quality, such as COD, BOD_5, TN, TP, NH_3-N, DO, etc., cannot be directly monitored by remote sensing. The indirect monitoring accuracy is low, and the mechanism is unclear. Hence, introducing a new technology that makes up for the deficiency of remote sensing water quality monitoring is essential.

In fact, the process of water pollution follows a nonlinear regression that fluctuates with many factors, and the accuracy of the water quality inversion result is limited by the traditional linear inversion model [11]. However, machine learning has a good nonlinear approximation ability, and the application of machine learning in water quality monitoring provides a new idea to improve the accuracy of water quality monitoring. Alves simplified the input variables of the feed forward neural network through principal component analysis, thus accurately inverting the water quality index (WQI) [12]. Gogu proved that there is a good potential in using a neural network to invert the salt content of river water through experiments [13]. Wang [11] estimated the WQI of water quality in the Ebinur Lake basin based on the support vector machine (SVR) model by using near-surface spectroscopy technology, and found that the nonlinear model has great potential in water quality observation.

Although the water quality parameter estimation model provides relatively highly accurate data, the result is uncertain due to the complex and changeable water environment. The reason is that the water spectrum shows the entire water environment rather than a single water quality parameter. Many scholars have developed a single water quality parameter estimation model based on water spectral data [14–16]. Therefore, the estimation model of individual water parameters introduces a certain degree of uncertainty. At this point, the establishment of the water quality index reflecting the whole water environment to evaluate the whole water environment is necessary. Moreover, a good water quality evaluation method should not only accurately reflect the spatial change of the water quality but also conveniently monitor the water quality level. Data on the Water Quality Index (WQI) is compiled by the Ministry of Water Resources and the Water Environment Monitoring and Evaluation Center to evaluate the quality of drinking water [17,18]. The WQI was originally proposed by Horton and Brown [19,20]. Scholars have devised various methods to calculate the water quality Index (WQI) [21,22], which is a mathematical tool of converting large amounts of water quality data into a single value that represents the water environment and reflects the overall water quality level [23]. However, it is impossible to identify the temporal and spatial variation of water quality, which is crucial for the comprehensive evaluation and management of water quality, even though the WQI method can provide reasonable accuracy of the water quality of a single sample.

In this paper, the relationship between the water quality index, water optical characteristics, and water reflectance is quantitatively analyzed. The specific research objectives in this paper include: (1) to better mine the information of remote sensing data by using a series of technologies, such as the remote sensing image differential algorithm, which are introduced to realize the extraction of water remote sensing information; (2) to construct a remote sensing spectral index (DI, RI and NDI) and fluorescence baseline height for monitoring water quality in arid areas; and (3) to establish a WQI model based on machine learning technology (particle swarm optimization algorithm) to achieve water quality monitoring. This study will provide an effective method for rapid, quantitative, and sustainable water quality management in arid areas, as well as a typical example for ecological conservation in arid areas, and it will also effectively contribute to the health of the ecological environment in arid areas.

2. Study Area

The Ebinur Lake watershed (43°380–45°520 N and 79°530–85°020 E) is located in northwest Xinjiang, China (Figure 1). The study area is 50,621 km^2, comprising Bortala River Valley, Jinghe oasis, Wusu Oasis, Dandagai desert, and the Mutetaer desert zone of the lower reaches of the Akeqisu-Kuitun River. The Ebinur Lake is in the lowest elevation of the watershed and is the largest saltwater lake in Xinjiang. It has all the typical characteristics that all other lakes do in the arid region of Central Asia. The area experiences a typical arid continental climate in the middle temperate zone and is characterized by drought, low rainfall, drastic temperature variations, and severe soil salinization. The average lake depth is merely 1.4–1.6 m, with a water density of about 1.079 g/cm^3, pH 8.49, and mineralization of 112.4 g/L. The watershed is one of the key areas of China's Silk Road Economic Belt, and can be divided into three sub-basins, namely, the Jinghe River basin, Boltala River basin, and Kuitun River basin. The Ebinur Lake basin consists of a varied landscape of mountain, desert, and oasis, where land is mainly use for agricultural. The annual average temperature is 7.2 °C, with the highest 9.1 °C and the lowest 5.3 °C. The annual extreme high and low temperature is 41 °C and −34.7 °C, respectively. The annual average precipitation is only 149 mm, but the potential evapotranspiration reaches up to 2281 mm.

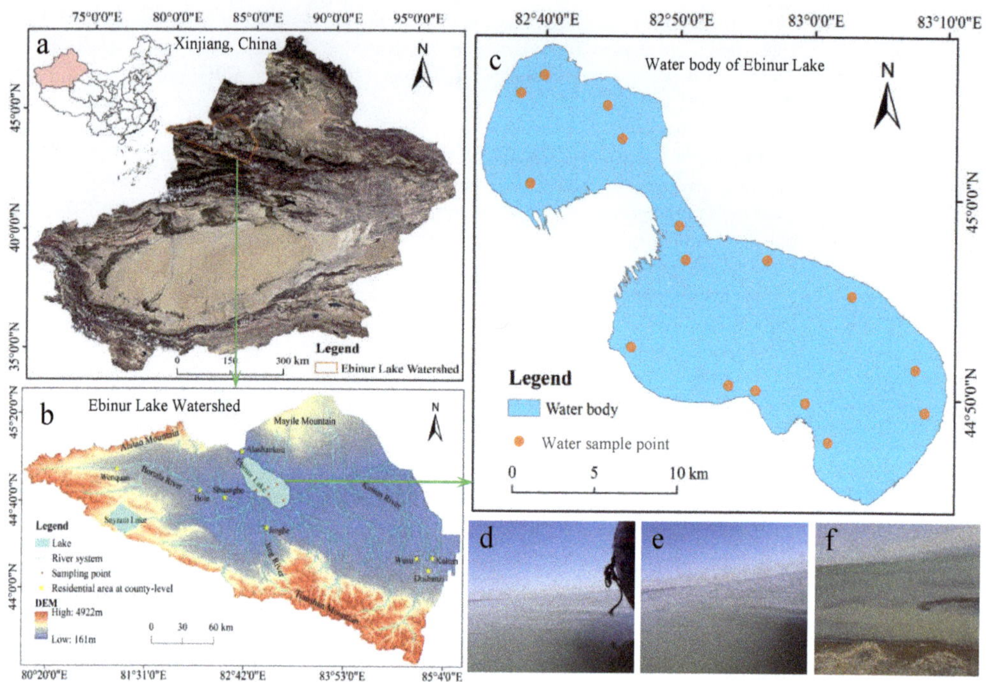

Figure 1. The study area: (**a**) The Xinjiang Uyghur Autonomous Region in northwestern China; (**b**) The Ebinur Lake Watershed with elevation and drainage information; (**c**) The water body and 16 sampling points at Ebinur lake extracted on October 2017; (**d**) A typical view of Ebinur Lake with sunny weather; (**e**) Ebinur Lake surface water landscape; and (**f**) The unique visual effect of Ebinur Lake.

3. Data and Methods

3.1. Data Collection

Water Quality Data Collection

The field investigations and water quality sample collections for this study were conducted in October 2015. They were integrated with the main body of the experiment, which included three parts of water sampling, including water surface spectral measurement, GPS record location, and other auxiliary information. Spectroscopy was measured by a FieldSpec®ProFR (wavelength range: 350–2500 nm), a portable ASD spectrometer (Analytical Spectral Devices, Boulder, CO, USA). Water samples were sent to the laboratory for analysis within a specified time frame.

The researchers collected a total of 16 water samples. For each sampling point, a water sample collector was used to collect water samples at 0.5 cm depth, just below the water surface, with 1000 mL of water samples collected at each sampling point. Samples were stored in Teflon plastic bottles (for standard and easy transportation). Teflon plastic bottles were washed several times with collected water before each collection. After the samples were collected, they were immediately put into the benzene board incubator with ice and transported to the laboratory where the water quality index was determined as soon as possible.

3.2. Remote Sensing Data Collection

The European Space Agency (ESA) recently launched the Copernicus Project, which is expected to improve the monitoring of forest conditions and land use, as well as enhance disaster management through the launch of Sentinel satellites. The Sentinel-2 Satellite

Multispectral Imager covers 13 spectral segments (443–2190 nm), a width of 290 km, with a spatial resolution of 10 m (4 visible spectral segments and 1 near infrared spectral segment), 20 m (6 red edge spectral segments and short-wave infrared spectral segments), and 60 m (3 atmospheric correction spectral segments). The Sentinel-3 was launched on 16 February 2016 [24]. The Sentinel-3-3 satellite has two payloads: one is the OLCI (Sea-Land Colorimeter) and the other is the SLSTR (Sea-Land Surface Temperature Radiometer). The OLCI is an optical instrument designed to provide data continuity for ENVISAT's MERIS. The OLCI is a push-sweep imaging spectrometer that measures solar radiation reflected from the Earth in 21 spectral bands with a ground-based spatial resolution of 300 m [25]. Multispectral remote-sensing data of the Sentinel-2 MSI and Sentinel-3 OLCI data were obtained from the ESA (2 October 2021, https://Sentinel.esa.int/web/Sentinel/home). In this study, only ENVI (ENVI5.4.1) soft data were used for preprocessing, including radiometric calibration and FLAASH atmospheric correction.

3.3. Methods

3.3.1. Construction of Spectral Index

The information from the ground objects observed by remote sensing data is mainly displayed by the difference and change of the spectral characteristics of the ground objects [26]. The ground features obtained by the different spectral channels have different correlations with different elements or some characteristic states of ground features. However, complex remote sensing data can only be represented by a single channel or multi-channel spectral combination [11]. Therefore, further mining with very limited remote sensing signals is necessary to represent ground object information through remote sensing data. In this study, the combination of multi-spectral remote sensing data (such as linear and non-linear combination, subtraction, multiplication, and division) was selected to achieve the effective expression of spectral information and to lay a foundation for the qualitative and quantitative evaluation of water body information. The optimal remote sensing indices (RI, DI, and NDI) were selected for the estimation of WQI, in which multiband remote sensing data were used as variable factors. Subsequently, a combined operation was conducted for various bands and the sensitivity of WQI information, which was obviously better than that of the single-band models, highlighting the advantages of using band combinations. The remote sensing index of water quality in arid area was constructed by Formulas (1)–(3):

$$RI(i,j) = \frac{R_i}{R_{j_i}} \tag{1}$$

$$RI(i,j) = \frac{(R_i - R_j)}{(R_i + R_j)} \tag{2}$$

$$DI(i,j) = R_i - R_j \tag{3}$$

where RI (i, j) is the ratio remote sensing index, NDI (i, j) is the water body normalized remote sensing index, DI (i, j) is the water body difference remote sensing index, and i, j is any band of the data of any two bands of the 350–2500 nm band.

3.3.2. Fluorescence Line Height

The statistical algorithm, based on the correlation between fluorescence line height (FLH) and chlorophyll concentration, is called the fluorescence baseline height method. The general algorithm is derived based on three wavelengths, including the central wavelength which is the maximum value of chlorophyll fluorescence (around 685 nm, which varies with the concentration of water components), and the other two baseline bands which are located on both sides of the fluorescence peak, as shown in Figure 2 [27]. The fluorescence line height (FLH) was calculated as follows: where C was the concentration of chlorophyll on the water surface (unit: mg/m^3); and FLH is the fluorescence baseline height (unit: mW/(cm^{2*} Sr *nm)). a, b, and k are the coefficients.

$$FLH = K + \frac{a \times C}{1 + b \times C} \tag{4}$$

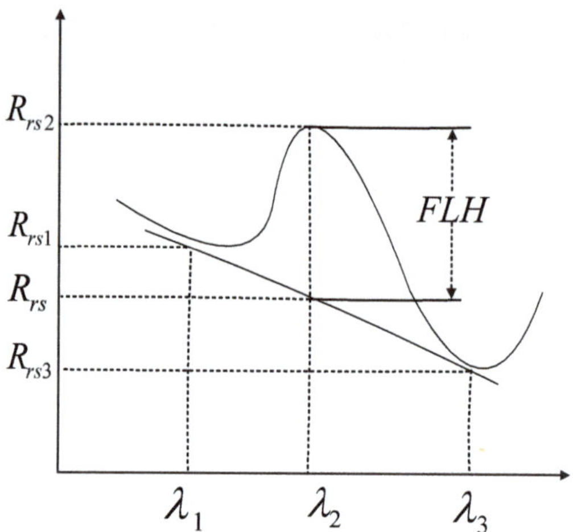

Figure 2. The principle of fluorescent line height.

The calculation formula of FLH is shown in Formula (4), where $\lambda2$ is the central wavelength, and $\lambda1$ and $\lambda3$ are the selected baseline wavelengths. L1, L2, and L3 are the radiance values of corresponding wavebands (unit: mW/(cm^{2*} Sr *nm)). The fluorescence channel designs are 665, 681.25, and 709 nm.

$$FLH = L2 - \left[L1 + (L1 - L3) \times \frac{\lambda2 - \lambda1}{\lambda1 - \lambda3} \right] \tag{5}$$

3.3.3. Water Quality Index (WQI)

The WQI is a comprehensive water environment index, which can reasonably quantify the degree of water pollution [28–30]. The method was first proposed by Horton and Brown [19,20], leading to the development of many water quality indices thereafter [21,22]. WQI can effectively reflect the water quality according to research objectives. Consequently, the WQI has been widely used in water environment assessments [31,32]. The smaller the WQI, the better the water quality. The researchers chose the water quality index constructed by Wang [11] for calculation. The index is constructed by using the measured water quality data of the Ebinur Lake basin, which meet the needs of water quality evaluation in arid areas. The water quality index scale is shown in Table 1.

Table 1. Water Quality Index scale.

Class	Threshold Value	Water Quality
I	$\geq$50	Excellent water
II	(50–100)	Good water
III	[100–200)	Poor water
IV	[200–300)	Very poor water
V	$\geq$300	Unsuitable for drinking

3.3.4. SVM Model

The Support Vector Machine (SVM) is a kind of machine learning technology based on the principle of structural risk minimization. It can solve the problems of small sample, nonlinear, high dimension, and local minimum well. It has an excellent prediction and generalization ability. The penalty factor C and the kernel function parameter σ in a support vector machine directly affect the prediction accuracy of the model. According to previous studies, the following three optimization algorithms can improve the accuracy of the SVM algorithm: Cross-validation selecting the optimal parameter (CV_cg); Genetic Algorithm (GA); and Particle Swarm Optimization (PSO) [33,34]. In this study, particle swarm optimization was selected for parameter optimization, as Wang proved that particle swarm optimization was more suitable for Ebinur Lake [11].

3.3.5. Estimate the Evaluation Index of the Model

In the establishment of the estimation model and the evaluation of accuracy, the fitting coefficient R^2, standard deviation SD, and root mean square error RMSE were selected in this study. R^2 is the determination coefficient. RPD refers to relative analysis error. RPD < 1.4 indicates that the model is unreliable; 1.4 < RPD < 2 indicates that the model has a general accuracy; and RPD > 2 indicates that the model has a high prediction ability [11].

4. Results and Analysis

4.1. Analysis of Spatial Variation Trend of WQI

Figure 3 shows the spatial distribution pattern of the WQI in Ebinur Lake, whereby the maximum value of WQI is 5678.35 and the minimum value is 1066.65. Overall, the degree of water pollution of Ebinur Lake is very high, and the salt content in Ebinur Lake is at a high level as well. However, different parts of Ebinur Lake are polluted at differing degrees. Specifically, the northwestern part of Ebinur Lake is the most polluted area. Similarly, the water environment and ecological environment safety of the Junggar Basin in northern Xinjiang are threatened by water quality issues. Therefore, efficient digital management of water quality is particularly important to ensure water sustainability in these areas.

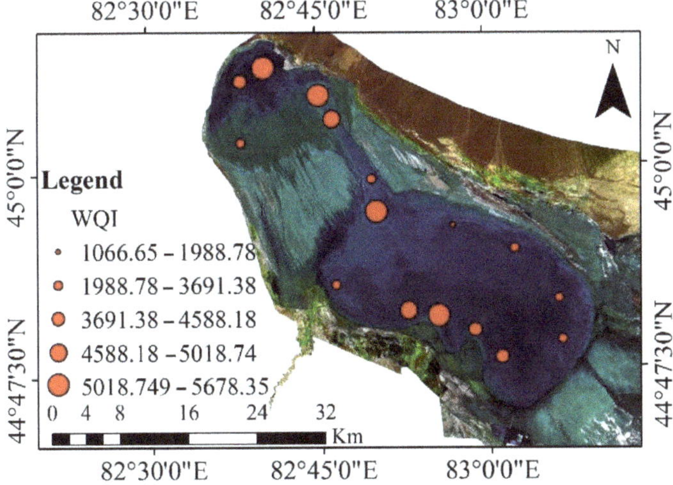

Figure 3. The spatial distribution of water quality index (WQI) in Ebinur Lake.

4.2. Study on Reflectance Spectral Characteristics of Water in Ebinur Lake

4.2.1. Spectral Characteristics of Water Based on Sentinel 3 Data

To obtain the most sensitive and effective water quality monitoring information, Sentinel 3 images were processed with the 1st, 2nd, and 3rd derivatives. However, the pixel reflectance value obtained by the 3rd derivative processing was the same due to the coarse spatial scale resolution. Thus, the pixel reflectance value was not considered in this study. The fluorescence baseline height (FHL) of the watercolor sensor was one of the main parameters examined in this study. The FHL calculated values in this paper are shown in Figure 4.

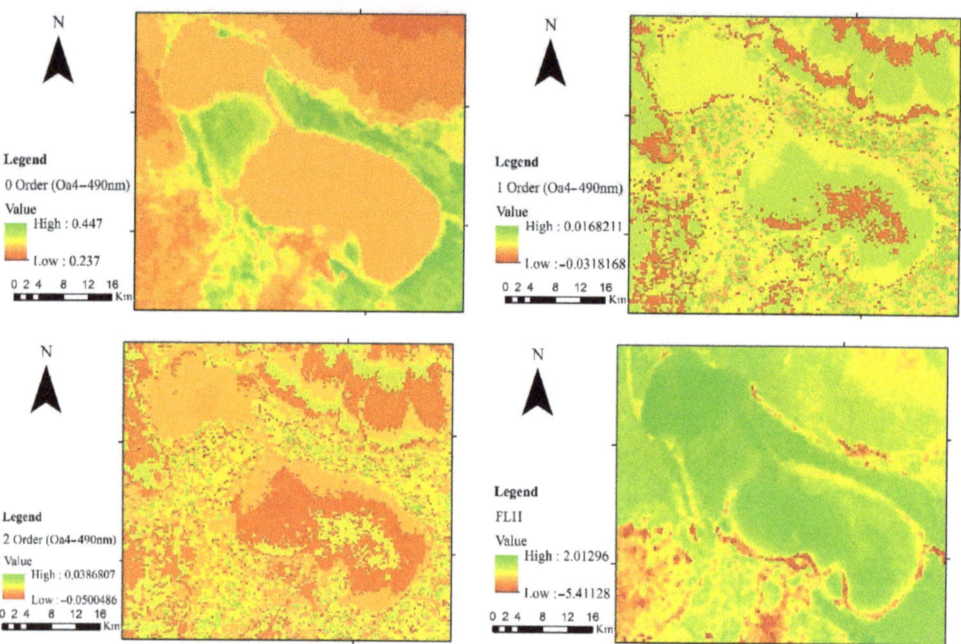

Figure 4. Spectral characteristics of water of Sentinel 3 (0 order, 1 order, 2 Order, and FHL).

To intuitively study the remote sensing data mining by image derivatives algorithm, we demonstrate the images of the fourth band (Oa4) of Sentinel 3 data with a central wavelength of 490 nm, shown in Figure 3. The raw data show that the lowest reflectance is 0.237 and the highest is 0.447. In the wetlands around the lake, salt spills out and forms salt shells on the surface due to the high degree of salinization. Therefore, the maximum reflectivity is in the salt crust around the lake, and the land-water boundary is very clear, but the difference between the land-water boundary is not as clear in the shallow lake depth. In the first derivative data, the lowest reflectance is −0.0318 and the highest is 0.0168. The boundary between land and water disappears. In terms of color, the reflectance of the surrounding mountains is in the same range as that of the center of the lake, but for the lake as a whole, the spectrum of the water body is different. In the second derivative data, the lowest reflectance is −0.05 and the highest is 0.0386. In terms of reflectance values, the second derivative amplifies the difference in reflectance values better than the first derivative. Although the boundary between water and land is blurred, it is still distinguishable. The reflectance of the surrounding mountains is in the same range as that of the center of the lake, but the spectral of the water body is different for the whole lake. In the fluorescence line height (FLH) image, the lowest value is −5.41128 and the highest value is 2.01296, where the reflectance value increases several times, the land-water boundary is clear, the color difference in the lake is obvious, and the spectral difference of water body is distinguishable. The results show that the derivative algorithm can amplify the reflectivity difference, but it cannot separate the land-water boundary.

4.2.2. Spectral Characteristics of Water Based on Sentinel 2 Data

We also used the derivative method to process Sentinel 2 data and showed the data of the fourth band (B2) with a central wavelength of 490 nm in Figure 5. The raw data show the reflectivity ranged from 0.0003 to 0.6848. Furthermore, the maximum reflectivity is in the salt crust around the lake. The land-water boundary is very clear, but the difference is not distinguishable in the shallow water around the lake. In the first derivative data, the lowest reflectance is 0.00605 and the highest reflectance is 0.1872. The land-water boundary is clear, with the surrounding mountains and land almost distorted, but the land-water boundary cannot be clearly distinguished. In the second derivative data, the lowest reflectance is -0.3296 and the highest reflectance is 0.3591. The second derivative over the first derivative and the original image data magnify the difference in reflectivity values. The boundary between land and water is very clear, and the small lakes in the southwest can also be distinguished. The spectral difference between the surrounding plain land and vegetation cover area is clear, but the spectral difference between the water body in the lake is not significant. The third derivative image data shows that the lowest reflectivity is -0.209225 and the highest reflectivity is 0.1361. In the reflectance value, the difference of the reflectance value can be reduced by the third-order derivative image data compared with the first-order derivative and second-order derivative image data. The boundary between water and land is very clear.

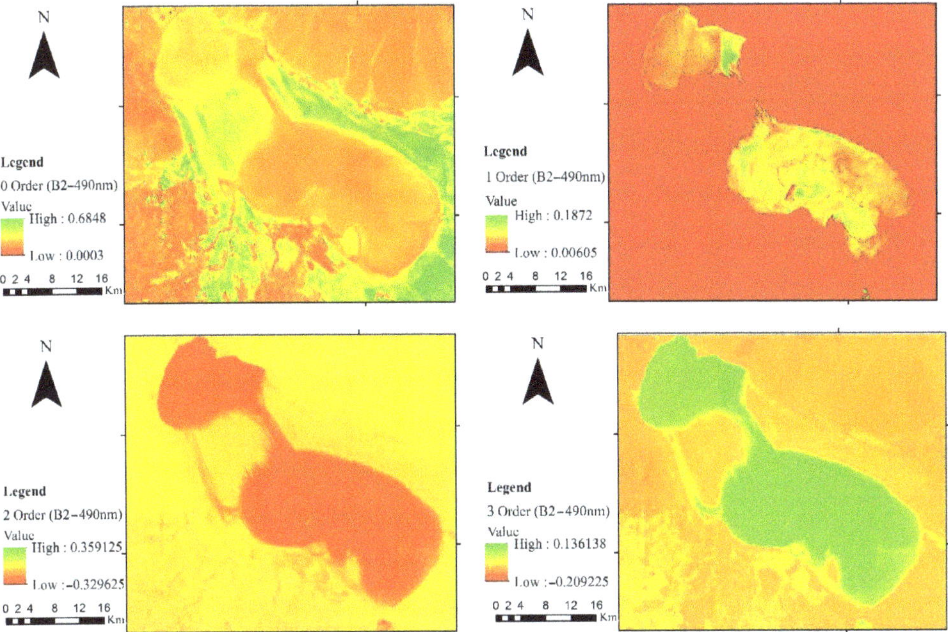

Figure 5. 0 order, 1 order, 2 Order, and 3 Order image of B14.

4.3. Relationship between Spectral Parameters and WQI

4.3.1. Relationship between WQI and Spectral Parameters from Sentinel 3 Data

(1)   Relationship between Single Band Reflectance and WQI

The correlation coefficients between the WQI and the spectral reflectance form the raw image, and the first and second order derivative spectral values of the Sentinel-3 OLCI image data were calculated in this study. The results are shown in Figure 6. These correlation coefficients were tested at the 0.01 significance level. As the derivative order increases, the number of bands passing the significance test also increases, and the correlation coefficient also increases. The bands Oa4, Oa5, and Oa21 in the first-order differential passed the significance test, with the bands Oa3, Oa4, Oa5, Oa11, and Oa21 in the second-order differential also passing the significance test. The results further show that the differential method is helpful in remote sensing spectral data mining.

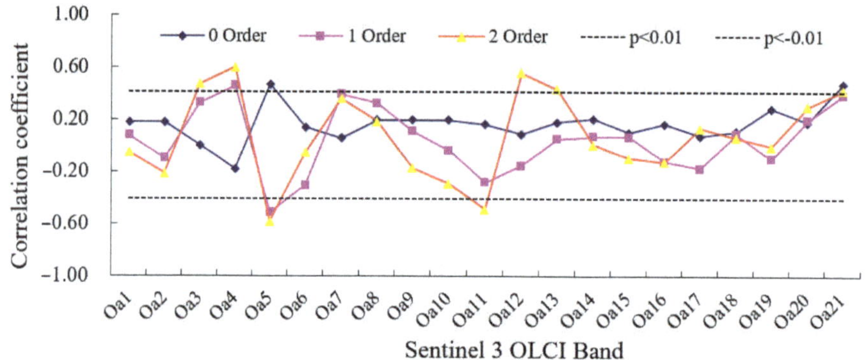

Figure 6. Relationship between single band reflectance and water quality WQI in Sentinel 3 data.

(2)   Relationship between spectral index from Sentinel 3 data and WQI

To enhance the spectral difference between a water body and other ground objects, we have constructed the water spectral index. In this study, NDI, DI, and RI were selected as the combination methods of spectral indexes, and the relationship between the WQI and spectral indexes was studied through Sentinel 3 data, as shown in Figure 7, providing a basis for the further construction of a water quality evaluation model. The correlation coefficient between the WQI and spectrum index of water is shown in Table 2.

We found that DI and NDI chose the same band in the same derivative order, such as the 0 derivative. For RI, the highest correlation coefficient between the raw spectral reflectance, derivative spectral reflectance value, and WQI is 0.701 at the 0-order derivative. The combined band is Oa13 and Oa17, the lowest correlation coefficient is 0.602, and the combined band is Oa5 and Oa20 at the second order derivative. For DI, the highest correlation coefficient between the raw spectral reflectance, derivative spectral reflectance value, and WQI is 0.705 at the 0-order derivative. The combined band is Oa3 and Oa8, with the lowest correlation coefficient 0.602, and the combined band is Oa5 and Oa21 at the second order derivative. For NDI, the highest correlation coefficient between the raw spectral reflectance, derivative spectral reflectance value, and WQI is 0.701 at the 0-order derivative. The combined band is Oa4 and Oa5, the lowest correlation coefficient is 0.592, and the combined band is Oa5 and Oa21 at the second order derivative. The study found that the derivative algorithm for Sentinel 3 data did not significantly improve the relationship between the spectral index and WQI, because the relationship between the spectral index and water quality index (WQI) constructed from Sentinel 3 raw data was the best.

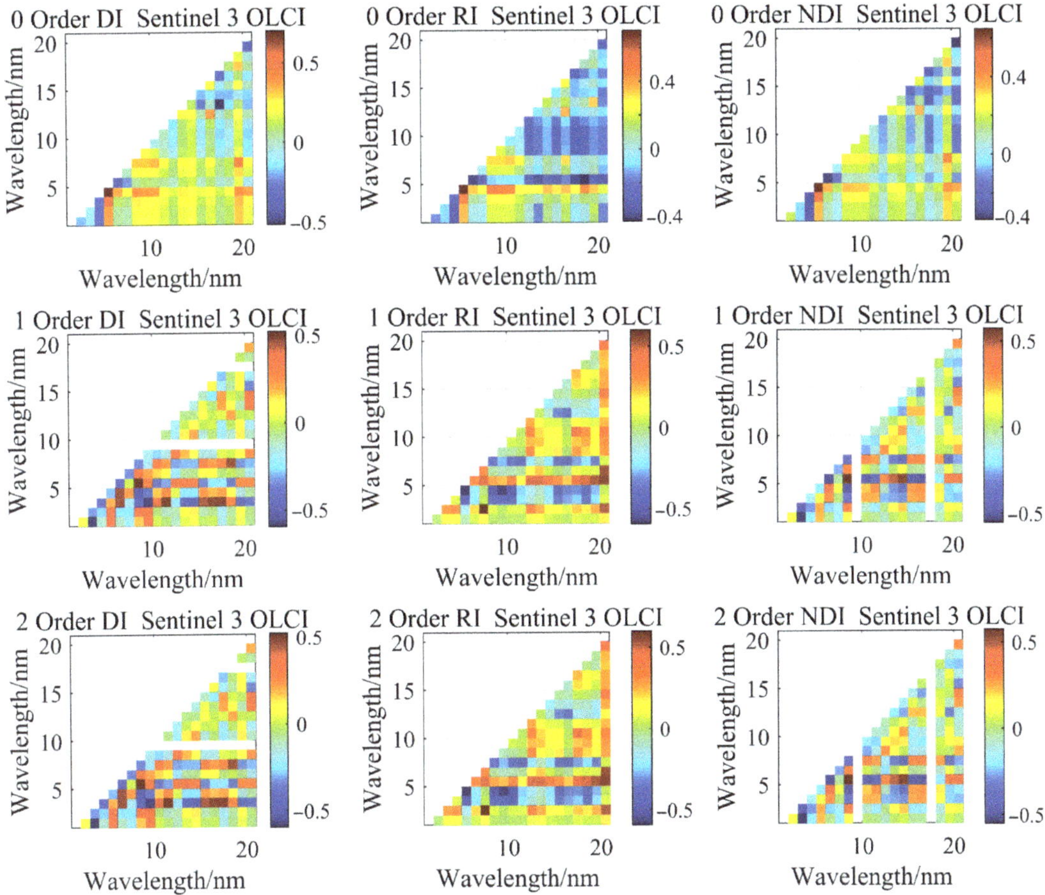

Figure 7. Relationship between WQI and spectrum index from Sentinel 3.

Table 2. Correlation coefficient between WQI and spectrum index of water.

Derivative Order	RI		DI		NDI	
	Band	R	Band	R	Band	R
0	Oa13/Oa17	0.701	Oa4 − Oa5	0.705	(Oa4 − Oa5)/(Oa4 + Oa5)	0.701
1	Oa5/Oa20	0.695	Oa3 − Oa8	0.662	(Oa3 − Oa8)/(Oa3 + Oa8)	0.622
2	Oa1/Oa3	0.602	Oa5 − Oa21	0.602	(Oa5 − Oa21)/(Oa5 + Oa21)	0.592

4.3.2. Relationship between WQI and Spectral Parameters from Sentinel 2 Data

(1)   Relationship between single band reflectance and WQI

The correlation coefficients between the WQI and the spectral reflectance form the raw image, and the first and second order derivative spectral values of Sentinel-2 MSI image data were calculated in this study. The results are shown in Figure 5, in which correlation coefficients were tested at the 0.01 significance level. The correlation coefficient curves of Sentinel-2 MSI original spectral reflectance, derivative spectral values of order 1, 2, and 3, and water quality index WQI calculation are shown in Figure 8.

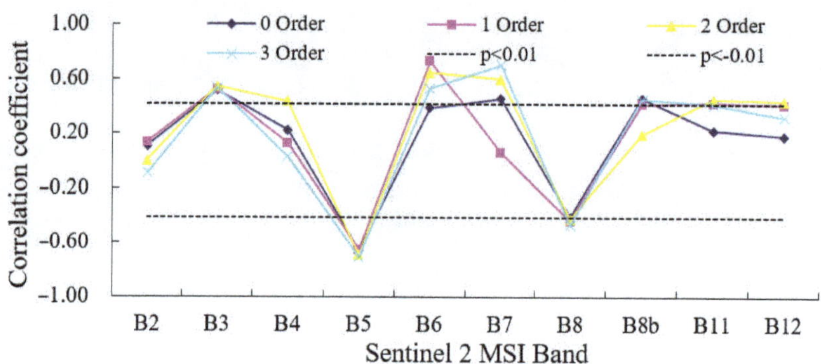

Figure 8. Relationships between WQI and spectral from sentinel-2 MSI.

The relationships between the raw reflectance of Sentinel-2 MSI image data and WQI was significantly correlated in four bands: B3, B5, B7, and B8b. The number of bands passing the significance test increased and the correlation coefficient also increased, with the derivative order increasing as well. The first order derivative was significant in the bands B3, B5, B7, and B8b, and the second order derivative was significant in the bands B3, B4, B5, B7, and B8b. The Order 2 derivative was significant in the bands B3, B5, B6, B7, and B8b. Although the bands' number of significance tests passed varied, the trend of the curves from the phase values in Figure 8 was consistent.

(2)   Relationship between Spectral Index of Sentinel 2 Data and WQI

In this study, NDI, DI, and RI were selected as spectral indices, and the relationship between the spectral index and WQI was explored, as shown in Figure 9 and Table 3. For RI, the relationship between the RI spectral indices and the WQI was significant at the first-order derivative, with a R value of 0.763. For DI, the relationship between the DI spectral indices and the WQI was significant at the second-order derivative, with a R value of 0.778. For NDI, the relationship between the NDI spectral indices and the WQI was significant at the first-order derivative, with a R value of 0.776. We found that the derivative algorithm of Sentinel 2 MSI data improves the relationship between the spectral index and WQI.

4.4. Verification and Precision Analysis of Water Quality Estimation Model
4.4.1. Validation of WQI Estimation Model by Sentinel 2 Data

We used 15 groups of field sample data to train the SVR model, input images of Ebinur Lake to calculate WQI, and then extract the WQI of sampling points as the predicted WQI for model precision analysis. The predicted WQI is represented by WQI_P, and the measured WQI is represented by WQI_M. The relationship between the two is shown in Table 4. We found that the optimal model was Sentinel 2 MSI data based on the third derivative data. The R^2 and RPD of the model were 0.81 and 1.86, respectively. These results indicate that the model has a strong stability.

Table 3. Correlation coefficient between WQI and spectrum index of water.

Derivative Order	RI		DI		NDI	
	Band	R	Band	R	Band	R
0	B2/B4	0.706	B5 − B6	0.741	(B2 − B4)/(B2 + B4)	0.704
1	B3/B5	0.763	B3 − B6	0.763	(B3 − B5)/(B3 + B5)	0.776
2	B3/B4	0.741	B4 − B11	0.778	(B3 − B4)/(B3 − B4)	0.731
3	B5 − B8	0.736	B5 − B7	0.741	(B4 − B5)/(B4 − B5)	0.735

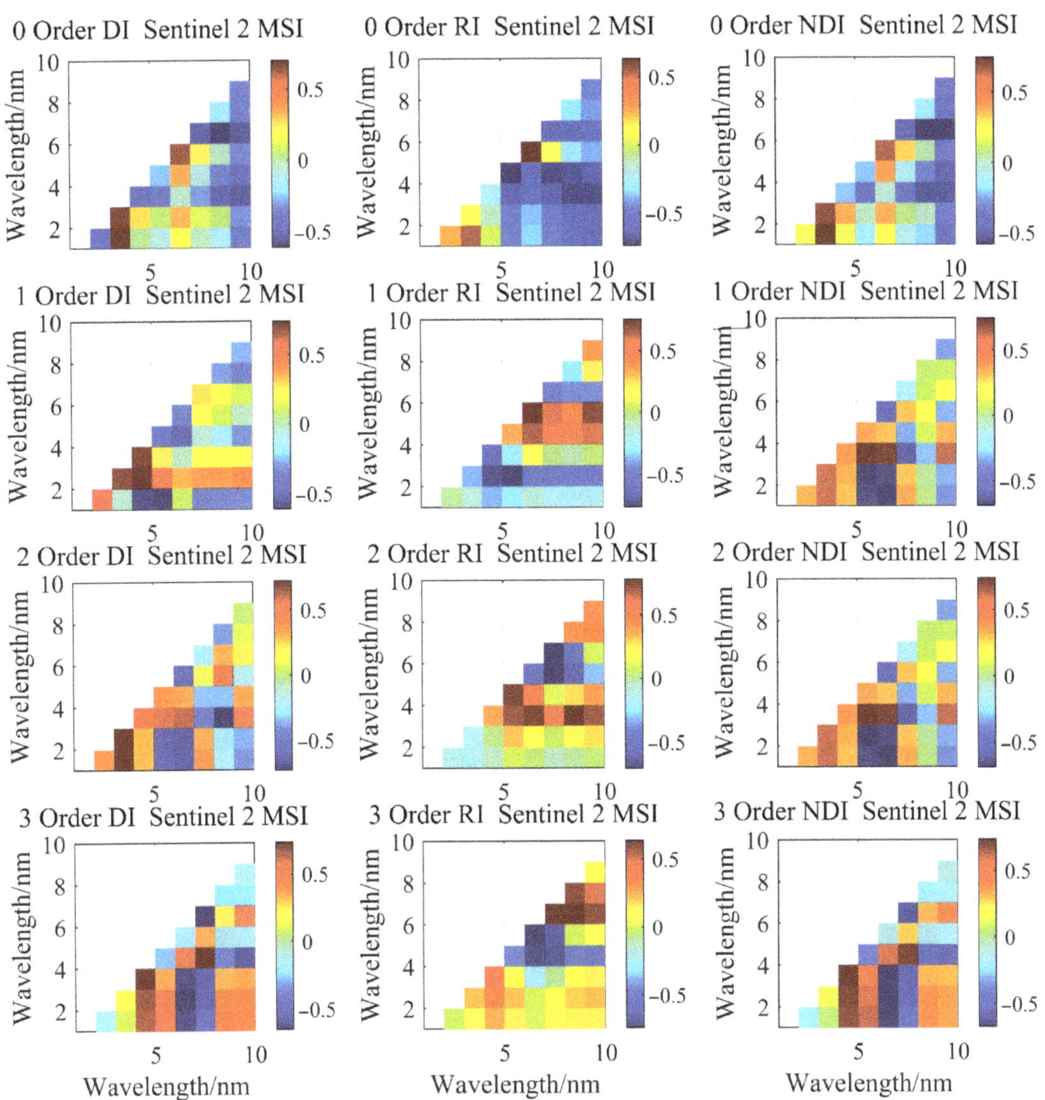

Figure 9. Correlation coefficient between WQI and spectrum index of water.

Table 4. Summary of relationship between the measured values and predicted values.

Order	X	Y	PSO-SVR					
			R^2	RMSE	SD	RPD	Slope	N
0	WQI_M	WQI_P	0.69	344.67	503.07	1.45	0.73	16
1	WQI_M	WQI_P	0.73	302.18	492.36	1.62	0.77	16
2	WQI_M	WQI_P	0.79	245.69	398.06	1.62	0.81	16
3	WQI_M	WQI_P	0.81	213.41	398.72	1.86	0.84	16

4.4.2. Validation of WQI Estimation Model Supported by Sentinel 3 Data

Similarly, we used 15 groups of field sample data and corresponding Sentinel 3 OLCI data for SVR model training, to input images of Ebinur Lake to calculate WQI, and then to extract the WQI of sampling points as the predicted WQI for precision analysis. The predicted WQI is represented by WQI_P, and the measured WQI is represented by WQI_M. The relationship between the two is shown in Table 5. The best model was the fluorescence baseline data of Sentinel 3 OLCI data. The R^2 and RPD of the model were 0.80 and 1.79, respectively, showing that the model has a strong stability.

Table 5. Summary of relationship between the measured values and predicted values.

Order	X	Y	PSO-SVR					
			R^2	RMSE	SD	RPD	Slope	N
0	WQI_M	WQI_P	0.76	233.14	412.38	1.76	0.76	16
1	WQI_M	WQI_P	0.73	342.72	521.09	1.52	0.72	16
2	WQI_M	WQI_P	0.69	354.47	519.84	1.46	0.71	16
FLH	WQI_M	WQI_P	0.80	200.78	359.28	1.79	0.84	16

4.5. Spatial Distribution Map of WQI in Ebinur Lake

A spatial distribution map of WQI based on an optimal model constructed from Sentinel 2–3 derivative data is presented, showing that the water quality in thenorthwest of Ebinur Lake is the lowest in that region. The northwest of Ebinur Lake is eroded by the Alashan Pass gale, and the water depth is less than 1 m. The water quality in the northeast of Ebinur Lake was the second highest, but the water quality was deteriorated by the salinization of large saline-alkali land and soil around the lake. The deterioration of water quality in the northeast of Ebinur Lake is closely related to human activities in the north, which is one of the largest halogen insect production bases in China. The distribution of WQI in Ebinur Lake is shown in Figure 10.

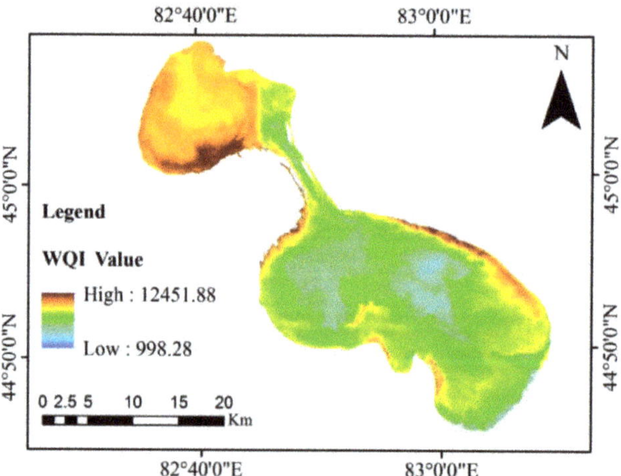

Figure 10. Inverted water quality index (WQI) in Ebinur Lake.

5. Discussion

5.1. Water Quality Index (WQI) as a Potential Proxy for Water Environment

Overall, the results of this study are very indicative, and in agreement with [11] our prediction, proving that remote sensing is a very useful potential tool for water quality monitoring. However, it should be noted that the uncertainty of the WQI remote sensing monitoring model for lake water quality was analyzed from the perspective of time and space. (1) In terms of time, this experiment was limited to the Ebinur Lake watershed during the dry season, aiming to clarify the relationship between WQI and spectral. Although WQI has seasonal variability, WQI also has great variability in the same period and within the same watershed. Therefore, the precision of the WQI model is not limited by season and has its portability in time. (2) The spatial WQI is mainly affected by water in the watershed, whereby spectral also reflects the integration of the whole water environment. The WQI estimation model was established based on the relationship between the spectral index and WQI. In the space under the influence of the study area, portability needs further validation for the model. However, the Ebinur Lake watershed is a typical area of arid area in Central Asia, and its model has certain portability in Central Asia. The extension of a wider range needs further verification. In short, it should be noted that the WQI estimation model is spatially uncertain.

5.2. Spectral Derivative Method and Spectral Indices as Useful Tools for Remote Sensing Modeling of Water Quality

To better mine the information of spectral data from remote sensing, we introduced the spectral derivative method to realize the extraction of spectral information of a water body. The results show that the spectral derivative method can improve the relationship between water body spectral and WQI, whereby the R^2 value of 0.6 is at the most sensitive wavelengths. The derivative technology is not only a powerful tool for analyzing spectra, but also improves multiple collinearity problems considerably [35]. The derivative technology has a strong effect on the peak of the micro spectrum; therefore, it can be used to improve the spectral resolution and sensitivity of the analysis. To some extent, it has the function of removing noise. Fractional derivatives can reduce the intense peak deformation and effectively retain the structure of the original curve, which is more advantageous than other integer derivatives.

Spectral indices are useful for remote sensing modeling of water quality: the optimal remote sensing indices (RI, DI and NDI) were selected for the estimation of WQI, in which multiband remote sensing data were used as variable factors; a combined operation was conducted for various bands, and the sensitivity of WQI information, which was obviously better than that of the single-band models, highlights the advantages of using band combinations. Fernández-Buces et al. used a combined spectral response index to map the soil salinity of bare soil and vegetation. They found a correlation between the normalized difference vegetation indices (NDVI) and electrical conductivity [36]. Therefore, we applied this method, as well as a formula that uses the DI, RI, and NDI of the reflectance values, to establish a new spectral index for estimating WQI.

6. Conclusions

In this paper, the Ebinur Lake basin was selected as the study area, with the aims of revealing the response between water quality index and water body reflectivity, as well as to describe the relationship between water quality index and water reflectivity. A remote sensing monitoring model of WQI was further established, and the water quality of the lake was evaluated by remote sensing. The results indicate:

(1) A Water Quality Index (WQI), based on remote sensing techniques, effectively evaluated the water environment in Ebinur Lake. The Water quality of Ebinur Lake is the lowest, with a WQI value as high as 4000;

(2) To better mine the information of spectral data from remote sensing, we introduced the spectral derivative method to realize the extraction of spectral information from a water body. The results show that the spectral derivative method can improve the relationship between the water body spectral and WQI, whereby the R^2 value of 0.6 is at the most sensitive wavelengths;

(3) When multi-source spectral data were integrated through the spectral index (DI, RI, and NDI) and fluorescence baseline, the correlation between the spectral sensitivity index and WQI was found to be greater than 0.6 at the significance level of 0.01;

(4) The distribution map of WQI in Ebinur Lake was obtained by the optimal model, which was constructed based on the third derivative data of Sentinel 2 data. Results indicate that the water quality in the northwest of Ebinur Lake was the lowest in the

region. In conclusion, remote sensing techniques were found to be highly effective and lay a foundation for water quality detection in arid areas.

Author Contributions: Conceptualization, F.Z. and X.W.; methodology, N.W.C.; software, X.W.; validation, C.L., T.G. and X.L.; formal analysis, X.W.; investigation, C.L.; resources, F.Z.; data curation, X.W.; writing—original draft preparation, X.W.; writing—review and editing, H.-T.K. and J.S.; visualization, C.L., W.W. and N.C.; supervision, F.Z.; project administration, F.Z.; funding acquisition, F.Z. All authors have read and agreed to the published version of the manuscript.

Funding: This research was funded by the National Natural Science Foundation of China (U1603241), the National Natural Science Foundation of China (Xinjiang Local Outstanding Young Talent Cultivation) (Grant No. U1503302), and the Tianshan Talent Project (Phase III) of the Xinjiang Uygur Autonomous Region. We want to thank the editor and anonymous reviewers for their valuable comments and suggestions to this paper.

Data Availability Statement: The data presented in this study are available on request from the corresponding author. The data are not publicly available due to confidentiality.

Conflicts of Interest: The authors declare no conflict of interest.

References

1. Cardinale, B.J. Biodiversity improves water quality through niche partitioning. *Nature* **2011**, *472*, 86–89. [CrossRef] [PubMed]
2. Pekel, J.F.; Cottam, A.; Gorelick, N.; Belward, A.S. High-resolution mapping of global surface water and its long-term changes. *Nature* **2016**, *540*, 418–422. [CrossRef] [PubMed]
3. Michalak, A.M. Study role of climate change in extreme threats to water quality. *Nature* **2016**, *535*, 349–350. [CrossRef] [PubMed]
4. Feyisa, G.L.; Meilby, H.; Fensholt, R.; Proud, S.R. Automated Water Extraction Index: A new technique for surface water mapping using Landsat imagery. *Remote Sens. Environ.* **2014**, *140*, 23–35. [CrossRef]
5. Kumar, V.; Sharma, A.; Chawla, A.; Bhardwaj, R.; Thukral, A.K. Water quality assessment of river Beas, India, using multivariate and remote sensing techniques. *Environ. Monit. Assess.* **2016**, *188*, 137. [CrossRef] [PubMed]
6. Dörnhöfer, K.; Oppelt, N. Remote sensing for lake research and monitoring—Recent advances. *Ecol. Indic.* **2016**, *64*, 105–122. [CrossRef]
7. Cheng, P.; Wang, X.L. The design and implementation of Remote-sensing Water Quality Monitoring System based on SPOT-5. In Proceedings of the Second IITA International Conference on Geoscience & Remote Sensing, Qingdao, China, 28–31 August 2010.
8. Pulliainen, J.; Kallio, K.; Eloheimo, K.; Koponen, S.; Hallikainen, M. A semi-operative approach to lake water quality retrieval from remote sensing data. *Sci. Total Environ.* **2001**, *268*, 79–93. [CrossRef]
9. Su, T.C. A study of a matching pixel by pixel (MPP) algorithm to establish an empirical model of water quality mapping, as based on unmanned aerial vehicle (UAV) images. *Int. J. Appl. Earth Obs. Geoinf.* **2017**, *58*, 213–224. [CrossRef]
10. Burns, P.; Nolin, A. Using atmospherically-corrected Landsat imagery to measure glacier area change in the Cordillera Blanca, Peru from 1987 to 2010. *Remote Sens. Environ.* **2014**, *140*, 165–178. [CrossRef]
11. Wang, X.P.; Zhang, F.; Ding, J.L. Evaluation of water quality based on a machine learning algorithm and water quality index for the Ebinur Lake Watershed, China. *Sci. Rep.* **2017**, *7*, 12858. [CrossRef]
12. Alves, E.M.; Rodrigues, R.J.; Corrêa, C.S.; Fidemann, T.; Rocha, J.C.; Buzzo, J.L.L.; Note, P.O.; Núñez, G.F. Use of ultraviolet–visible spectrophotometry associated with artificial neural networks as an alternative for determining the water quality index. *Environ. Monit. Assess.* **2018**, *190*, 319. [CrossRef]
13. Gogu, R.; Carabin, G.; Hallet, V.; Peters, V.; Dassargues, A. GIS-based hydrogeological databases and groundwater modelling. *Hydrogeol. J.* **2001**, *9*, 555–569. [CrossRef]
14. Gholizadeh, M.H.; Melesse, A.M.; Reddi, L. A comprehensive review on water quality parameters estimation using remote sensing techniques. *Sensors* **2016**, *16*, 1298. [CrossRef] [PubMed]
15. Kutser, T.; Paavel, B.; Verpoorter, C.; Ligi, M.; Soomets, T.; Toming, K.; Casal, G. Remote sensing of black lakes and using 810 nm reflectance peak for retrieving water quality parameters of optically complex waters. *Remote Sens.* **2016**, *8*, 497. [CrossRef]
16. Giardino, C.; Bresciani, M.; Cazzaniga, I.; Schenk, K. Evaluation of multi-resolution satellite sensors for assessing water quality and bottom depth of lake garda. *Sensors* **2014**, *14*, 24116–24131. [CrossRef] [PubMed]
17. Farhad, M.; Fatemeh, D.; Mostafa, R.; Baden, M. Introducing a water quality index for assessing water for irrigation purposes: A case study of the ghezel ozan river. *Sci. Total Environ.* **2017**, *589*, 107–116.
18. Nuraslinda, A.; Pauzi, A.M.; Bakar, A.A.A. Methodology of water quality index (WQI) development for filtrated water using irradiated basic filter elements. *Math. Sci. Its Appl.* **2017**, *1799*, 040010.
19. Horton, R.K. An index number system for rating water quality. *J. Water Pollut. Control Fed.* **1965**, *37*, 300–306.
20. Brown, R.M.; McClelland, N.I.; Deininger, R.A.; Tozer, R.G. A water quality index-do we dare? *Water Sew. Work.* **1970**, *117*, 339–343.

21. Debels, P.; Fıgueroa, R.; Urrutia, R.; Barra, R.; Niell, X. Evaluation of water quality in the Chilla'n river (Central Chile) using physicochemical parameters and a modified water quality index. *Environ. Monit. Assess.* **2005**, *110*, 301–322. [CrossRef]
22. Saeedi, M.; Abessi, O.; Sharifi, F.; Maraji, H. Development of groundwater quality index. *Environ. Monit. Assess.* **2009**, *163*, 327–335. [CrossRef] [PubMed]
23. Şener, Ş.; Şener, E.; Davraz, A. Evaluation of water quality using water quality index (WQI) method and gis in Aksu river (Sw-turkey). *Sci. Total Environ.* **2017**, *584–585*, 131–144. [CrossRef]
24. D'Amico, F. Prompt-Particle-Events in ESA's Envisat/MERIS and Sentinel-3/OLCI Data: Observations, Analysis and Recommendations. Master's Thesis, University of Pisa, Pisa, Italy, 2015.
25. Sibanda, M.; Mutanga, O.; Rouget, M. Examining the potential of Sentinel-2 MSI spectral resolution in quantifying above ground biomass across different fertilizer treatments. *ISPRS J. Photogramm. Remote Sens.* **2015**, *110*, 55–65. [CrossRef]
26. Liu, C.J.; Duan, P.; Zhang, F.; Jim, C.Y.; Tan, M.L.; Chan, N.W. Feasibility of the Spatiotemporal Fusion Model in Monitoring Ebinur Lake's Suspended Particulate Matter under the Missing-Data Scenario. *Remote Sens.* **2021**, *13*, 3952. [CrossRef]
27. Hoge, F.E.; Lyon, P.E.; Swift, R.N.; Yungel, J.K.; Abbott, M.R.; Letelier, R.M.; Esaias, W.E. Validation of Terra-MODIS phytoplankton chlorophyll fluorescence line height. I. Initial airborne lidar results. *Appl. Opt.* **2003**, *42*, 2767. [CrossRef] [PubMed]
28. Brown, R.M.; McClelland, N.I.; Deininger, R.A.; O'Connor, M.F. A Water Quality Index -Crashing the Psychological Barrier. In *Indicators of Environmental Quality*; Thomas, W.A., Ed.; Springer: Boston, MA, USA, 1972. [CrossRef]
29. Dede, O.T.; Telci, I.T.; Aral, M.M. The Use of Water Quality Index Models for the Evaluation of Surface Water Quality: A case study for Kirmir Basin, Ankara, Turkey. *Water Qual. Expo. Health* **2013**, *5*, 41–56.
30. Harkins, R.D. An Objective Water Quality Index. *J. (Water Pollut. Control. Fed.)* **1974**, *46*, 58591.
31. Wu, Z.S.; Wang, X.L.; Chen, Y.W.; Cai, Y.J.; Deng, J. Assessing river water quality using water quality index in Lake Taihu Basin, China. *Sci. Total Environ.* **2018**, *612*, 914–922. [CrossRef]
32. Mukate, S.; Wagh, V.; Panaskar, D.; Jacobs, J.A.; Sawant, A. Development of new integrated water quality index (IWQI) model to evaluate the drinking suitability of water. *Ecol. Indic.* **2019**, *101*, 348–354. [CrossRef]
33. Wiley, E.O.; Mcnyset, K.M.; Peterson, A.T.; Robins, C.R.; Stewart, A.M. Niche modeling and geographic range predictions in the marine environment using a machine-learning algorithm. *Oceanography* **2003**, *16*, 120–127. [CrossRef]
34. Marjanović, M.; Kovačević, M.; Bajat, B.; Voženílek, V. Landslide susceptibility assessment using SVM machine learning algorithm. *Eng. Geol.* **2011**, *123*, 225–234. [CrossRef]
35. Peón, J.; Fernández, S.; Recondo, C.; Calleja, J.F. Evaluation of the spectral characteristics of five hyperspectral and multispectral sensors for soil organic carbon estimation in burned areas. *Int. J. Wildland Fire* **2017**, *26*, 230–239. [CrossRef]
36. Fernández-Buces, N.; Siebe, C.; Cram, S.; Palacio, J.L. Mapping soil salinity using a combined spectral response index for bare soil and vegetation: A case study in the former Lake Texcoco, Mexico. *J. Arid Environ.* **2006**, *65*, 644–667. [CrossRef]

MDPI
St. Alban-Anlage 66
4052 Basel
Switzerland
Tel. +41 61 683 77 34
Fax +41 61 302 89 18
www.mdpi.com

Water Editorial Office
E-mail: water@mdpi.com
www.mdpi.com/journal/water